ARC WELDING

Edited by **Wladislav Sudnik**

INTECHOPEN.COM

Arc Welding
Edited by Wladislav Sudnik

Published by InTech

Janeza Trdine 9, 51000 Rijeka, Croatia

Edition 2014

Publishing Process Manager Marija Radja

Technical Editor Teodora Smiljanic

Cover Designer InTech Design Team

Image Copyright pmakin, 2011. DepositPhotos

Additional hard copies can be obtained from orders@intechopen.com

Arc Welding, Edited by Wladislav Sudnik
p. cm.
ISBN 978-953-307-642-3

INTECH
open science | open minds

Contents

Preface

Ever since the invention of arc technology in 1870s and it's early use for welding lead during the manufacture of lead-acid batteries, advances in arc welding throughout the twentieth and twenty-first centuries have seen this form of processing applied to a range of industries and progress to become one of the most effective techniques in metals and alloys joining.

The objective of this book is to introduce relatively established methodologies and techniques which have been studied, developed and applied in industries or researches. State-of-the-art development aimed at improving technologies will be presented covering topics such as weldability, technology, automation, modelling, and measurement. This book also seeks to provide effective solutions to various applications for engineers and researchers who are interested in arc material processing.

This book is divided into 4 independent chapters corresponding to recent advances in this field.

The editor expresses thankfulness to all authors for the presented materials and their timely design, and also to the technical editor and to the book manager Mrs. Marija Radja - for the big work on preparation and the edition of this book.

Editor

Prof. Dr. Wladislav Sudnik
R & E Center 'Computer Hi-Tech in Materials Joining'
Welding Department
Tula State University,
Russian Federation

Part 1

Arc Welding Technology

1

Hardfacing by Plasma Transferred Arc Process

Víctor Vergara Díaz[1], Jair Carlos Dutra[2] and Ana Sofia Climaco D'Oliveira[3]
[1]*University of Antofagasta, Mechanical Engineering Department*
[2]*University Federal de Santa Catarina, Mechanical Engineering Department*
[3]*University Federal do Paraná, Mechanical Engineering Department*
[1]*Chile*
[2,3]*Brasil*

1. Introduction

According to the literature, the plasma transferred arc welding process which employs the filler metal in wire form is known as Plasma Arc Welding (PAW) while that which employs powder filler material is generally referred to as Plasma Transferred Arc (PTA), Dai et al., 2008.

The PTA process can be considered a derivation of the PAW process. The similarities between the two processes can be observed in Figure 1. Both welding processes employ a non-consumable tungsten electrode located inside the torch, a water-cooled constrictor nozzle, shield gas for the protection of the molten pool, and the plasma gas. The difference between the two welding processes lies in the nature of the filler material, powder instead of wire, which requires a gas for its transport to the arc region. The diagram in Figure 1 shows the two processes with their differences and similarities.

The equipment required to carry out the deposition through the PTA plasma process is very similar to that used in PAW. When PAW is employed the equipment must be able to drive spooled wires of various gages and different materials, at constant or pulsed velocities. In the PTA plasma welding process, the filler material is used in the form of a powder, and specific powder feeding equipment is required to transport it to the voltaic arc to produce the coating. With respect to its application for coating, the PTA process is appropriate since it produces dilution values of the order of 6 to 10 % (Gatto, et al., 2009), much lower than those obtained with other arc soldering process which are around 20 to 25 %. The low distortion, the small zone affected by the heat and the refined microstructure are also features of this technique (Zhang, et al., 2008; Liu, et al., 2008).

In the PTA and PAW processes an inert gas is used as the plasma gas, which is forced to pass through the orifice of the constrictor nozzle, where the electrode is concentrically fixed. The shield gas passes through an external opening, concentric to the constrictor nozzle, effectively protecting the weld against contamination from atmospheric air (active or inert). On the other hand, in the PTA process a carrier gas is used to transport the filler material through flexible tubes to the constrictor nozzle, allowing its entrance into the plasma arc in a convergent form. The gas used for this purpose is generally argon.

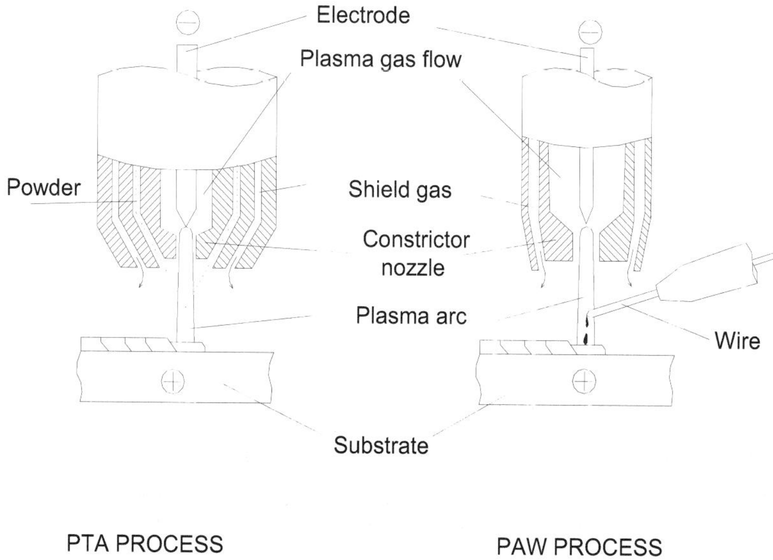

Fig. 1. Comparison of Plasma Transferred Arc processes PTA and PAW.

Given that the tungsten electrode lies within the constrictor nozzle of the welding torch, it is difficult to open the arc by contact, and thus equipment called a plasma module must be used to establish the arc opening. An electronic igniter provides voltage peaks between the tungsten electrode and constrictor nozzle, generating a small spark in this region. Thus, with the passage of the plasma gas a low intensity electric arc appears between the tungsten electrode and constrictor nozzle, called the pilot arc (non-transferred arc). The pilot arc forms a pathway of low electrical resistance between the tungsten electrode and the workpiece to be welded facilitating the establishment of the main arc when a power source is added.

In practice, the parameters which control the quality of the weld are the rate at which the material is added, the gas flow rate (shield gas, plasma gas, carrier gas), the weld current, the nozzle to workpiece distance (see below) and the welding speed.

The basic configuration of the constrictor nozzle is shown in Figure 2, where the parameters employed in the process are indicated. The distance from the external face of the constrictor nozzle to the substrate is called the nozzle to workpiece distance (NWD).

The recess (Rc) of the electrode is measured from the electrode tip to the external face of the constrictor nozzle. Alterations in the arc characteristics are influenced by this factor, which defines the degree of constriction and the rigidity of the plasma jet (Oliveira, 2001).

Oliveira (2001) studied the influence of the electrode recess of the plasma transferred arc process fed by wire in order to identify whether the degree of arc constriction influences the arc voltage. The results showed that, on average, a 2.4 V/mm variation in the voltage occurred as a function of the electrode recess.

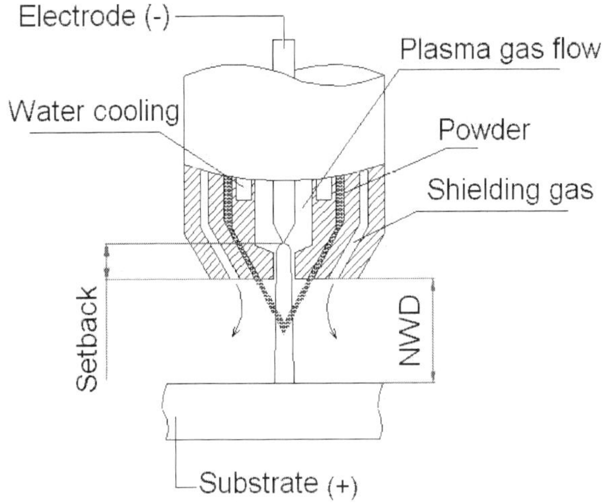

Fig. 2. Nozzle to workpiece distance (NWD) and electrode setback (Rc) (Vergara, 2005).

In general, the maximum and minimum values for the adjustment of the electrode recess vary according to the welding torch. The electrode recess of the welding torch PWM-300, manufactured by Thermal Dynamics Corporation, for instance, has a range of adjustment of 0.8 to 2.4 mm.

As the electrode recess is reduced, the weld bead width increases and weld beads with lower penetration depth are obtained. This variation in the geometric characteristics of the weld bead is due to a reduction in the constriction effect producing a larger area of incidence of the arc on the substrate.

The constrictor nozzle (made of copper), where the electrode is confined, has a central orifice through which the arc and all of the plasma gas volume pass. The diameter of the orifice of the constrictor nozzle has a great influence on the quality of the coating since this relationship is directly related to the width and penetration of the weld bead produced. An insufficient plasma gas flow rate affects the useful life of the constrictor nozzle since it leads to its wear. The weld current reduces as a function of the decrease in the diameter of the constricting orifice, due to an increase in the weld arc temperature.

The extent to which the nozzle to workpiece distance influences the coating is strongly dependent on the electrode recess in relation to the constrictor nozzle and the diameter of the constrictor orifice. The larger the electrode recess adopted and the smaller the constrictor orifice diameter the greater the effect of the arc constriction, making it more concentrated.

In the "melt-in" technique small electrode recess values are used, the arc being submitted to a low degree of collimation, assuming a conical form. In this situation, a variation in the nozzle to workpiece distance, even within normal limits, results in a change in the characteristics of the weld bead, in the same way as occurs in the GTAW process. Thus, the greater the nozzle to workpiece distance the lower the penetration and wider the width of the weld bead due to the increase in the area of incidence of the arc on the substrate.

Hallen et al. (1991) reported that to obtain a good deposition yield, the nozzle to workpiece distance should not be greater than 10 to 15 mm. At values higher than this range the efficiency of the shield gas is significantly reduced.
The authors of this paper have also reported results in relation to the nozzle to workpiece distance, for two values: 15 and 20 mm. The study showed that as the nozzle to workpiece distance increases the degree of dilution decreases.
The general objective of this study was to investigate the PAW and PTA welding processes with a view to their application in surface coating operations, particularly on hydraulic turbine blades worn by cavitation. This research was motivated by the observation that information is scare in relation to the benefits offered by the plasma welding process using powder instead of wire filler material in the application of coatings. The geometric characteristics of the weld beads, degree of dilution, hardness and microstructure were evaluated.

2. Materials and Methods

2.1 Test bench

Initially, a test bench was assembled based on equipment previously developed at LABSOLDA (Oliveira, 2001; Vergara, 2005) which allowed tests to be carried out on the plasma transferred arc welding process fed by wire. On the same test bench, a similar process fed by powder was assembled. The welding source was equipment which, via an interface, was connected to a PC. By way of a very versatile software program almost all of the process variables could be controlled.
Of the three gas circuits, that which received most attention was the plasma gas given its considerable relevance in terms of the quality of the deposits. A mass flow controller was used, in which the control is carried out electronically and the command signal is a reference voltage. The other gas flow circuits are simply monitored by electronic flow meters, however these are volumetric.
One of the fundamental parts of the equipment is the device known as the plasma module, which enables any version of plasma welding to be carried out based on conventional welding sources for GTAW or coated electrode. For the displacement of the welding torch an electronic device (Tartílope) was used. The system component which was integrally designed for this specific development was the powder feeding device, which functions through a combination of an endless screw and a gas flow as the powder carrying mechanisms. The weld torch was developed based on the plasma torch for keyhole welding. The great advantage of this lies in its multiprocess aspect which allows it to work with plasma employing powder or with conventional plasma. Also, the design adaptation allows the use of constrictor nozzles with different angles of convergence for the powder feeding. Initially, analysis was carried out on the torches to be used in this research. It was observed that the PTA torch had a nozzle with a constrictor diameter of 4.8 mm. In the case of the PAW torch, the manufacturer provides three nozzles with constrictor diameters of 2.4, 2.8 and 3.2 mm, which are designed according to the welding current to be applied.
In this case, the nozzle with the largest constrictor diameter available for the PAW torch was selected, that is, 3.2 mm.
Figure 3 shows a general view of the equipment developed, that which forms part of the test bench for the PAW and PTA welding processes being shown in the upper part of the figure.

In this study argon with a purity of 99.99 % was used as the plasma, shield and carrier gases. A tungsten electrode with 2% thorium oxide (EWTh-2) and with a diameter of 4.8 mm was used. The angle of the electrode tip was maintained at 30° for all of the experiments.

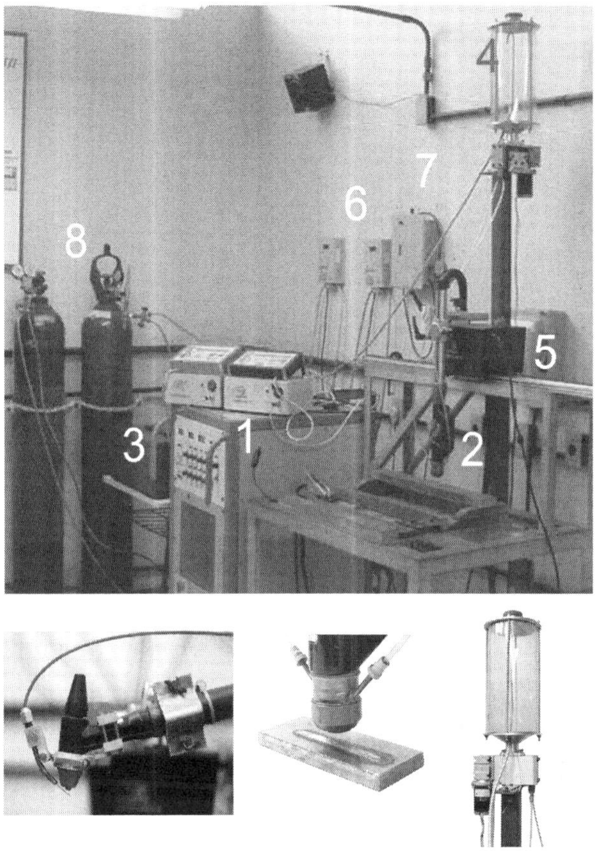

Adapted PAW torch | Adapted PTA torch | PTA Powder Feeder

Fig. 3. Test bench assembled at the welding laboratory 1-Welding source; 2-Adapted plasma torch; 3-Plasma module; 4-Powder feeder; 5-Torch displacement system; 6-Digital gas meters; 7-Electronic gas valve; 8-Gases

2.2 Constrictor nozzle in PTA process

The configuration of the constrictor nozzle developed in this study included two conduits for the passage of the carrier gas, the role of which is to feed the powder to the plasma arc in a convergent form. Figure 4 shows a cross-section of the constrictor nozzle. At 60° the constrictor nozzle allows the entrance of powder directly into the molten pool, when a nozzle to workpiece distance of 10 mm is used.

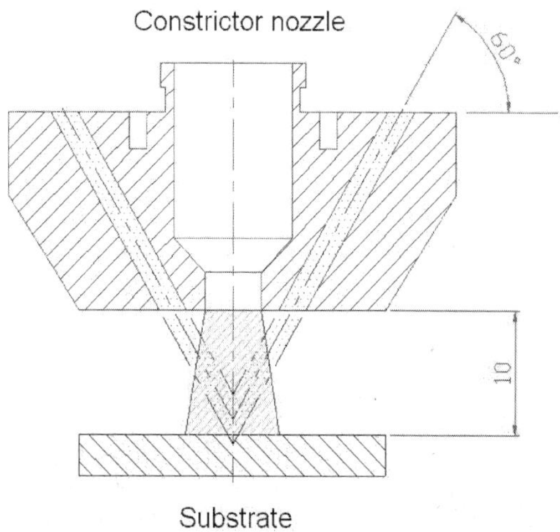

Fig. 4. Cross-section of constrictor nozzle showing the entrance of the powder flow into the plasma arc. (Vergara, 2005).

2.3 Characterization

Deposits of the atomized alloy Stellite 6, Figure 5, were processed on carbon steel plates (class ABNT 1020; dimensions 12.5 x 60 x 155 mm), using a constant continuous current. Table 1 shows the chemical composition of the substrate. The chemical analysis of the different filler materials was carried out by optical emission spectrometry and the results are shown in Tables 2 and 3.

Single weld beads were deposited with the parameters indicated in Table 4 and samples were removed for their characterization. This table gives the operational parameters for the PTA and PAW plasma welding processes, in which there are parameters which could not remain constant in the two process, for example: nature of the filler material (in PAW wire and in PTA powder); wire speed (not required in PTA); carrier gas (not required in PAW); constrictor nozzle diameter (in PTA 4.8 mm and in PAW 3.2 mm).

Initially, the weld beads were submitted to visual inspection for the presence of welding defects, the degree of dilution was determined by the areas method using micrographs of the cross-sections of the deposits, etched with 6% nital. Profiles of the Vickers microhardness, with a load of 500g, enabled the evaluation of the uniformity of the weld beads processed, according to the procedure of the standard ABNT6672/81. The determination of the microhardness profiles, average of three measurements, was carried out at the center of the weld beads and in the region where they overlap. To determine the microstructure by optical microscopy a cross-section was prepared following standard procedures, the microstructure being revealed after electrolytic attack with oxalic acid.

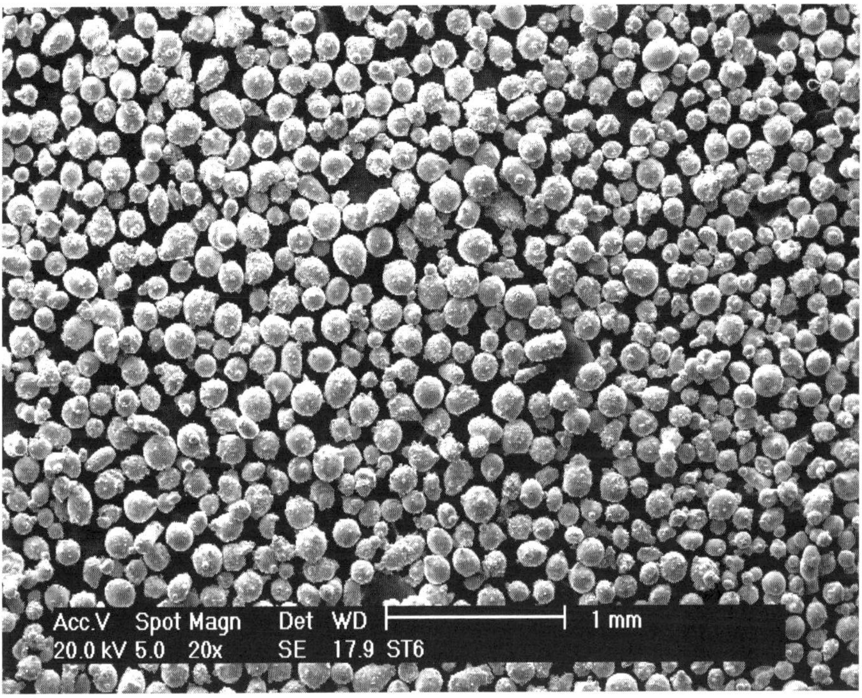

Fig. 5. Morphology of powder deposited by the PTA process (Stellite 6).

C	Si	Mn	P	S	Cr	Mo	Ni	Al
0.11	0.22	0.74	0.021	0.008	0.027	0.024	0.011	0.06
Cu	V	W	Sn	Fe				
0.016	0.015	0.026	0.065	98.6				

Thickness: 12.7 mm

Table 1. Chemical composition of the low carbon steel substrate.

C	Si	Mn	Cr	Mo	Ni	Co	W	Fe
1.32	1.30	0.028	30.01	0.24	2.45	Bal	5.21	2.05

Hardness: 38-47 Rc; Particle size: 45 to 150 μm; Density: 8.3 g/cm^3

Table 2. Chemical composition of the filler material Stellite 6 in the form of a powder (BT-906)

C	Si	Mn	Cr	Mo	Ni	Co	W	Fe
0.9-1.4	2.0	1.0	26-32	1.0	3.0	Bal	3.0-6.0	2.0

Table 3. Chemical composition of filler material Stellite 6 in the form of steel (BT-906T).

PTA Process		
Welding current	A	160
Welding speed	cm/min	20
Plasma gas flow rate	l/min	2.2; 2.4; 3.0
Shield gas	l/min	10
Carrier gas	l/min	2
Feed rate	kg/h	1.4
Constrictor nozzle diameter/ convergence angle	mm/°	4.8/30
Nozzle to workpiece distance	mm	10
Setback	mm	2.4
PAW Process		
Wire diameter (tubular)	mm	1.2
Wire speed	m/min	3.0
Deposition rate	kg/h	1.4
Constrictor nozzle diameter	mm	3.2
Welding current	A	160
Welding speed	cm/min	20
Plasma gas flow rate	l/min	2.2; 2.4; 3.0
Shield gas	l/min	10
Feed rate	kg/h	1.4
Nozzle to workpiece distance	mm	10
Setback	mm	2.4

Table 4. Welding variables and parameters.

3. Results and discussion

3.1 General characteristics

Figure 6 shows the external aspect of the beads where significant differences between them can be observed. The PTA process produced a better surface finish, better dilution, better wetting and wider width.

Figures 7 and 8 show cross-sections of the beads obtained using the two processes (PAW and PTA) where considerable differences in the penetration profile of the welds can be noted and Figure 9 shows the results for the geometric parameters of the beads, for the three levels of plasma gas flow rate tested in this study: 2.2; 2.4 and 3.0 l/min. On comparing the deposits obtained from the two processes it can be observed that the reinforcement and the penetration are always smaller in the PTA process (Figure 9). In the PTA process there was a significantly wider cord width, which is due to the use of a constrictor nozzle with a wider diameter.

The data shown in Figure 9 together with an analysis of the variance in Tables 5, 6 and 7, indicate that the welding process and plasma gas flow rate have significant effects on the geometric parameters of the bead.

In relation to the convexity index (CI = 100*r/W), Silva et al. (2000) establishes that values close to 30% are desirable for the relation between the width (W) and reinforcement (r) of the weld bead. Figure 10 shows the convexity index of the weld bead for the PAW and PTA processes as a function of the plasma gas flow rate.

Analysis of Figure 10 shows that for the three plasma gas flow rates tested the PTA process provided acceptable convexity of the weld beads (less than 30%), a highly desirable condition. In the case of the PAW process, the convexity index was acceptable only for low plasma gas flow rates.

The average values for the areas of the metal deposited varied for the two welding processes studied, as expected, due to the difference in the diameters of the constriction orifices used in each case and the material loss according to the efficiency of the deposition process.

Figure 11 shows that in the PTA process there was loss of material. Lin (1999) observed that losses occur mainly due to vaporization and also dispersion of the particles after making contact with the substrate.

Vergara (2005), reports that the carrier gas flow rate influences the dispersion of the particles. In many cases it is possible, at the end of the finishing operation, to observe unmolten powder particles adhered to the sides of the finish. On the other hand, when the deposition rate is very high (1.5 kg/h) in relation to the welding current (160 A) unmolten power can be seen spread over the substrate. Vergara [9] observed that the PTA process has a deposition efficiency of the order of 87% when a constrictor nozzle of 30° is used. Similar results have been reported by Davis (1993), who demonstrated a range of 85 to 95 % deposition yield for the PTA process.

The graph in Figure 12 shows the effect of the plasma gas flow rate on the degree of dilution using the wire Stellite 6, 1.2 mm tubular diameter. The results indicate that the dilution increases with the plasma gas flow rate possibly due to the greater pressure of the plasma jet. Similar results were found for the PTA process, with dilution values being lower than those achieved with the PAW process, as expected, due to the difference in the diameters of the constrictor orifice. Vergara (2005) reports that the diameter of the constrictor nozzle orifice has a considerable influence on the quality of the finish since it is directly related to the width and penetration of the weld bead produced. The data in Figure 12 together with the analysis of variance in Table 8 indicate that, in general, the welding process and the plasma gas flow rate significantly affect the dilution. Similar conclusions have been reported by Silvério (2003) for the alloy Stellite 1.

The good results obtained for the PTA process are associated with:

- Wider weld beads ⇒ greater area of covering
- Lower dilution ⇒ deposits with composition closer to that of the filler alloy
- Better wetting, lower convexity ⇒ reduced risk of lack of penetration/ fusion between weld beads.

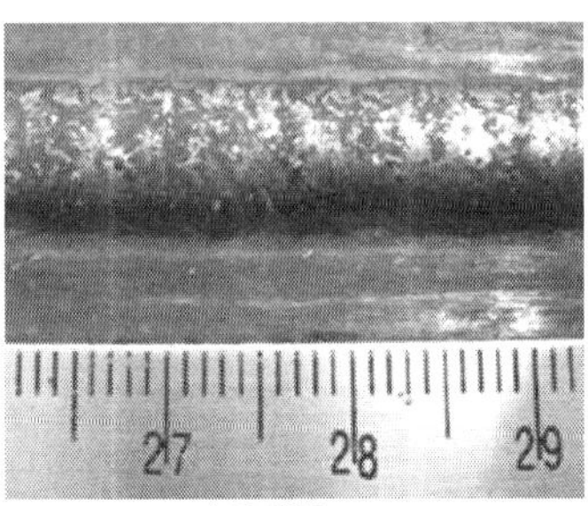

a) PAW

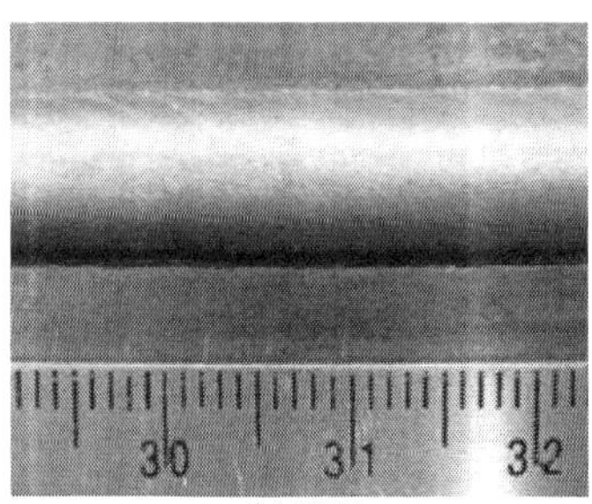

b) PTA

Fig. 6. Superficial aspect of Stellite 6 deposited by: a) PAW and b) PTA. Welding current = 160 A, Welding speed = 20 cm/min, Feed rate =1.4 kg/h, Plasma gas flow rate = 2.4 l/min.

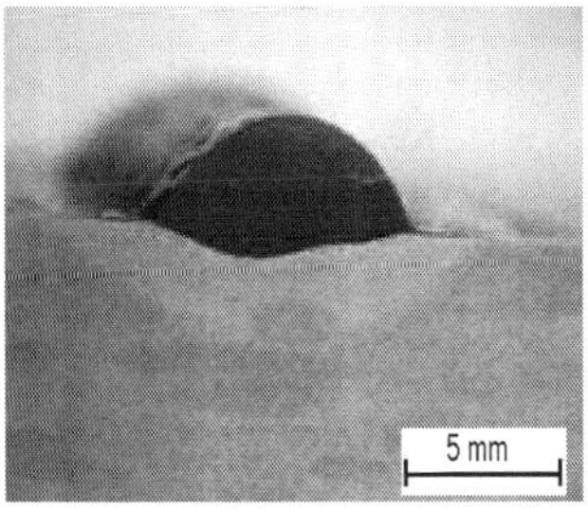

(a)

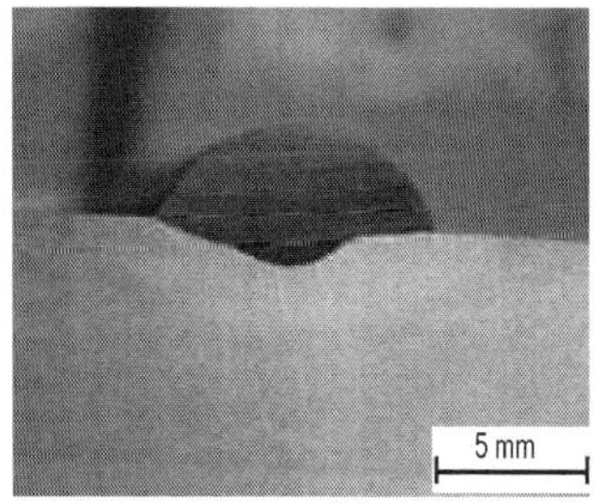

(b)

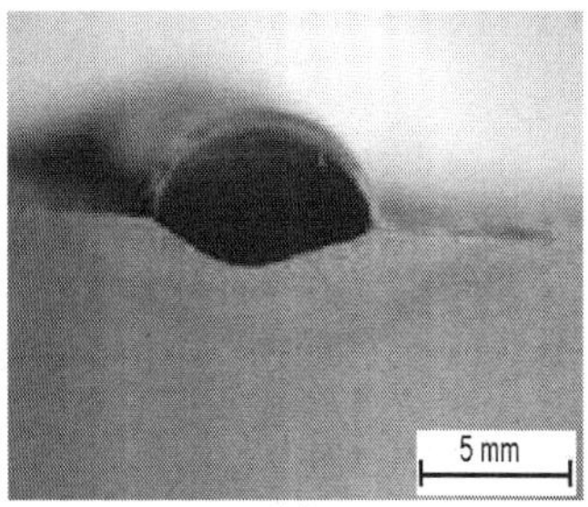

(c)

Fig. 7. Cross-section of weld beads processed via PAW. Plasma gas flow rate: (a) 2.2 (l/min); (b) 2.4 (l/min); and (c) 3.0 (l/min)

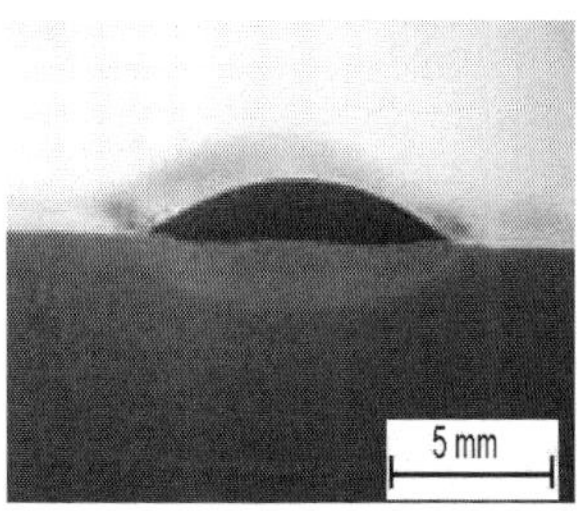

(a)

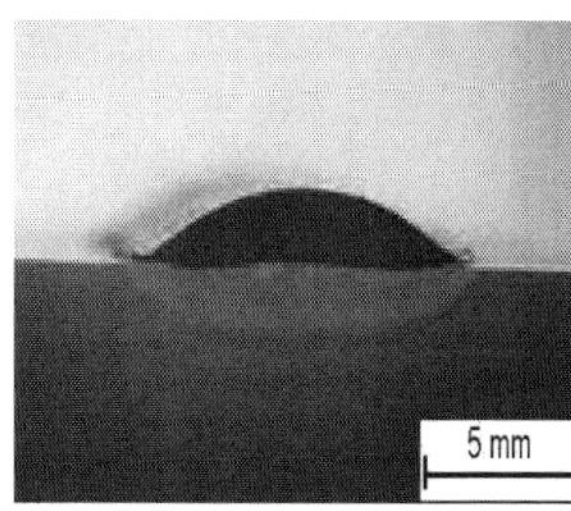

(b)

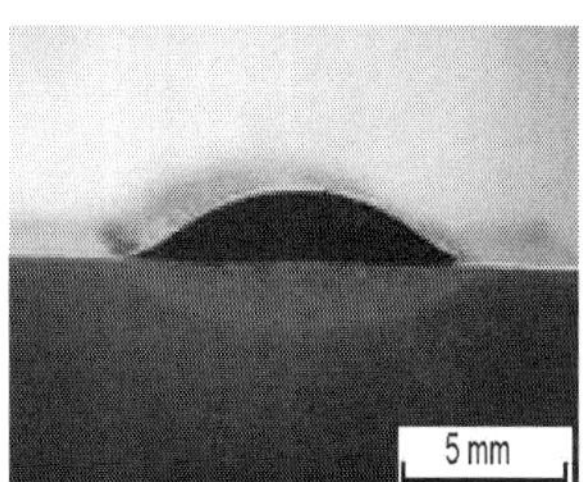

Fig. 8. Cross-section of weld beads processed via PTA. Plasma gas flow rate: (a) 2.2 (l/min); (b) 2.4 (l/min); and (c) 3.0 (l/min).

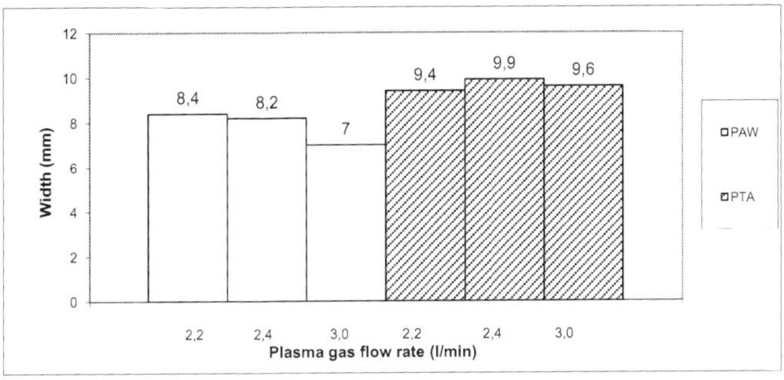

a) Width

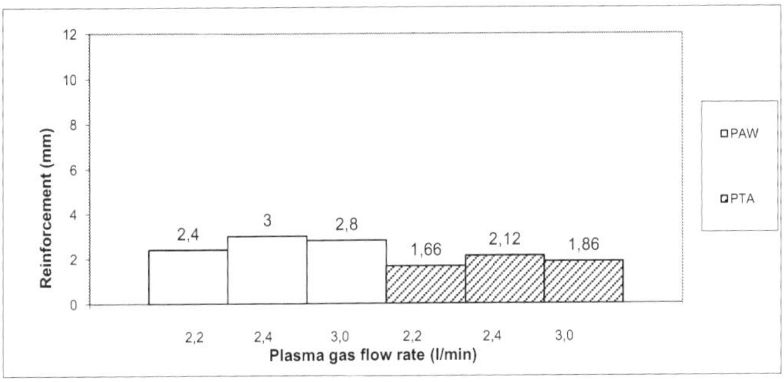

b) Reinforcement

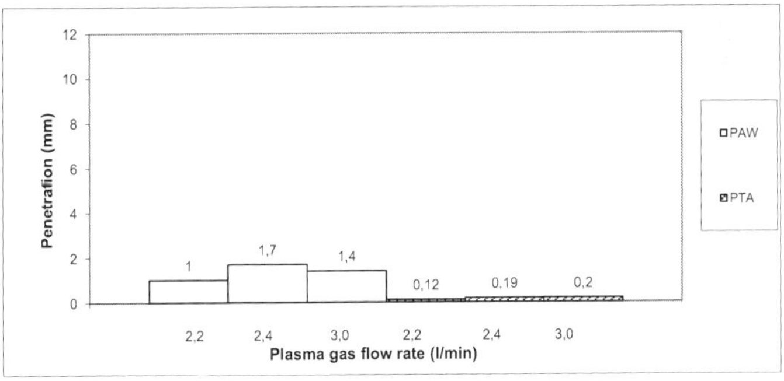

c) Penetration

Fig. 9. Effect of plasma gas flow rate on geometric parameters (Width, reinforcement, penetration).

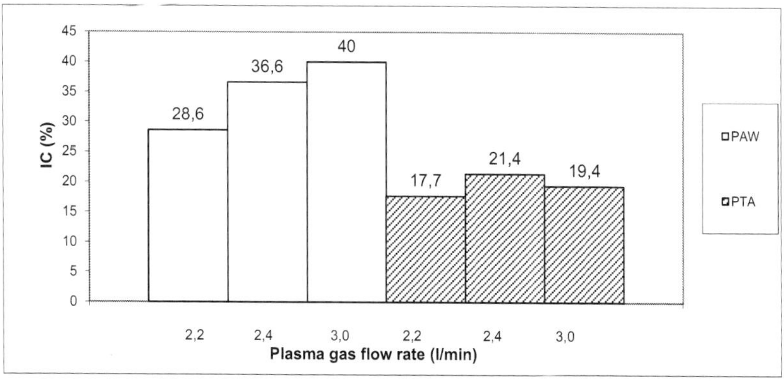

Fig. 10. Effect of plasma gas flow rate on convexity index.

Source of variation	*Sum of squares*	*Degrees of freedom*	*Average of squares*	*F observed*	*F critical*
Welding process	17.85	1	17.85	1444.35	
Plasma gas flow rate	2.316	2	1.16	93.67	
Interaction	2.33	2	1.16	94.14 >	3.55
Residual	0.22	18	0.0124		
Total	22.72	23			

Obs.: Index of significance (α) = 5%

Table 5. Results of the analysis of variance for width.

Source of variation	*Sum of squares*	*Degrees of freedom*	*Average of squares*	*F observed*	*F critical*
Welding process	4.29	1	4.29	1353.78	
Plasma gas flow rate	1.33	2	0.66	209.016	
Interaction	0.098	2	0.049	15.45 >	3.55
Residual	0.057	18	0.0032		
Total	5.77	23			

Obs.: Index of significance (α) = 5%

Table 6. Results of analysis of variance for reinforcement.

Source of variation	*Sum of squares*	*Degrees of freedom*	*Average of squares*	*F observed*	*F critical*
Welding process	8.35	1	8.354	5323.15	
Plasma gas flow rate	0.58	2	0.288	183.74	
Interaction	0.37	2	0.185	118.06 >	3.55
Residual	0.02825	18	0.00157		
Total	9.33	23			

Obs.: Index of significance (α) = 5%

Table 7. Results of analysis of variance for penetration.

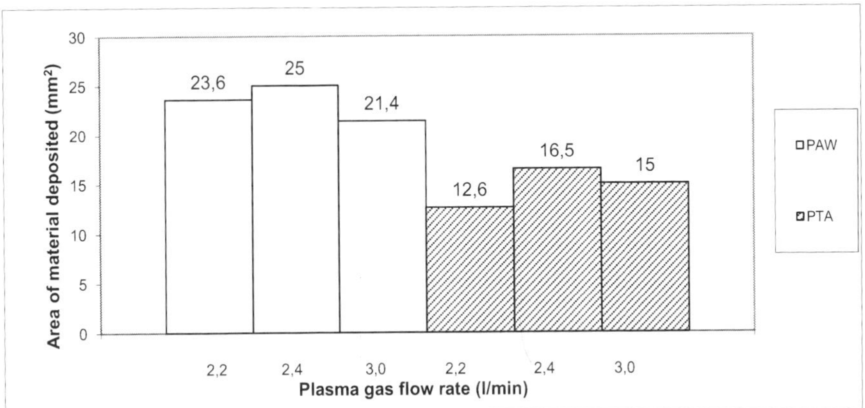

Fig. 11. Area of material deposited in PAW and PTA processes.

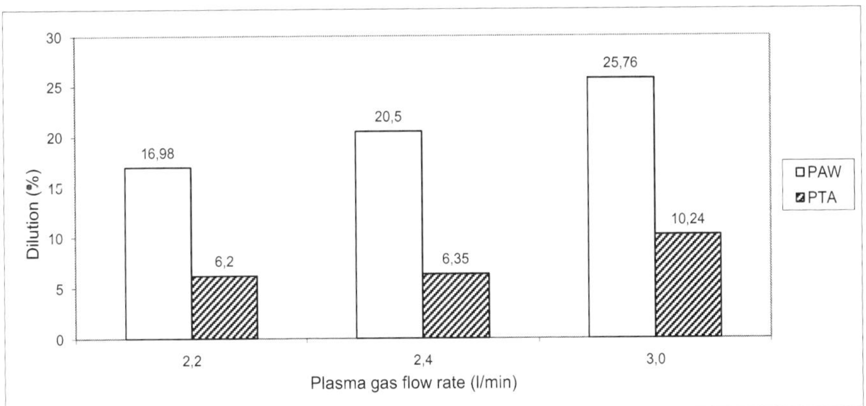

Fig. 12. Effect of plasma gas flow rate on degree of dilution in PAW and PTA processes.

Source of variation	Sum of squares	Degrees of freedom	Average of squares	F observed	F critical
Welding process	1102.43	1	1102.43	25289.88	
Plasma gas flow rate	182.16	2	91.08	2089.39	
Interaction	25.93	2	12.96	297.4 >	3.55
Residual	0.785	18	0.044		
Total	1311.305	23			

Obs.: Index of significance (α) = 5%

Table 8. Results of analysis of variance for dilution

3.2 Microhardness and microstructure

Figure 13 shows the typical microstructures of the solidification in the center of the weld bead. When a plasma gas flow rate of 2.2 l/min was used in the PAW and PTA processes the microstructure was more refined. For a plasma gas flow rate of 3.0 l/min for both welding processes the microstructure was less refined.

The microhardness profiles evaluated along the cross-section of the deposits are shown in Figures 14 and 15 for the PAW and PTA processes, respectively.

The data in Figure 14 together with the analysis of variance in Table 9, related to the PAW process, indicate that, in general, the plasma gas flow rate significantly affects the hardness. On the other hand, the data in Figure 15 together with the analysis of variance in Table 10, which relate to the PTA process, indicate that the plasma gas flow rate does not significantly affect the hardness. Deposits obtained with the PAW process have lower hardness values, which is to be expected given the less refined structures and higher degrees of dilution.

Source of variation	Sum of squares	Degrees of freedom	Average of squares	F observed	F critical
Plasma gas flow rate	18214.93	2	9107.463	151.9637 >	3.2381
Residual	2337.341	39	59.93183		
Total	20552.27	41			

Obs.: Index of significance (α) = 5%

Table 9. Results of analysis of variance for average hardness of microstructure - PAW.

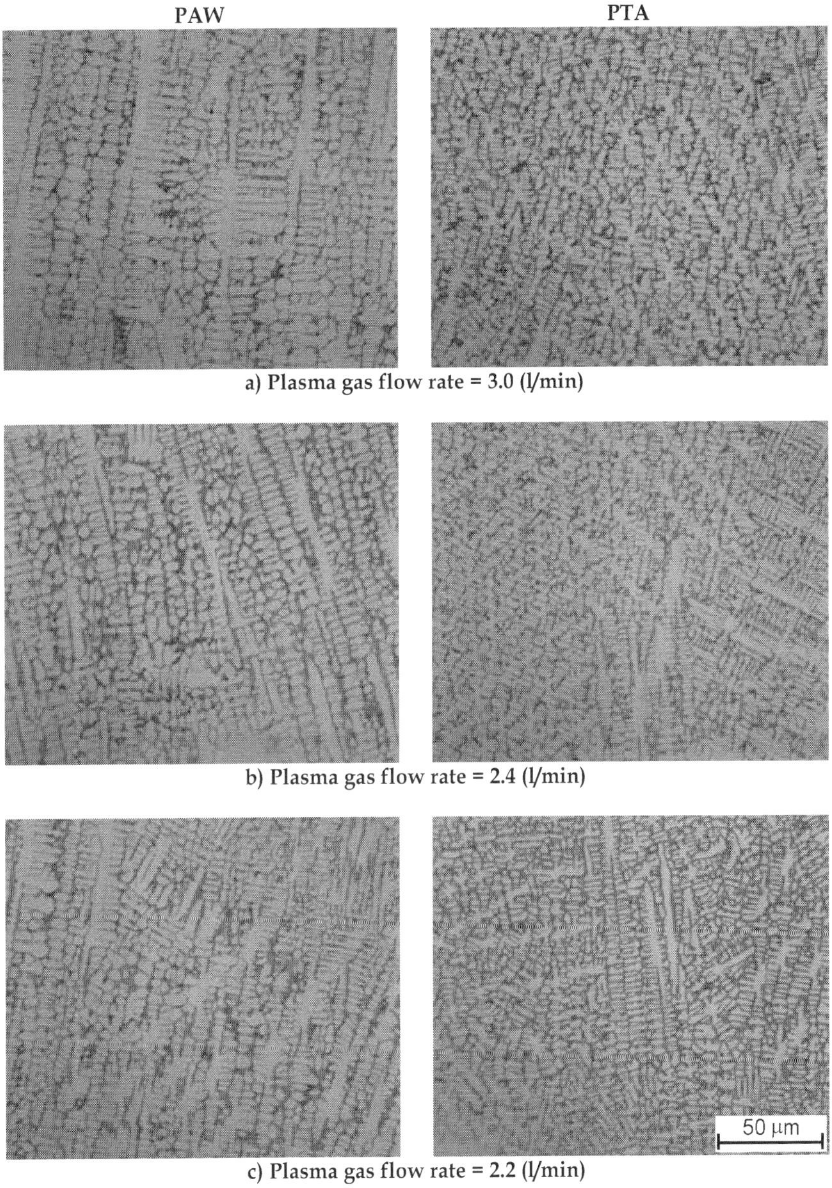

Fig. 13. Micrographs of the samples of Stellite 6 for the PAW and PTA processes. Centre of weld bead.

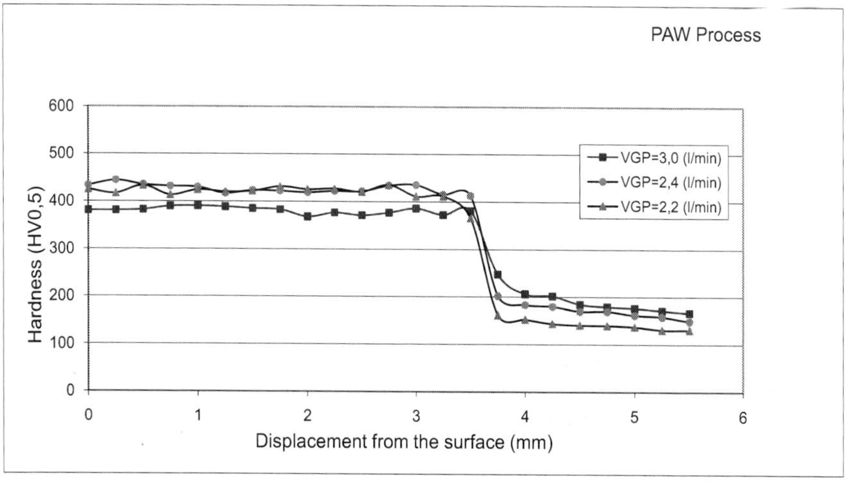

Fig. 14. Effect of plasma gas flow rate on hardness in PAW process.

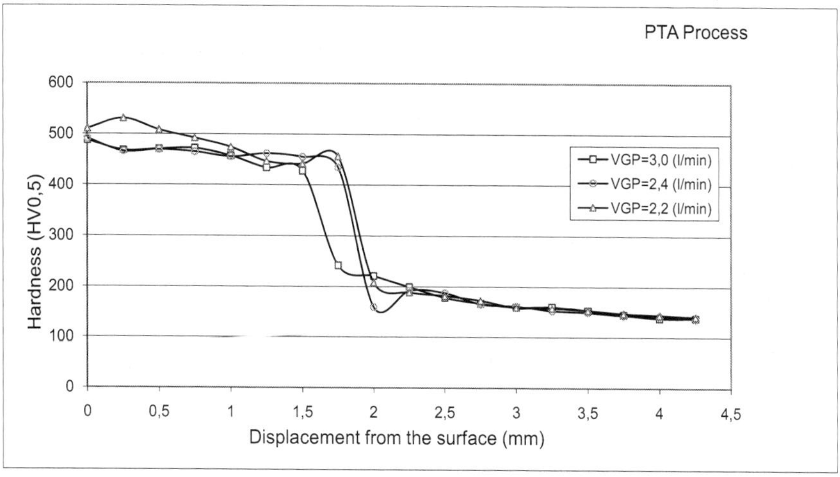

Fig. 15. Effect of plasma gas flow rate on hardness in PTA process.

Source of variation	*Sum of squares*	*Degrees of freedom*	*Average of squares*	*F observed*	*F critical*
Plasma gas flow rate	2729.185	2	1364.593	2.388627 <	3.554561
Residual	10283.17	18	571.2875		
Total	13012.36	20			

Obs.: Index of significance (α) = 5%

Table 10. Results of analysis of variance for average hardness of microstructure -- PTA.

It was verified that the PTA process generates a more refined microstructure and consequently greater hardness than the PAW process, as also observed by Silvério (2003).

4. Conclusions

Based on the experimental results obtained in this study the conclusions are as follows:

- The PTA process produced a better surface finish and better wetting. Due to the deposition efficiency and the difference in the orifice diameter of the constrictor nozzle used in the welding processes studied the main results are:
- In the PTA process lower dilution values were achieved in comparison with the PAW process.
- Greater weld bead width was obtained using the PTA process.
- On comparing the deposits obtained through the two processes it could be observed that the reinforcement and penetration are always lower in the PTA process.
- Deposits obtained with the PAW process had lower hardness values as expected due to the less refined structures and higher degrees of dilution.

5. References

Dai, W. S.; Chen, L. H. & Lui, T. S. (2001). *SiO_2 particle erosion of spheroidal graphite cast iron after surface remelting by the plasma transferred arc process.* Available at: <http://www.sciencedirect.com> Accessed in: Nov. 2008.

Gatto, A.; Bassoli, E. & Fornari, M. Plasma Transferred Arc deposition of powdered high performances alloys: process parameters optimisation as a function of alloy and geometrical configuration. Available at:<http://www.sciencedirect.com> Accessed in: Jun. 2009.

Zhang, L.; Sun, D. & Yu, H.(2008). *Effect of niobium on the microstructure and wear resistance of iron-based alloy coating produced by plasma cladding.* Available at: <http://www.elsevier.com/locate/msea> Accessed in: Nov. 2008.

LIU, Y. F.; Mu, J. S. & Yang, S. Z. (2007). *Microstructure and dry-sliding wear properties of TiC-reinforced composite coating prepared by plasma-transferred arc weld-surfacing process.* Available at:<http://www.elsevier.com/locate/msea> Accessed in: Nov. 2008.

Oliveira, M. A. (2001). Estudo do processo plasma com alimentação automática de arame: 78p. Dissertação (Mestrado em Engenharia Mecânica)-Programa de Pós-Graduação em Engenharia Mecânica, UFSC, Florianópolis.

Vergara, V. M. (2005). Inovação do equipamento e avaliação do processo plasma de arco transferido alimentado com pó (PTAP) para soldagem fora de posição: 2005. 174p. Doctoral Thesis, Mechanical Engineering Department - UFSC, Florianópolis.

Hallen, H.; Lugscheider, E.; Ait-Mekideche, A. *Plasma transferred arc surfacing with high deposition rates. In: Proceedings of conference on thermal spray coatings: properties, processes and applications*, Pittsburgh, USA, 4-10 May 1991. ASM International; 1992. p. 537-9.

SIlva, C. R.; Ferraresi, V. A & Scotti, A. (2000). *A quality and cost approach for welding process selection.* J. Braz. Soc. Mech. Sci., Campinas, v. 22, n. 3. Available from <http://www.scielo.br/scielo.php?script=sci_arttext&pid=S0100-73862000000300002&lng=en&nrm=iso>. Accessed on 29 Nov. 2009. doi: 10.1590/S0100-73862000000300002.

LIN, J. A. (1999). *Simple model of powder catchment in coaxial laser cladding.* Optics & Laser Technology, 233-238.

Davis, J. R. - Davis and Associates. (1993). *Hardfacing, Weld Cladding and Dissimilar Metal Joining.* In: ASM Handbook - Welding, Brazing and Soldering, Vol. 6. 10th ed. OH: ASM Metals Park. p. 699-828.

Silvério, R. B. & D'Oliveira , A.S. C. M. Revestimento de Liga a Base de Cobalto por PTA com Alimentação de Pó e Arame. In: Congresso Brasileiro de Engenharía de Fabricação, Uberlândia-MG, Maio. 2003.

2

Fusion Welding with Indirect Electric Arc

Rafael García[1], Víctor-Hugo López[1], Constantino Natividad[2],
Ricardo-Rafael Ambriz[3] and Melchor Salazar[4]
[1]*Instituto de Investigaciones Metalúrgicas-UMSNH,*
[2]*Facultad de Química-UNAM*
[3]*Instituto Politécnico Nacional CIITEC-IPN,*
[4]*Instituto Mexicano del Petróleo*
México

1. Introduction

The indirect electric arc technique (IEA) is a welding process that was initially developed to weld aluminum metal matrix composites (MMCs) reinforced with high content of TiC particles. Later on, its use was extended to weld MMCs reinforced with low contents of SiC and Al_2O_3 particles and monolithic materials such as carbon steels, aluminum and aluminum alloys. This technique is based on using the gas metal arc welding process (GMAW). In this instance, however, fusion of the base metal is not realized by the direct contact between the electric arc and the work pieces. Instead, the application of the electric arc is on thin plates of feeding metal placed on top of the work pieces and aligned with the groove of the joint. The filler wire, fed in a spray transfer mode, forms a weld pool with the plates of feeding metal and the molten metal is instantaneously fed, at high temperature, into the groove of the joint. The heat input supplied with the molten metal melts the side walls of the work pieces enabling welding upon solidification. The IEA technique allows using feeding material with the same chemical composition of the base metal. It has been found that the microstructure obtained in the weld metal with this technique, in carbon steels, improves the resistance to stress corrosion in hydrogen sulfide. The IEA technique has proved to be effective in welding MMCs with low and high content of ceramic particles, aluminum and its alloys as well as carbon steels such as API X-65 employed for transport and storage of hydrocarbons.

The design of the IEA joint enables welding of plates, 12.5 mm thick, in a single welding pass with a reduced heat input and thereby a reduction in the thermal affection of the base metal. Trials to weld materials such as aluminum and MMCs with a thickness of 12.5 mm in one welding pass without joint preparation, i.e. square edges, resulted in deficient welds with partial penetration. Successful welding of these plates demands 3 or 4 welding passes using a single V joint design. Conversely, the use of the IEA technique with preheating of the joint led to welds with full penetration and without lack of fusion in the side walls in a sole welding pass. The multipass welding procedure required for the single V groove joint means a larger heat input which inevitably has an impact on the microstructure of the different regions of the welded joint and of course on its mechanical performance. A thermal balance of the IEA process revealed a larger thermal efficiency as compared to the traditional use of the GMAW welding process. This increase is ascribed to the fact that the electric arc is not openly exposed to the atmosphere. Instead, it is established in a hidden fashion within the groove formed by the feeding plates, reducing thus heat losses. This characteristic of the IEA technique has outstanding repercussions on the microstructure and mechanical properties of the welds. For example, degradation of the reinforcement during fusion welding of aluminum matrix composites reinforced with SiC particles is a common occurrence. Welding of this type of MMC with IEA did not show signs of reaction and a larger fraction of SiC particles were incorporated into the weld metal increasing the tensile properties of the welded joint. Also, in carbon steels, aluminum and its alloys the use of the IEA leads to a different solidification mode and grain refined microstructures. In particular, for an API X-65 this refinement has a profound positive effect in terms of sulfide stress cracking (SSC) behavior. Regarding heat treatable aluminum alloys, it is well known the overaging effect in the heat affected zone (HAZ) which weakens the strength of the alloy and predisposes failure in this region with a very low stress. The reduced heat input of the IEA technique also reduces the loss of strength in the HAZ of this type of alloys.

The main disadvantage of the IEA technique is that it leaves the residual feeding plates on top of the welded joint. This would be unacceptable in most of the applications. Thus, an additional step in the process demands removing these strips from the weld. To overcome this inconvenient, a modification of the design of the IEA joint was proposed and tested in heat treatable aluminum alloys. The use of the strips of feeding material on top of the work pieces was omitted and in the upper part of the work pieces a lash was machined, simulating the original feeding plates. The modified indirect electric arc (MIEA) technique drew similar results than the original IEA and when compared with the conventional single V groove joint, the behavior of the MIEA welds was better in both static and dynamic testing.

So far, the IEA technique and its evolution into the MIEA has emerged as an attractive alternative to weld a number of materials with peculiar microstructural characteristics that have a positive impact in the mechanical and stress corrosion cracking (SCC) behaviors. This chapter details a broad description of the process, emphasizing its advantages with respect to the conventional practice of fusion welding. An overview of the findings and benefits observed in different materials as well as the evolution of the original idea throughout ten years of research are provided.

2. Overview of the IEA welding process

The indirect electric arc (IEA) technique is a novel welding process that has been successfully used to join MMCs (Garcia et al, 2002a, 2002b, 2003). It is a variation of the metal inert gas (MIG) process in which fusion of the base metal is not caused by direct contact with the electric arc. Instead, the electric arc is established between the filler metal and feed metal, in plate form, placed over the base metal (Fig. 1), where the feed metal plates, base metal and filler metal all have similar chemical compositions. The plates are prepared with square edges and with a small single-V preparation with an angle of 45° in the upper part. The IEA technique allows using feeding material with the same chemical composition. The resultant droplets, in the form of a spray, produce a molten pool on the plates and the liquid is fed instantaneously at high temperature into the groove formed between the workpieces (Lu & Kou, 1989a, 1989b). The high temperature of the liquid metal melts the parent materials, producing the welded joint. Due to the increase of the thermal efficiency of the IEA process, complete penetration and uniform weld beads, in a single pass, were obtained, as well as a reduction in the heat input and thereby a reduction in the thermal affection of the base metal compared with that provoked by the direct application of the electric arc (Garcia et al., 2002a, 2002b, 2003).

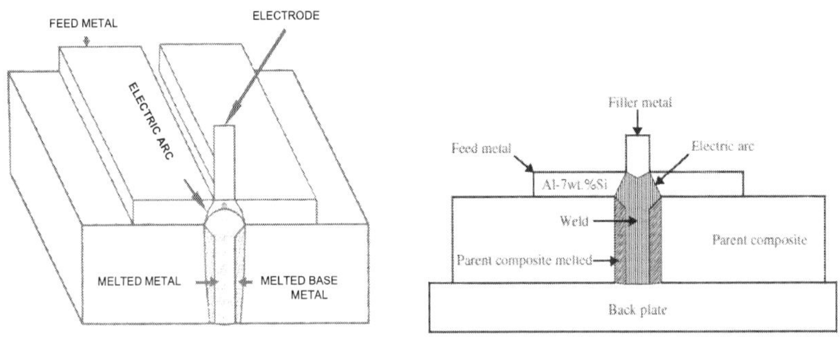

(a) (b)

Fig. 1. IEA welding process: (a) General set-up (Garcia et al., 2002); (b) Set-up for application to 359 aluminum MMCs reinforced with 20%SiC (Garcia et al., 2007).

The fusion process of the base material is carried out by means of a liquid diffusion process, similar that the shown in Fig. 2a, where the liquid material diffuses through the grain boundaries, because these are areas of lower melting point than the matrix of the grains. This phenomenon is favored due to segregation of impurities and alloying elements in metals of high purity and alloying elements in alloys, respectively. Fig. 2b-c illustrates the difference between the indirect electric arc and the traditional electric arc welding processes. In the majority of the electric arc welding processes, the electric arc is established between the electrode and the base metal. The high energy developed by the electric arc is in direct contact with base metal and the forces generated in the weld pool affect the weld form and solidification mode. As illustrated in Fig. 2b, the temperature gradient within the weld pool induces a corresponding density gradient that enhances the flow and generates radial forces (RF) and circular forces (CF). Fig. 2b shows that when materials of a higher temperature and lower density are moving toward the bottom of the puddle, the buoyancy forces tend to rise up the liquid flow across the center of the pool. The flow moves radially outward, the molten metal is forced along the surface and then down the side of the weld pool toward the bottom (Domey et al., 1995). In the IEA technique, the electric arc is established within the feeding plates and as soon as the filler metal and feed metal are melted, liquid metal with low density is supplied into the joint geometry at high temperature. Thus, in the IEA process the buoyancy-driven flow is interrupted due to the presence of a deep groove and as a result, the radial forces instead of becoming circular forces, are transformed into "drag forces", a combined effect of the pressure exerted by the electric arc and the gravity action, drives the molten pool downwards into the groove formed by the workpieces, as can be seen in Fig. 2c. In addition, the hidden arc in the IEA method suggests that the loss of energy by radiation is suppressed and as a consequence, the efficiency of the electric arc in melting the feed metal is increased.

Along with the stir forces generated in the weld pool, the extent of supercooling also accounts for the mode of solidification. Although there are some exceptions, in electric arc welding epitaxial growth is a typical occurrence, wherein the first grains of the weld pool nucleate directly from randomly oriented grains in the HAZ and grow toward the greatest thermal gradient within the puddle (Domey et al., 1995). On the other hand, the solidification in the IEA welding method is different due to the distinct generation of forces and flow patterns in the weld pool. Fusion of the base metal is realized by the high energy of the melt supplied into the joint geometry. The improvement of the efficiency of the MIG welding process to about 95% using IEA may be ascribed to the better use of heat generated by the electric arc, which is established in a hidden form and the contact with the environment is minimum reducing thus heat losses.

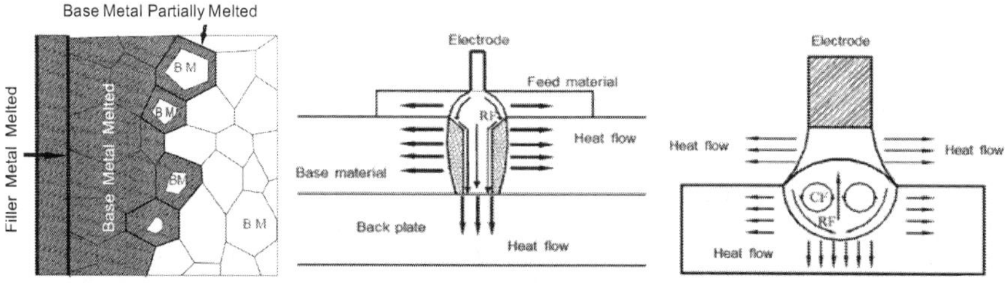

Fig. 2. (a) Fusion process by liquid diffusion; Schematic of the MIG welding process using (b) direct electric arc and (c) indirect electric arc (Garcia et. al, 2002).

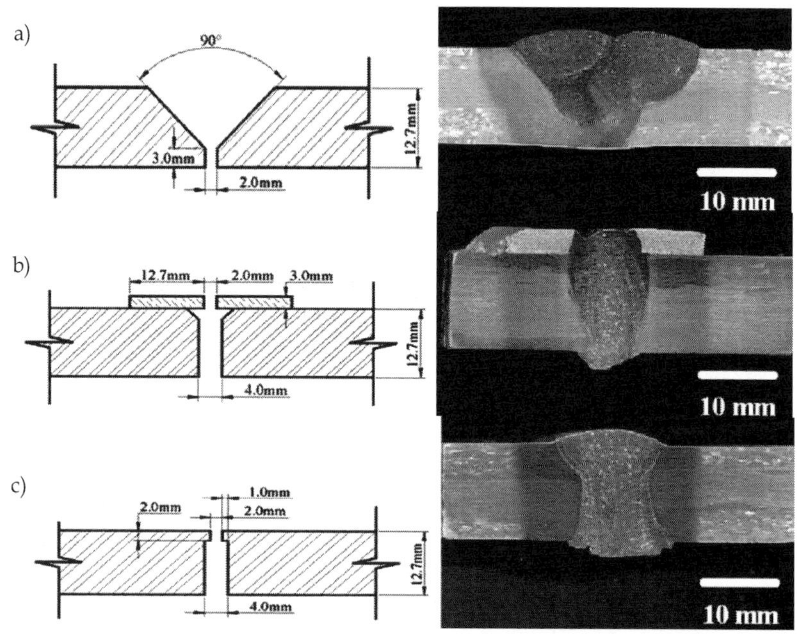

Fig. 3. Joint designs and typical geometry of the welds obtained; a) single V groove joint, b) IEA joint and c) MIEA joint.

The Fig. 3 shows different joint designs and dimensions as well as the typical geometries of the welds obtained by a) typical single V groove joint, b) IEA joint and c) MIEA joint. Whilst for the single V groove joint three or more welding passes are required to fill the groove, the IEA and MIEA joints only need one welding pass. The macrographs of the transverse weld profiles depict the geometries of the weld beads. Macroetching of the welds also revealed the HAZ. Roughly, it can be seen the larger thermal affection in the single V groove joint as compared to the IEA and MIEA welds.

3. Applications of the IEA welding process

In this section, an overview of the findings and benefits observed in different materials are provided. Between these materials are the MMCs, aluminium and its alloys and carbon steels.

3.1 Composites

The development of MMCs was a breakthrough in materials technology in the 80´s. Acceptation and use of new materials rely on their readiness to be joined. Sorting out the challenges of incorporating ceramic reinforcements into molten metals and alloys was not enough for spreading the use of MMCs. A major problem was also encountered when trying to join this type of materials with conventional fusion welding processes. Exposure of the Al/SiC-type composite to temperature above the liquidus of the aluminium alloys, as typically experienced in welding, results in a severe lose of mechanical properties due to the formation of brittle and hygroscopic aluminium-carbon compounds, mainly aluminium carbide. In addition, fusion welding processes produce a weld pool that has poor fluidity and solidifies with large volumes of porosity in the weld, because of the realization of hydrogen from the melted aluminium powder, which is used to make many MMCs (Ahearn et al. 1982). A reduction in the porosity in welds deposited by gas tungsten arc welding (GTAW) was achieved by previous vacuum degassing for long periods of time before welding. Nevertheless, both Al_4C_3 and Al-Si eutectic were detected in degassed composites. A number of authors (Cola & Lundin, 1989; Devletian 1987; Fukumoto & Linert, 1993; J Ahearn, et al, 1982, Ellis, et al, 1995; Urena et al, 2000, Lundin et al, 1989, Lienert, et al, 1993) reported that Al_4C_3 is always formed in the weld metal in MMCs reinforced with SiC no matter which of the fusion welding processes is employed to weld the MMCs (laser, electron beam, TIG, MIG and so on), and the formation of this compound occurs according to Eq. (1).

$$3SiC_{(s)} + 4Al_{(l)} \rightarrow Al_4C_3 + 3Si_{(s)} \qquad (1)$$

This reaction is not reversible and the Al_4C_3 is formed as plates in the microstructure. The presence of the plates has two deleterious effects. First, the material becomes extremely brittle, and second, it becomes very prone to corrosion in presence of water, leading to the release of acetylene gas. In an extreme case, this has led to total corrosion of the weld within a few days. In response to the problematic issue of welding MMCs, the idea of the indirect electric arc was conceived (Garcia et al 2002) with the metal inert gas (MIG) welding process in order to overcome the difficulties of welding MMCs. The concept is based on the fact that experimental measurements indicate that the temperature of the droplets in the MIG welding process with spray transfer is between 2000 to 2327 °C for aluminium and its alloys according to (Lu & Kou, 1989, Kim et al, 1991). If a molten metal with a large overheating is casted into a "mould" shaped by the parent materials to be welded, the sensible heat of the weld pool formed is sufficient to melt the side walls (the matrix in MMCs) whilst instantly filling the groove, yielding the welded joint upon freezing. The indirect application of the electric arc reduces the degradation of the ceramic but still the temperature of the molten metal is large to induce spontaneous and instantaneous wetting so that continuity is seen between weld metal and the matrix of the composites when the content of reinforcement is large (Garcia et al, 2002, 2003) and significant incorporation of particles into the weld metal occurs for composites with low fraction of reinforcement (Garcia et al, 2002, 2007).

The MIG welding process with IEA is a novel fusion welding method, which was developed to join MMCs with a reduced HAZ in the base metal. In the IEA welding method, the fusion of the base metal is not caused by direct contact with the electric arc, instead, the electric arc is established between the solid electrode and a plate of the same base metal, placed over the parent material. The resultant droplets, in the form of a spray, produce a molten pool on the plates and the liquid is fed instantaneously at high temperature into the groove formed by the base metal; Fig. 4 depicts the experimental set-up in a dissimilar join. Experimental measurement indicate that the droplets temperature in the MIG welding process with spray transfer is between 2000 to 2327 °C for aluminium and its alloys according to (Lu & Kou, 1989, Kim et al, 1991). As a result of its elevated temperature, the liquid metal melts the parent materials (the matrix in MMCs), yielding the welded joint upon freezing.

Profiles of MMCs welds using IEA are shown in Fig.4. Irrespective of the reinforcement content (high, medium or null), full penetration was attained in one welding pass and uniform weld profiles are obtained with little fusion of the base materials and a minimized heat input. The contour of the weld at the top depicts the configuration of the electric arc, which does not impinge on the surface of the parent plates; rather, it strikes inside the channel formed by them, as illustrated previously in Fig.1. Thus, during welding the electric arc is hidden and the typical flashing and sputtering of the normal MIG welding process is no longer observed when welding is carried out in any kind of material. It is well known that in order to weld 9, 10 mm thick MMCs and 12.5 mm thick aluminium plates, conventional MIG welding practice demands more than one welding pass with low travel speed, which leads to a large HAZ. It is worthy bearing in mind that most of the attempts to weld MMCs have been performed rather in thin sections or depositing bead on plate welds.

It has been reported during MMCs welding by different welding processes, that the high energy developed by the electric arc produces a wide HAZ accompanied by dissociation of the ceramic particles and the formation of hygroscopic compounds (Al_4C_3). This was confirmed by (Garcia et al, 2007) when welding an A359/SiC/20p commercial composite. Fig. 5a shows jagged SiC particles within the weld metal, this feature was not seen when the composite was welded with IEA. Fig 5b shows also the particles incorporated into the weld metal but they retain their initial angularity meaning that significant degradation (according to the resolution of the optical microscope) did not occur during welding with IEA. Tensile testing of the welded joints drew a tensile strength of 234 MPa for the IEA weld (one welding pass) as compared to 209 MPa for the weld with direct application of the electric arc (three welding passes). This behaviour is related to a larger incorporation of SiC particles into the weld metal and the reduced porosity for the IEA weld. The authors stated that the degradation of the SiC particles observed in the plain weld played a minor role during mechanical testing.

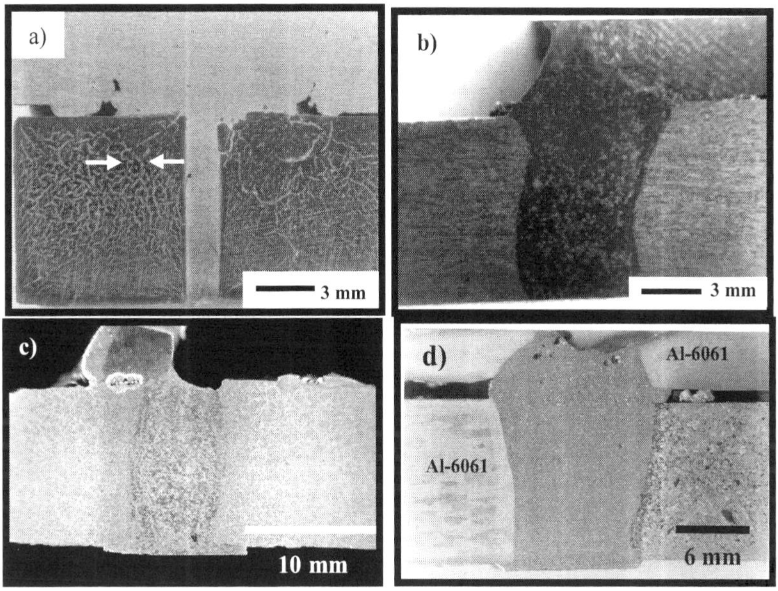

Fig. 4. Weld profiles obtained with the MIG-IEA technique in different MMCs.
a) Al-1010/TiC/50p (Garcia et al, 2003), b) Al-6061/Al_2O_3/20p (Garcia et al, 2002), c) A359/SiC/20p and d) dissimilar joint.

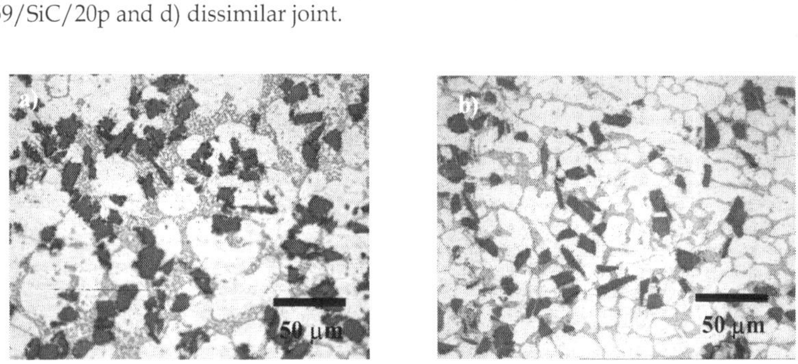

Fig. 5. Effect of the a) direct and b) indirect application of the electric arc on the SiC particles during welding an A359/SiC/20p commercial composite (Garcia et al, 2007).

3.2 Carbon steels

Pipelines of low carbon steel welded by electric arc have been used for many years and are widely used in the petroleum industry. However, frequent failures during operation over the years have prompted several studies of the design, construction, operation and maintenance of equipment and metallic structures used in this industry (Craig, 1997). Oil and gas from Mexico contain high concentrations of H_2S and CO_2, these constituents induce failures. In the oil industry, the welding processes commonly used for joining pipeline are electrical resistance (ERW) and submerged arc welding (SAW). With these processes, wide

HAZs are created and as a result, a high probability of cracking due to the heterogeneous weldment exists. Studies of weld bead failures have demonstrated that these occur mainly in the HAZ because of the variation in microstructure (grain growth), residual stresses and a higher susceptibility to embrittlement by hydrogen. These microstructures are produced by the thermal gradients experienced in the joint during welding. Therefore, it is very important to develop welding processes with a narrow HAZs, this is possible with the IEA process. The welding process has an important influence in the SSC susceptibility of materials. The different welding processes promote different changes in the microstructure of the welded zone; these changes affect the SCC resistance and the yield strength of the pipeline steel.

A comparative study of SSC resistance between IEA, SAW and MIG was carried out (Natividad et. al., 2007) through slow strain rate tests (SSRT), electrochemical tests and hydrogen permeation measurements. The base metal used was API grade X-65 pipeline steel. Cylindrical tensile specimens with a gauge length of 25 mm and gauge diameter of 2.50 mm were machined from the pipeline perpendicular to the rolling direction. The specimens were subjected to conventional slow strain rate tests in air (as an inert environment) and in the standard NACE solution (5% NaCl, 0.5% acetic acid, saturated with hydrogen sulphide (H_2S)) at a strain rate of $1x10^{-6}$ s^{-1} at room temperature (25°C) and at 37°C and 50°C. All of the tests were performed at the open circuit potential (OCP). The loss in ductility was assessed in terms of the percentage reduction in cross-sectional area (% RA), as follows

$$\%RA = \frac{A_1 - A_f}{A_f} x100 \tag{2}$$

where A_i and A_f are the initial and final cross-sectional areas, respectively. The index of susceptibility to SSC (I_{SSC}) was calculated as follows

$$I_{SSC} = \frac{\%RA_{AIR} - \%RA_{NACE}}{\%RA_{AIR}} \tag{3}$$

where % RA_{AIR} and % RA_{NACE} are the percentage reduction in area values in air and in the H_2S-saturated NACE solution, respectively. An I_{SSC} value close to unity indicates high susceptibility towards SSC whereas a value close to zero indicates immunity. The fracture surfaces were then examined using scanning electron microscopy (SEM).

Fig. 6 shows the macro and microstructures of the weldments obtained by the three welding processes. This figure shows clearly the different zones: base metal (BM), weld bead (WB), HAZ and fusion zone (FZ). Full penetration and a narrow HAZ are observed in microstructure obtained by IEA process. The weld bead and HAZ is very different from that obtained by both SAW and MIG processes. In general, the microstructure obtained with the IEA welding process is more homogeneous than the obtained by SAW and MIG processes. The corrosion and the SSC susceptibility of the welds are affected by the differences in composition, microstructure, and electrochemical potentials among the different zones. A lower electrochemical potential of the weld bead is commonly related to composition, microstructure and distribution of inclusions (Dawson et. al., 1997). Similarly, in a study performed by Turnbull and Nimmo, about SCC susceptibility (Turnbull & Nimmo, 2005), a direct relation to OCPs with mechanical properties like microstructure or hardness of phases was reported.

The limit of hardness recommended for avoiding cracking in a sour environment is 22 Rc (248 Hv) (NACE/ISO, 2009). Although susceptibility to SSC generally increases with increasing hardness, some microstructures are more susceptible to cracking than others at the same hardness. Fig. 7 shows hardness measurements obtained from a transverse section of the weld for each welding process and it is observed that the values of hardness in the weld bead made by the SAW process are the lowest. For the MIG process, there is no significant difference between the HAZ and the WB hardness values. On the other hand, for the IEA process, the hardness value of the HAZ decreases by nearly 35 HV with respect to weld bead values. These values are within the recommended limits to avoid the fracture and cracking of the weld bead.

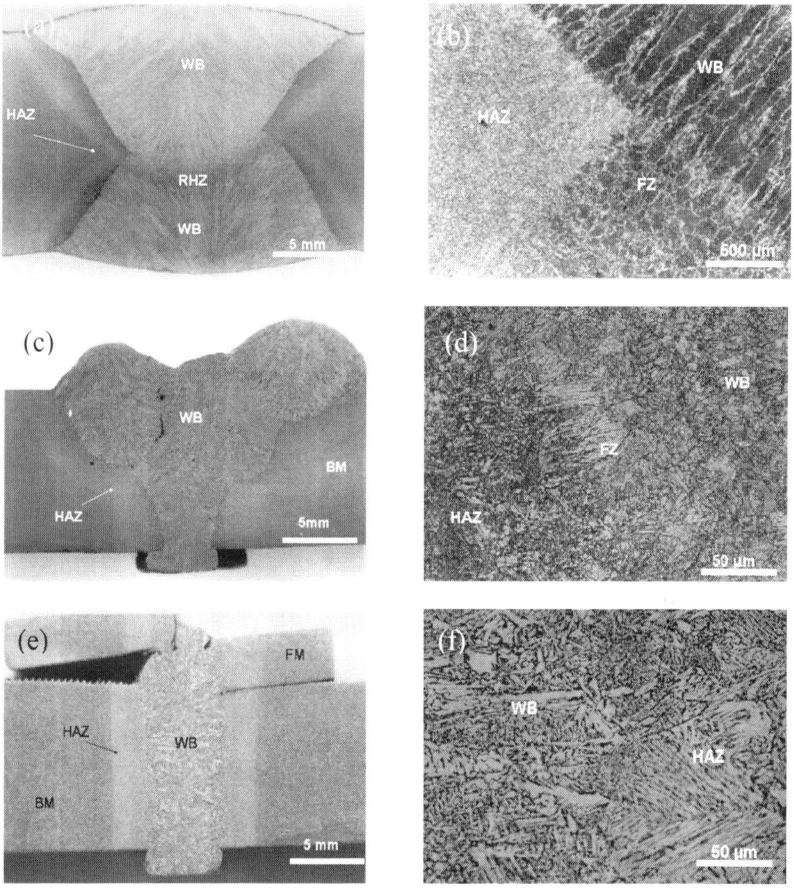

Fig. 6. Macro and microstructures of the weldments obtained by (a-b) SAW, (c-d) MIG and (e-f) IEA, (Natividad et al., 2007).

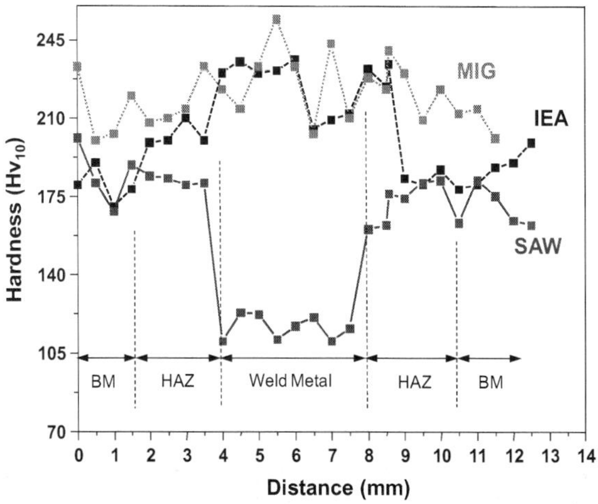

Fig. 7. Hardness values in weldments obtained by the different welding processes. (Natividad et al., 2007).

Fig. 8 shows the I_{SCC} measurements on the three weldments at the three bath temperatures. The welded joint obtained by SAW shows the highest I_{SCC} values, which was correlated with the hydrogen permeation results. Similarly, the I_{SCC} measurements in the IEA case indicate a lower concentration of both the trap sites in the metal and the susceptibility to hydrogen embrittlement, because here the most hydrogen flux passed to the anodic cell side. Although the high electrochemical activity generates a higher atomic hydrogen concentration, a smaller concentration of this atomic hydrogen was trapped in the bulk. In addition, the I_{SCC} results were affected by the change of welded microstructure, where the IEA presented a refined higher concentration of bainite compared with the grain coarse ferrite+pearlite microstructure obtained by the SAW process as reported before (Turnbull & Nimmo, 2005), and the consequent change in hardness due to the modified grain size during welding process (Omweg et al, 2003). Some evidence of the hydrogen diffusion effect into the welded joints is presented in the SCC fractographs illustrated in Fig. 9. The SAW weldments (Fig. 9b) show a more brittle fracture than the MIG and IEA weldments. The IEA weldment presented a less brittle fracture (Fig. 9a) and the I_{SCC} values show this behaviour. However, the three welded joints do not show a completely brittle behaviour, but in general, the fracture behaviour was closer to a quasi-cleavage fracture. The hydrogen diffusion was low in quantity and low in permanence time into the electrolyte generating hydrogen embrittlement, but there is evidence of the hydrogen damage to the weldments. Additionally in this work, the IEA material was the least susceptible to hydrogen embrittlement damage, of course, the SCC resistance was higher, and was related to the lowest OCP activity promoted by the change in microstructure of the weldment.

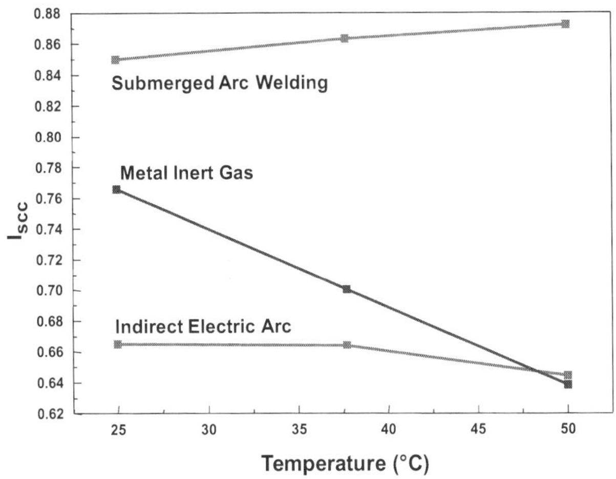

Fig. 8. Variation of I_{scc} as function of temperature on the weldments (Natividad et al., 2007).

With the features of the IEA process, a better SSC resistance at 25 °C was obtained in comparison to the SAW and MIG processes. For specimens obtained by the IEA process subjected to SSR tests, the failure occurred on the base metal and for specimens produced by SAW and MIG processes it occurred in the weld bead and HAZ respectively. At 37 °C and 50 °C, the SCC resistance of all processes show a similar behaviour (failed in the base metal). The higher atomic hydrogen permeation flux presented by the IEA process was promoted by the ferrite phase from the base metal, which is comparable to the fracture which occurred from SSRT.

A significant increase in the corrosion current (more positive OCP) was presented by the IEA material, and was attributed to the greatest galvanic cell formation between the welded and base metal, but this material shows superior resistance to the SCC susceptibility than that for the MIG and SAW processes. The hydrogen permeation results and the difference in microstructure presented by the IEA process corroborate this behaviour.

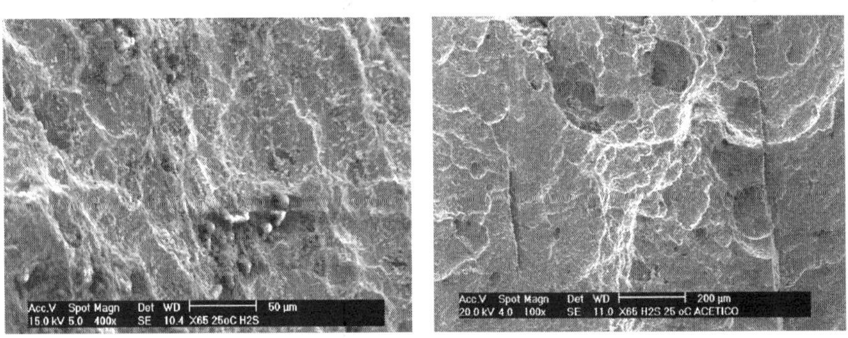

Fig. 9. SEM images of the SCC fractographies of (a) IEA weldment material and (b) SAW weldment material tested at 25 °C into the NACE solution saturated with H_2S, (Natividad et al., 2007).

3.3 Aluminium and Al-alloys

Aluminium and its alloys are important engineering materials with a vast number of applications as structural and functional components. The shiny and attractive appearance of these materials is due to the native oxide layer that always envelops the bulk of the material. The aluminium oxide layer is highly compact and this characteristic prevents further thickening of the oxide and permeation of aggressive media, making of Al and its alloys corrosion resistant materials in a number of environments. Besides, their lightness means a large strength/weight ratio for heat treatable Al-alloys (Davies, 1993 & Heinz et al, 1990). The high technological relevance of Al and its alloys demands quite often joining operations. The first barrier to overcome for successful joining is the native oxide layer which avoids coalescence between faying surfaces. In order to succeed fusion welding with the electric arc, direct current and reverse polarity are used to generate an ionic striking effect that dissolves the oxide and enables mixing between molten metals. Another aspect to take into consideration is the high solubility of hydrogen in aluminium melts. If care is not taken in preventing sources of hydrogen during welding, this effect may lead to welds with a large level of porosity in the weld metal. Besides, Al-alloys are prone to solidification and liquation cracking owing to their relatively high thermal expansion, large change in volume upon freezing as well as wide solidification temperature range (Gittos et al, 1981; Enjo & Kuroda, 1982; Kerr & Katoh, 1987; Miyazaki et al, 1990; Malin, 1995; Huang & Kou, 2004). All the above mentioned defects may have a profound impact on the mechanical performance of the welded joint and therefore they must be prevented. Furthermore, exposure of heat treatable Al-alloys to the welding thermal cycles gives rise to a soft zone in the HAZ caused by overaging where the mechanical strength may decrease up to 50% (Malin, 1995). Partial or almost full recovery of the mechanical properties can be achieved by post weld heat treating (solutioning, quenching and aging) but this procedure might be restricted in many instances and the welded joint must be on service in the as-welded condition. On this context, contributions aimed to improve the mechanical performance of aluminium alloys are desirable and valuable.

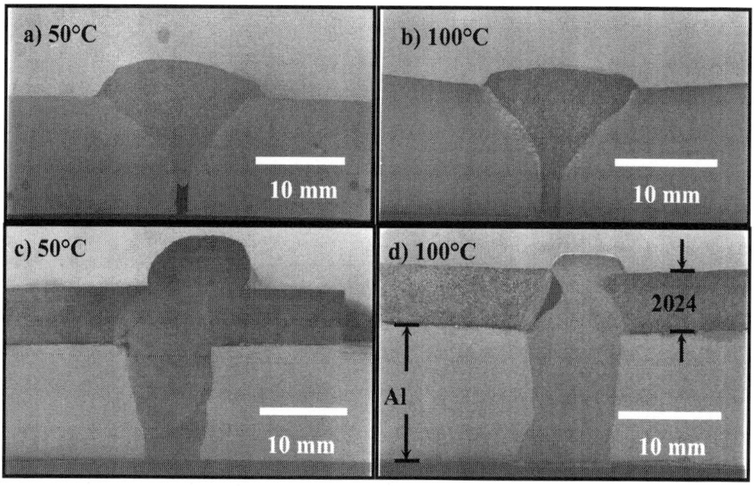

Fig. 10. Comparison of the weld profiles obtained with the direct (a-b) and indirect (c-d) application of the electric arc with preheating of the joint (Garcia et al, 2007).

The IEA technique developed by (Garcia et al, 2002) was applied to study the feasibility of welding Al and its alloys by using an ER4043 filler wire. Preliminary welding trials (Garcia et al, 2007), in plates of aluminium of commercial purity, revealed that the IEA technique has the capability of joining plates of 12.5 mm in thickness in a single welding pass with full penetration by preheating the plates to 50 or 100°C as shown in Fig. 10. Conversely, with the same preparation of the parent plates but without the feeding plates, welds with partial penetration are obtained as seen in Fig. 10. Needless to say, this problem shows why traditionally welding of thick sections is performed by a multi-pass procedure either with a single or double V preparation of the plates.
Microstructural characterization of the welds disclosed for the welds with direct application of the electric arc the typical epitaxial and columnar grain growth from the partially melted grains of the base metal whereas for the IEA welds, partially melted grains of the base metal were found trapped within the weld metal as shown in Fig. 11. The mechanism of this phenomenon was previously mentioned in section 2. Dragging and survival of partially melted grains modified the mode of solidification in the weld metal by blocking columnar growth. Thermal analysis of the IEA technique showed that heat losses by radiation and convection are reduced since the electric arc is established in a hidden fashion within the channel formed by the feeding plates. The authors stated that the low ionization potential of the aluminium vapour, 5.986 eV, plays also an important role in the IEA technique.

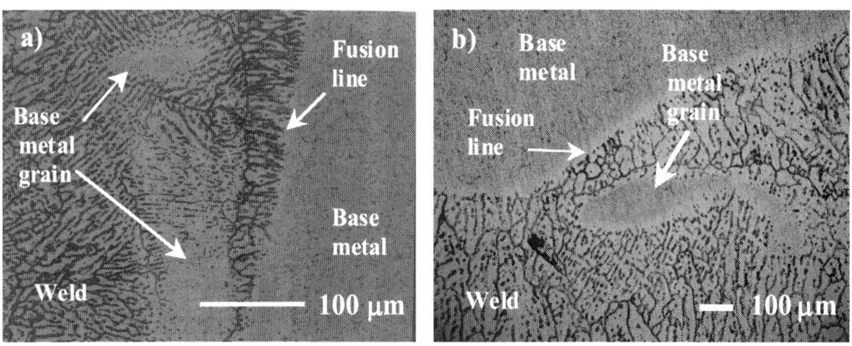

Fig. 11. Optical micrographs of the base metal/weld interface: (a) in the transversal and (b) longitudinal axes of a MIG-IEA weld (Garcia et al, 2007).

The preliminary findings suggested that if the number of welding passes are reduced, so it is the heat input. This advantage means also a reduction in the thermal affection of the base metal and therefore a minimization in the loss of strength in heat treatable aluminium alloys. To prove this, plates of an Al-6061-T6 were welded with the IEA technique and the results were compared with a single V joint (Ambriz et al, 2009). Table 1 shows the results of the mechanical properties under tension of the welded joints in the as-welded condition. At first sight, it is clear that the mechanical properties are lower than the base metal independently of the joint design employed. The worst mechanical properties were, however, exhibited by the single V-groove joint, which has approximately 37.4 and 61% mechanical efficiency with respect to base material and filler wire respectively. In comparison to the single V groove joint, the IEA joints exhibited a notable increase in the mechanical efficiency with values ranging between 48 to 55%. Nevertheless, those efficiencies are still below the permissible limit of 57% for the 6061-T6 alloy according to the ASME Pressure Vessel Code, Section VIII. Irrespective of the joint design and preheating condition, the mechanical failure under tensile testing occurred in the base metal, at different distances from the fusion line, depending on joint design and preheating condition. The nearest failure from the fusion line occurred for the single V joint.

Joint type	Preheating temperature (°C)	Ultimate strength (MPa)	Yield strength (MPa)	Elongation (%)	Failure from fusion line (mm)	Efficiency from base metal (%)	Efficiency from filler metal (%)
ER-4043	--	190	164	--	--	--	--
6061-T6	--	328	300	14	--	--	--
DEA	25	116	57	17.6	3	35.36	61.05
IEA	50	148.9	71.6	14	11-13	45.39	78.36
IEA	100	172.1	76.8	14.8	13-15	52.46	90.57
IEA	150	168	86.7	15.2	13-16	51.21	88.42

Table 1. Mechanical properties of the weldments in the as-welded condition (Ambriz et al, 2009).

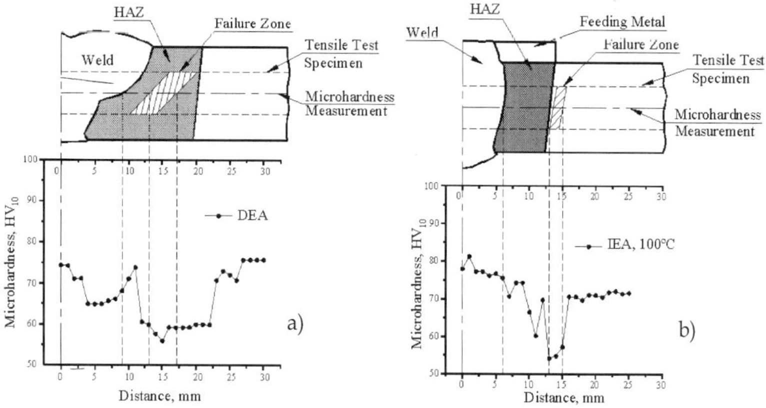

Fig. 12. Microhardness profiles for; a) the single V joint and b) the IEA joint preheated to 100°C (Ambriz et al, 2009).

Fig. 12 shows microhardness profiles along the cross sections of the welds, they corroborate the failure zone with the variation of mechanical properties, as a result of the thermal cycle of welding. The plots of these figures reveal, in each weld, a soft zone which matches well with the failure zone of the tensile specimens. This behavior, which nevertheless was expected, agrees with others (Gutierrez et al, 1996; Liu et al, 1991; Myhr et al, 2004; Shelwatkar, 2002; Kostrivas & Lippold, 2000). Worth to point out is the fact that the depth, in terms of microhardness, and width of the soft zone is related to the joint design and preheating temperature. These features are obviously reflected on the tensile strength of the welds. Note that although the IEA and single V groove are similarly "soft", the later fails with a lower stress due to the relaxation of the dislocations as a result of repeated exposure to heat. Another important aspect revealed by the microhardness profiles is the increase in hardness within the weld metal for the IEA joint as compared to the single V groove joint. This trend is related to the dilution rates for each joint, the larger the dilution the largest incorporation of alloying elements into the weld metal, and to the solidification mode inherent to every joint design. Whilst the single V groove joints exhibits the typical epitaxial and competitive-columnar grain growth, the IEA joints solidified with a grain refined microstructure.

4. Modified indirect electric arc (MIEA)

The main disadvantage of the IEA technique is that it leaves the residual feeding plates on top of the welded joint. This would be unacceptable in most of the applications. Thus, an additional step in the process demands removing these strips from the weld. To overcome this inconvenient, a modification of the design of the IEA joint was proposed and tested in heat treatable aluminum alloys. The use of the strips of feeding material on top of the work pieces was omitted and in the upper part of the work pieces a lash was machined, simulating the original feeding plates. This evolution was named MIEA (Ambriz at al., 2006, Ambriz et al. 2008). The MIEA provided several advantages with respect to the traditional arc fusion welding process, for instance:

i. The high thermal efficiency that allows welding plates by using a single welding pass. As a result, the thermal effect is reduced and the mechanical properties of the HAZ are improved as compared to a multi-pass welding procedure,
ii. The dilution percent of the weld pool is higher; which tends to improve the hardening effect after performing a post weld heat treatment (PWHT) (Ambriz et al., 2008),
iii. The solidification mode promotes an heterogeneous nucleation and jointly diminishes the micro-porosity formation,
iv. The geometry of the welding profile improves the fatigue performance of the welded joint (Ambriz et al., 2010a).

MIEA welding technique employs the same equipment that is required to weld by GMAW. General dimensions and the schematic representation of the MIEA joint along with the weld bead geometry obtained were shown in Fig. 3c. The dissipation of heat in this case is quite similar to the IEA joint so that large thermal efficiency is also obtained.

4.1 Microstructure in aluminum alloys welds

In a fusion welding process, the heat input produces a fusion-solidification phenomenon, which is different to that obtained in the solidification of an ingot. (i) In an ingot, solidification begins with heterogeneous nucleation at the chill zone meanwhile in a weld pool the liquid metal partially wets the grains of the parent metal and epitaxial growth takes place from the partially melted grains of the parent metal (Davies et al., 1975). (ii) The rate of solidification in a weld pool, which depends on the traveling speed as well as the welding process, is by far faster than in an ingot. (iii) The macroscopic profile of the solid/liquid interface in welds progressively changes as a function of the traveling speed of the heat source whereas it exclusively depends on the time for an ingot. (iv) The movement of the liquid metal in a weld pool is greater than in an ingot due to the Lorentz forces which create turbulence within the molten metal (Grong, 1997). Fig. 13 shows longitudinal views, which depict the direction of solidification of the welds, for a multi-pass welding and MIEA with preheating conditions at 50°C. The arrows indicate the displacing direction of the electric arc.

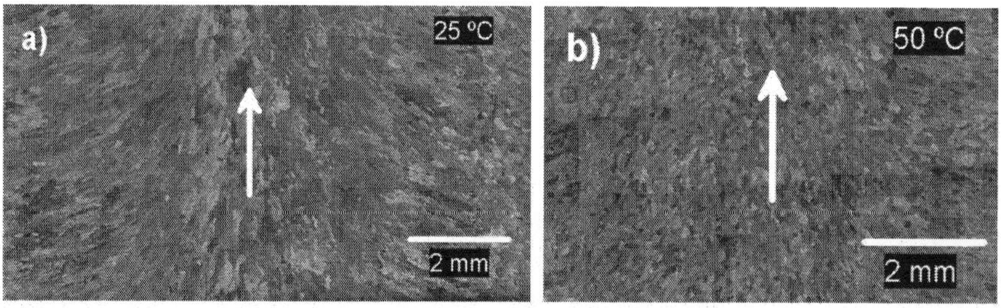

Fig. 13. Longitudinal top view of the weld metal grain structure at the mid plane for: a) single V groove, b) MIEA preheated at (50°C) (Ambriz, et al, 2010).

The longitudinal macrostructure for the MIEA joint, Fig. 13b, exhibit significant differences with respect to the multi-passes single V groove joint. Irrespective of the preheating condition, the local crystalline growth maintains an angle nearly constant in relation to the moving heat source. The virtual non existence of changes in growth direction means that the local and nominal rates of crystalline growth tend to be equal. This phenomenon yields a significantly different grain structure in the weld metal for the MIEA joint as compared to the structure observed for the single V groove joint. It leads, in fact, to a grain refining effect which is obviously affected by the initial preheating temperature of the joint.

The MIEA joint exhibits signs of heterogeneous nucleation which promotes grain refining. The levels of porosity in the MIEA joints decrease with preheating temperature and are comparatively lower than that obtained in the single V groove joint. Epitaxial solidification is also observed at the fusion line of the MIEA welds, however, competitive columnar growth was restricted instead grain structures alike those observed in the centre of the weld metal were present. The characteristics of solidification observed for the MIEA welds are the result of heterogeneous nucleation which is based on the principle of the formation of a critical radii needed to achieve the energy of formation from potential sites for nucleation such as inclusions, substrates or inoculants (Ti or Zr) (Rao et al., 2008; Ram et al., 2000; Lin et al., 2003). For the MIEA welds, these sites are principally the sidewalls of the joint in conjunction with the content of Ti in the filler and base metal since the significant dilution of base metal favors incorporation of Ti into the weld pool.

4.2 Mechanical properties in aluminum alloys welds

4.2.1 Microhardness

In order to determine the effect of the welding process in aluminum alloys, a common practice is to perform a microhardness profile in a perpendicular direction to the weld bead, as is showed in Fig. 14. Microhardness measurements give a general idea of the microstructural transformations and the variation of the local mechanical properties (Ambriz et al. 2011) produced after a welding process in aluminum alloys.

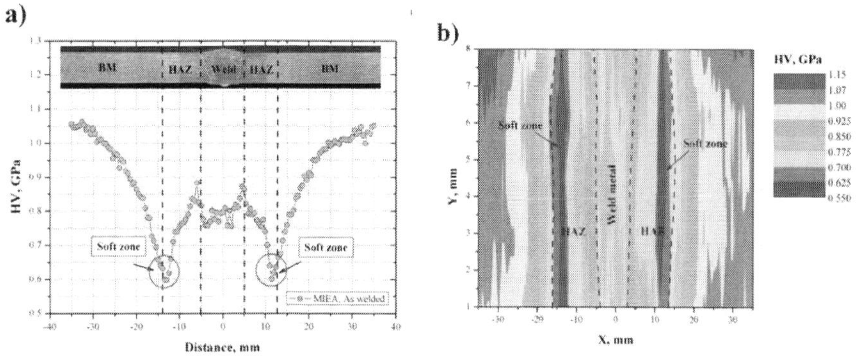

Fig. 14. a) Microhardness profile and b) microhardness map in 6061-T6 aluminum alloy welded by MIEA. Note that 1HV=9.8×10^{-3} GPa (Ambriz, et al, 2011).

Fig. 14a presents the Vickers hardness number profile in 6061-T6 aluminum alloy welds obtained by MIEA. A significant difference between the hardness number of the weld material and HAZ with respect to the base material is observed. Also, at the limit between the HAZ and the base metal, it is noted the presence of a soft zone which is formed nearly symmetrically in both sides of the welded joints. It should be note that the hardness obtained in this zone represents roughly 57 % of the hardness number of the base material. This seems to indicate that the tensile mechanical properties after welding process will be greatly different. Fig.14b visualizes the location of the soft zone highlighted by the Vickers hardness profile represented in Fig. 14a, by means of a hardness mapping. In this figure, the hardness values for each zone of the welded joint are well-defined. It is clear that in the soft zone (HAZ) the hardness number range is between 0.55 to 0.7 GPa. This soft zone results from the thermodynamic instability of the β″ needle-shaped precipitates (hard and fine precipitates) promoted by the high temperatures reached during a fusion welding process (Myhr et al., 2004). Indeed the temperatures reached during the welding process are favorable to transform the β′ phase, rod-shaped, according to the transformation diagram for the 6061 alloy.

4.2.2 Tensile properties

The individual mechanical behaviour of the base metal, weld metal, HAZ and welded samples in as welded condition for 6061-T6 aluminum welds by MIEA is shown in Fig. 15 as stress as function of strain graph. From Fig. 15, it can be observed that the experimental results for the base metal are in agreement with nominal values found in the literature for 6061-T6 alloy (American Society for Metals Fatigue and Fracture, 1996). The tensile properties of the sample obtained from the HAZ presents a 41% and a 19 % reduction of the ultimate strength with respect to the base metal and weld metal respectively. The loss of mechanical strength commonly referred to as over-aging, when welding a 6061-T6 alloy is a fairly well understood phenomenon and it is explained in terms of the precipitation sequence (Dutta & Allen, 1991). During welding, however, the base metal adjacent to the fusion line is subjected to a gradient of temperature imposed by the welding thermal cycle. At certain distance from the fusion line, the cooling curve crosses the interval of temperatures between 383 to 250 °C in which the β' phase, rod-shaped, is stable. It is thus the transformation of β'' into β' the responsible of the decrease in hardening of the α matrix due to the incoherence of the β' phase caused by the thermodynamic instability of β'' in a welding process. It is worthy to mention that the MIEA further increased the mechanical strength of this alloy as compared to the IEA joint (Ambriz et al, 2009).

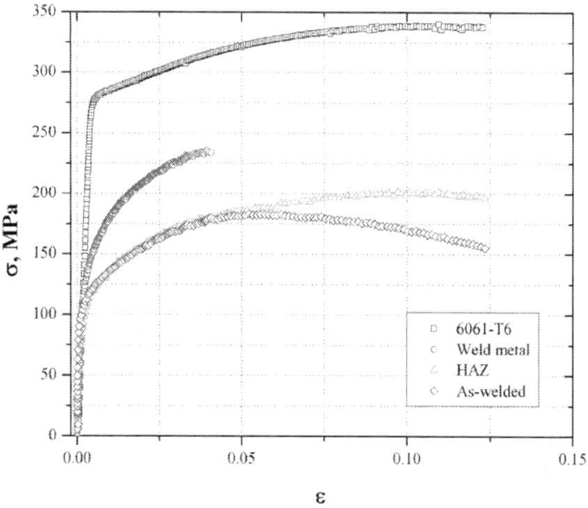

Fig. 15. True stress-strain curves for 6061-T6 plates, weld metal, HAZ and as welded (Ambriz et al, 2010).

4.2.3 Fatigue crack growth (FCG)

Fatigue behavior of aluminum alloys welded by conventional process and MIEA has been investigated (Ambriz et al. 2010b; Branza et al., 2009; Seto et al., 2004). Fig. 16a presents the crack length as a function of number of cycles for base metal, weld metal and HAZ in 6061-T6 welds by MIEA, for ΔP equal to 2.5 and 3.0 kN. In general, the *a-N* curves showed in Fig. 16a reveal a notable difference in terms of crack length for each material as a function of the number of cycles, nevertheless the small change in ΔP (Ambriz et al., 2010b). Experimental results for *a*, were plotted in *da/dN* versus ΔK graphs according to Paris law:

$$\frac{da}{dN} = C(\Delta K)^n \tag{4}$$

where C and *n* are constants obtained directly from the fitting curve. Fig. 16b-d, presents the FCG data obtained for the base metal, weld metal and HAZ in MIEA, as well as the comparison with Friction Stir Welding (FSW) data found in the literature (Moreira et al., 2008). Fig. 16b, shows the FCG for base metal in both directions. This graph shows that the microstructure aspect (anisotropy) does not have an important influence in terms of FCG as could be expected, taking into consideration that yield strength in the base metal parallel to rolling direction is higher than transverse direction. However this is not the case for the weld metal and HAZ (Figs. 16c-d), in which the crack tend to propagate faster than base metal. Under this scenario, the FCG behavior for base metal (L-T) was taken as a basis to perform a comparative table between the weld metal and HAZ of MIEA and FSW. Table 2,

presents the crack growth rate, *da/dN* and the stress intensity factor ΔK, for base metal, weld metal and HAZ corresponding to a critical crack length in MIEA welds. For comparison effects, values for *da/dN* in MIEA were taken to compute the ΔK in FSW.
The results presented in Table 2 indicate that, there is an important difference in ΔK for weld metal and HAZ, independently of the welding process. In this way, it should note that ΔK for weld metal in MIEA represents only 57% of the base metal, unlike the ΔK for weld metal in FSW, which reach a 79% with respect to base metal. This means that FCG rate are higher in MIEA weld metal than FSW, as can be seen in Fig. 16c. This behavior is totally related to the joining processes; it means that MIEA is a welding technique based on a fusion welding process that employs a high silicon content filler metal, which produces a self grain refining, but a brittle microstructure in the weld metal (Ambriz et al., 2010b). On the other hand, FSW is a solid-state joining process that does not use a filler metal (Nandan et al., 2008). Thus, chemical composition in weld metal is similar to the base metal and microstructural characteristics related to dynamic recrystalization tends to be better than MIEA. In contrast, Fig. 16d shows that FCG rate in MIEA and FSW is similar in the HAZ. The stress intensity factor relation was 64% with respect to base metal. It is noted that thermal effect produced by the microstructural transformation of very fine precipitates needle shape β'', to coarse bar shape β' precipitates, has a profound impact in the HAZ crack growth rate. It confirms that, independently of the welding process (MIEA and FSW), the crack growth conditions are directly influenced by the temperature within the HAZ, which is normally above of the aging temperature of the alloy, causing a hardening lost and important decrease in mechanical properties.

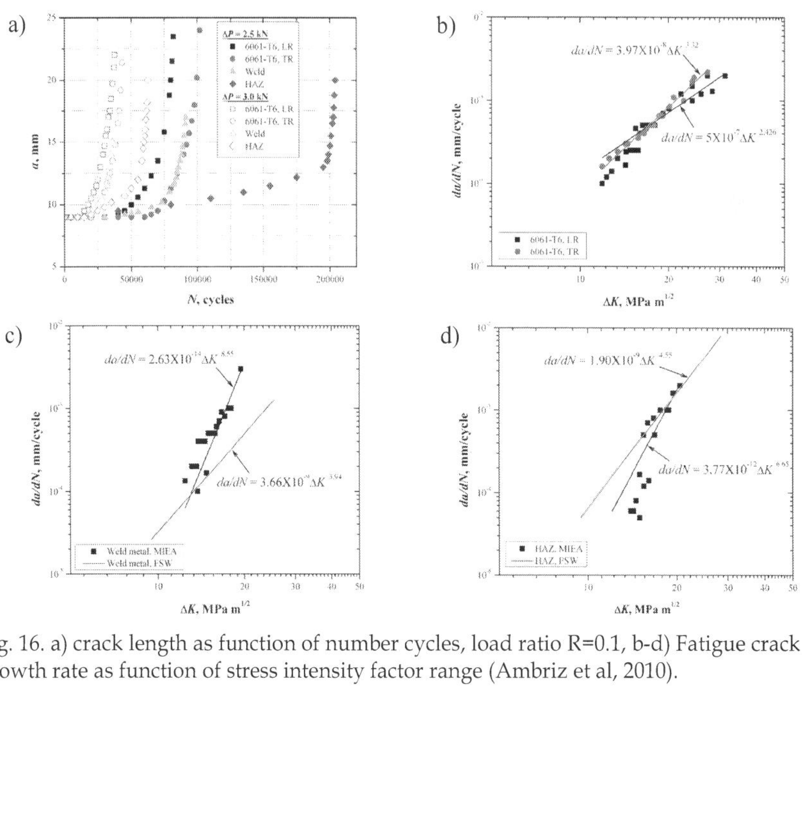

Fig. 16. a) crack length as function of number cycles, load ratio R=0.1, b-d) Fatigue crack growth rate as function of stress intensity factor range (Ambriz et al, 2010).

	Base metal	MIEA		FSW	
		Weld metal	HAZ	Weld metal	HAZ
da / dN (mm/cycle)	1.981×10^{-3}	1.0×10^{-3}	1.413×10^{-3}	1.0×10^{-3}	1.413×10^{-3}
ΔK (MPa $m^{1/2}$)	30.41	17.27	19.46	23.98	19.51
$(da/dN)_i/(da/dN)_{BM}$	1.0	0.50	0.71	0.50	0.71
$(\Delta K)_i/(\Delta K)_{BM}$	1.0	0.57	0.64	0.79	0.64

Table 2. Comparison between MIEA and FSW based on a critical crack length. BM = Base metal, *i* corresponds to weld metal or *HAZ for MIEA or FSW* (Ambriz et al, 2010).

5. Conclusions

The IEA technique and its evolution into the MIEA have emerged as an attractive alternative to weld a number of materials with peculiar microstructural characteristics that have a positive impact in the mechanical and stress corrosion cracking (SCC) behaviors. Between these material are aluminum metal matrix composites (MMCs) reinforced with TiC, SiC and Al_2O_3 particles and monolithic materials such as carbon steels (API X-60 and X-65), aluminum and its alloys, such as 6061, 2014 and 359.
With the IEA welding process, a columnar microstructure with a fine grain, more homogeneous structure and a small HAZ is obtained. In this process, the heat transfer developed does not affect the base metal as much as the direct electric arc process does, but the heat input is enough to partially melt the base metal yielding a weld profile with a high depth to-width ratio. In addition to the macroscopic features obtained, the use of this technique might lead to obtaining a fine microstructure in the weld for metallic materials and improved mechanical properties. The evolution into the MIEA has marked a significant progress regarding mechanical behavior as a result of the reduced thermal affection in heat treatable aluminum alloys.

6. References

Ahearn, J.; Cook, C. & Fishman, S. (1982). Fusion welding of SiC-reinforced Al composites. *Metal construction*, Vol. 14, Issue 4 (April 1982), pp. 192-197, ISSN 03077896

Ambriz, R.; Barrera, G. & García, R. (2006). Aluminum 6061-T6 welding by means of the modified indirect electric arc process. *Soldagem and Inspecao*, Vol. 11, Issue 1, pp. 10-17

Ambriz, R.; Barrera, G.; García, R. & López V.H. (2008). Microstructure and heat treatment response of 2014-T6 GMAW welds obtained with a novel modified indirect electric arc joint. *Soldagem and Inspecao*, Vol. 13, Issue 3, (Jul-sep 2008), pp. 255-263, ISSN 0104-9224

Ambriz, R.; Chicot, D.; Benseddiq, N.; Mesmacque, G. & de la Torre, S. (2011). Local mechanical properties of the 6061-T6 aluminium weld using micro-traction and instrumented indentation. *European Journal of Mechanics A/Solids*, Vol. 30, Issue 3, (May-Jun 2011), pp. 307-315, ISSN 0997-7538

Ambriz, R.; Mesmacque, G.; Ruiz, A.; Amrouche, A. & López, V.H. (2010). Effect of the welding profile generated by the modified indirect electric arc technique on the fatigue behavior of 6061-T6 aluminum alloy. *Materials Science and Engineering A*, Vol. 527, Issue 7-8, (Mar 2010), pp. 2057-2064, ISSN 0921-5093

Ambriz, R.; Mesmacque, G.; Ruiz, A.; Amrouche, A.; López, V.H. & Benseddiq, N. (2010). Fatigue crack growth under a constant amplitude loading of Al-6061-T6 welds obtained by modified indirect electric arc technique. *Science and Technology of Welding and Joining*, Vol. 15, Issue 6, (Aug 2010), pp. 514-521, ISSN 1362-1718

Ambriz, R.R.; Barrera, G.; Garcia, R.; López, V.H. (2009). A comparative study of the mechanical properties of 6061-T6 GMA welds obtained by the indirect electric arc (IEA) and the modified indirect electric arc (MIEA). *Materials and Design*, Vol. 30, pp. 2446-2453. ISSN 0261-3069

Branza, T.; Deschaux-Beaume, F.; Velay, V. & Lours, P. (2009). A microstructural and low-cycle fatigue investigation of weld-repaired heat-resistant cast steels. *Journal of Materials Processing Technology*, Vol. 209, Issue. 2, (Jan 2009), pp. 944-953, ISSN 0924-0136

Cicala, E.; Duffet, G.; Andrzejewski, H.; Grevey, D. (2005). Hot cracking in Al-Mg-Si alloy laser welding - Operating parameters and their effects. *Materials Science and Engineering A*, Vol. A395, pp. 1-9, ISSN 0921-5093

Cola, M.; Lienert, T.; Gouland, J. & Hurley, J. (1989). Laser welding of a SiC particulate reinforced aluminum metal matrix composite, *Proceedings of ASME l 1989, International Conference on Recent trends in welding science and technology*, pp. 1297-303, Gatlinburg, Tennessee, USA, May 14-18, 1989, ISBN 0-87170-401-3

Craig, B.D. (1998). Calculating the lowest failure pressure for electric resistance welded pipe. *Welding Journal*, Volume 77, Issue 1, (January 1998), pp. 61-63, ISSN 0043-2296

Davies, G. & Garland J. (1975). Solidification structures and properties of fusion welds. *International Metals Review*, Vol. 20, (Jun 1975), pp. 83-106

Davis, J. (1993). *Aluminium and aluminium alloys: ASM Specialty Handbook*, ASM International, New York, USA, ISBN 087170496X

Dawson, J.; Palmer, J.; Moreland, P. & Dicken G. (1999). Weld corrosion-chemical, electrochemical and hydrodynamic issues, inconsistencies and model, *Proceeding of Advances in Corrosion Control and Materials in Oil and Gas Production EFC 26*, pp. 155-169, ISBN 1-86125-092-4, Trondheim, Norway, Sept 22-25, 1997

Devletian, J. (1987). SiC/Al Metal matrix composite welding by a capacitor discharge process. *Welding Journal*, Vol. 66, Issue 6, (Jun 1987), pp. 33-39, ISSN 0043-2296

Domey, J.; Aidun, D.; Ahmadi , G.; Regel L. & and Wilcox W. (1995). Numerical - Simulation of the effect of gravity on weld pool shape. *Welding Journal*, Volume 74, Issue 8, (August 1995), pp. S263-S268, ISSN 0043-2296

Dutta, I. & Allen, S. (1991). A calorimetric study of precipitation in commercial Al alloys. *Journal of Materials Science Letters*, Vol. 10, Issue 6, (Mar 1991), pp. 323-326, ISSN 0261-8028

Enjo, T.; Kuroda, T. (1982). Microstructure in weld heat affected zone of Al-Mg-Si alloy. *Transactions of JWRI* , Vol.11, No.1, pp. 61-66, ISSN 0387-4508

Fukumoto, S; Hirose, A. & Kobayashi, K. (1993). Application of laser beam welding to joining of continuous fiber reinforced composite to metal. *Materials science and technology*, Vol.9, Issue 3, (March 1993), pp. 264-271, ISSN 0267-0836

Garcia, R.; Lopez, V. H.; Bedolla, E.; Manzano, A. (2007). Welding of aluminium by the MIG process with indirect electric arc (MIG-IEA), *Journal of Materials Science*, Vol.42, pp. 7956-7963, (Jun 2007), ISSN 0022-2461

García, R.; López, V.H.; Bedolla E. & Manzano A. (2002). MIG welding process with indirect electric arc. *Journal of Materials Science Letters*, Vol. 21, (2002), pp. 1965-1967, ISSN 0261–8028

García, R.; López, V.H.; Bedolla E. & Manzano A. (2003). A comparative study of the MIG welding of Al/TiC composites using direct and indirect electric arc processes. *Journal of Materials Science*, Vol. 38, (2003), pp. 2771– 2779, ISSN 0022–2461

García, R.; Manzano, A.; López, V.H. & Bedolla, E. (2002). Comparative welding study of metal matrix composites with the MIG welding process, using direct and indirect electric arc. *Metallurgical and Materials transactions B*, Vol. 33B, (December 2002), pp. 932 a 937, ISSN 1073-5615

Garcia, R; López, V.H.; Kennedy, A. & Arias, G. (2007). Welding of Al-359/20%SiC_P metal matrix composites by the novel MIG process with indirect electric arc (IEA). *Journal of Materials Science*, Vol. 42, pp. 7794-7800, ISBN 0022-2461

Gittos, N.; & Scott M. (1981). Heat-affected zone cracking of Al-Mg-Si alloys. *Welding Journal*, Vol.49, No.2, pp. 96-s-103-s, ISSN 0043-2296

Grong, O. (1997). *Metallurgical Modeling of Welding*, (Second Edition), Maney Materials Science, ISBN 1861250363, London.

Guiterrez, L.; Neye, G. & Zschech, E. (1996). Microstructure, hardness profile and tensile strength in welds of AA6013 T6 extrusions. *Welding Journal*, Vol.75, No.4, pp. 116s-121s, ISSN 0043-2296

Heinz, A.; Haszler, A.; Keidel, C.; Moldenhauer, S.; Benedictus, R. & Miller, W. (2000). Recent development in aluminium alloys for aerospace applications. *Materials Science and Engineering A*, Vol. A280, No.1, pp. 102-107, ISSN 0921-5093

Huang, C. & Kou, S. (2004). Liquation cracking in full-penetration Al-Mg-Si welds. *Welding Journal*. Vol. 83, No. 4, pp. 111s-122s, ISSN 0043-2296

Kerr, H. & Katoh, M. (1987). Investigation of heat-affected zone cracking of GMA welds of Al-Mg-Si alloys using the varestreint test. *Welding Journal*, Vol.66, No. 9, pp. 251s-259s, ISSN 0043-2296

Kluken, A. & Bjorneklett, B. (1997). A study of mechanical properties for aluminum GMA weldments. *Welding Journal*, Vol. 76, No.2, pp. 39-44, ISSN 0043-2296

Lienert, T.; Brandon, E. & Lippold, J. (1993). Laser and Electron beam welding of a SiC_P reinforced aluminum A-356 metal matrix composite, *Scripta metallurgical et materialia*, Vol. 28, Issue 11 (Jun 1993), pp. 1341-1346, ISSN 0956-716X

Lin, D.; Wang, G. & Srivatsan T. (2003). A mechanism for the formation of equiaxed grains in welds of aluminum-lithium alloy 2090. *Materials Science and Engineering A*, Vol. 351, Issue 1-2, (Jun 2003), pp. 304-309, ISSN 0921-5093

Liu, G.; Murr, L.; Niou, C.; McClure, J. & Vega, F. (1997). Microstructural aspects friction-stir welding of 6061-T6 aluminum. *Scripta Materialia*, Vol.37, No.3, pp. 355-361, ISSN 1359-6462

Liu, W. J.; Tian, X. & Zhang, X. (1996). Preventing weld hot cracking by synchronous rolling during weld. *Welding Journal*, Vol.75, No.9, pp. 297-s-304-s, ISSN 0043-2296

Lu, M. & Kou, S. (1989). Power input in gas metal arc welding of aluminum-Part 1. *Welding Journal*, Volume 68, Issue 9, pp. S382-S388, (Sep., 1989), ISSN 0043-2296

Lu, M. J. & Kou, S. (1989). Power input in gas metal arc welding of aluminum-Part 2. *Welding Journal*, Volume 68, Issue 11, pp. 452s - 456s, ISSN 0043-2296

Lundin, C.; Danko, J. & Swindeman, C. (1989). Fusion of a SiC reinforced aluminum alloy 2024, *Proceedings of ASME 1989 International Conference on Recent trends in welding science and technology*, pp. 303-307, ISBN ?, Gatlinburg, Tennessee, USA, May 14-18, 1989

Malin, V. (1995). Study of metallurgical phenomena in the HAZ of 6061-T6 aluminium welded joints. *Welding Journal*, Vol.74, No.9, pp. 305s-18s, ISSN 0043-2296

Miyazaki, M.; Nishio, K.; Katoh, M.; Mukae, S. & Kerr, H. (1990). Quantitative investigation of heat-affected zone cracking in aluminium alloy 6061. *Welding Journal*, Vol.69, Issue.9, pp.362s-71s, ISSN 0043-2296

Moreira, P.; De Jesus, A.; Ribeiro, A. & Castro, P. (2008). Fatigue crack growth in friction stir welds of 6082-T6 and 6061-T6 aluminium alloys: as a comparison. *Theoretical and Applied Fracture Mechanics*, Vol. 50, Issue 2, (Oct 2008), pp. 81-91, ISSN 0167-8442

Myhr, O.; Grong, O.; Fjaer, H. & Marioara, C. (2004). Modeling of the microstructure and strength evolution in Al-Mg-Si alloys during multistage thermal processing. *Acta Materialia*, Vol. 52, Issue 17, (Oct 2004), pp. 4997-5008, ISSN 1359-6454

NACE MR0175/ISO 15156 (2009). Petroleum and natural gas industries - Materials for use in H_2S-containing environments in oil and gas production Part 1: General principles for selection of cracking-resistant materials

Nandan, R.; DebRoy, T. & Bhadeshia H. (2008). Recent advances in friction-stir welding-process, weldment structure and properties. *Progress in Materials Science*, Vol. 53, Issue 6, (Aug 2008), pp. 980-1023, ISSN 0079-6425

Natividad, C.; Salazar, M.; Espinoza-Medina, M.A. & Pérez, R (2007). A comparative study of the SSC resistance of a novel welding process IEA with SAW and MIG. *Materials Characterization,* Volume 58, Issue 8-9, (August-September 2007), pp. 786-793, ISSN 1044-5803

Omweg, G.; Frankel, G.; Bruce, W.; Ramirez, J. & Koch G. (2003). Performance of welded high-strength low-alloy steels in sour environments. *Corrosion*, Volume 59, Issue 7, (July 2003), pp. 640-653, ISSN 0010-9312

Ram, G.; Mitra, T.K.; Raju, M. & Sundaresan, S. (2000). Use of inoculants to refine weld solidification structure and improve weldability in type 2090 Al-Li alloy. *Materials Science and Engineering A*, Vol. 276, Issue 1-2, (Jan 2000), pp. 48-57, ISSN 0921-5093

Rao, K.; Ramanaiah, N. & Viswanathan, N. (2008). Partially melted zone cracking in AA6061 welds. *Materials & Design*, Vol. 29, Issue 1, (2008), pp 179-186, ISSN 0261-3069

Seto, A.; Yoshida, Y. & Galtier, A. (2004). Fatigue properties of arc-welded lap joints with weld start and end points. *Fatigue & Fracture Engineering Materials & Structures*, Vol. 27, Issue 12, (Dec 2004), pp. 1147-115, ISSN: 8756-758X

Turnbull, A. & Nimmo, B. (2005). Stress Corrosion Testing of Welded Supermartensitic stainless gas steel for oil and gas pipelines. *Corrosion Engineering Science and Technology*, Volume 40, Issue 2, (June 2005), pp. 103-109, ISSN 1478-422X

Urena A.; Rodrigo P.; Gil L.; Escalera M. & Baldonedo J. (2001). Interfacial reactions in an Al-Cu-Mg (2009)/SiCw composite during liquid processing - Part II - Arc welding. *Journal of materials science*, Vol. 36, Issue 2 (Jan 2001), pp. 429-439, ISSN: 0022-2461

Urena, A.; Rodrigo, P.; Gil, L.; Escalera, M. & Baldonedo, J. (2000). Interfacial reaction in Al-Cu-Mg (2009)/SiCP composite during liquid processing Part 1, *Journal of materials science*, Vol.32, pp. 419-428, ISBN 0022-2461

Part 2

Arc Welding Automation

3

Arc Welding Automation

Eduardo José Lima II and Alexandre Queiroz Bracarense
Universidade Federal de Minas Gerais
Brazil

1. Introduction

It is very well known that the use of robotics in the industry increases productivity and quality in many aspects. It is also well known that some adjustments have to be made to grant payback for the investment and to reach the expected results. Today, the robotization is known as an alternate technique for production, increasing qualitative and quantitative competence of its industries. One point that has been observed in some applications is that inadequate procedure for robot application in welding process reflects in insatisfactory results. These procedures are excessive time expending for implementation, material loss, reworking and poor weld quality.

The main problem detected is that the number of experts in welding "and" robotics is still reduced and some industries are investing in robot without any planning or orientation, believing that the robot will solve all their problems. The results have been disastrous. In many cases even though the weld appearance is acceptable and the welding time is significantly reduced, the weld quality is poor. The experts in robotics know very well what they are doing. However, many of them do not have experience in welding to understand that many features related to welding physics and metallurgy have to be considered when a welding procedure has to be implemented. Because of this, some small and medium industries with large potential for robotization are holding investment and/or postponing it until its adaptation to implement the robot in their production line. It is believed however that very soon many small industries will have their own robot.

2. The paradigm of welding automation

Since the beginning, welding is a process that depends on the welder skills. This relation is so direct that the classification, according to the application methods, is based on the degree of control of the activities related to welding that depends on the human interference. These application methods are classified as manual, semi-automatic, mechanized, automatic, robotic and with adaptive control, according to American Welding Society (AWS).

This classification can be better understood when an agent is established (Table 1) to execute the normal activities involved to realize arc welding (Cary, 1994).

The manual welding is defined, according to the American Welding Society (AWS, 2001) as "welding where the torch or the electrode holder are carried or manipulated by human hands". In other words when the tasks, related with the execution and continuous control of the welding, are made by the hands of the human and are under responsibility of him.

	Manual	Semi-automatic	Mechanized	Automatic	Robotic	Adaptive control
Activities		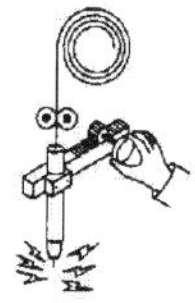	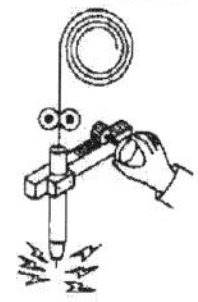		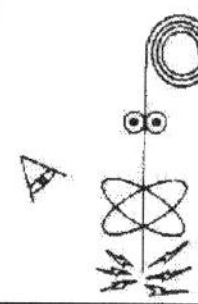	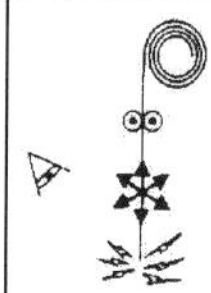
Arc start and maintenance	Human	Machine	Machine	Machine	Machine	Machine (with sensor)
Wire feeding	Human	Machine	Machine	Machine	Machine	Machine (with sensor)
Heat control to obtain penetration	Human	Human	Machine	Machine	Machine	Machine (with sensor)
Arc motion along the joint	Human	Human	Machine	Machine	Machine (robot)	Machine (with sensor)
Guide the arc along the joint	Human	Human	Human	Machine (with pre programed track)	Machine (robot)	Machine (with sensor)
Torch manipulation to direct the arc	Human	Human	Human	Machine	Machine (robot)	Machine (with sensor)
Arc corrections to compense erros	Human	Human	Human	Do not occours	Do not occours	Machine (with sensor)

Table 1. Application methods of welding process, adapted from (Welding Handbook, 2001).

Semi-automatic welding is defined as "manual welding with equipment that automatically control one or more welding conditions". The welder manipulates the torch while the wire/electrode is automatically fed by the machine.

Mechanized welding is defined as "welding with equipment that requires manual adjustments in response to visual observation during welding, with the torch or electrode holder carried by a mechanical system". The welder participation in this process consist in adjusting the parameters as he observe the operation.

Automatic welding is defined as "welding with equipment that only requires occasional observation and/or no observation of the weld and no manual adjustments". The function of the welder is only turn the machine on to begin the welding cycle and occasionally check the procedure.

Robotic welding is defined as "welding that is executed and controlled by a robotic equipment". In robotic welding and automatic welding the function of the welder is to guarantee the weld quality by performing periodic inspections of the results identifying weld discontinuities. When those are find, maintenance and programming must be done to fix such problems.

Welding with adaptive control is defined as "welding with equipment that has a control system that automatically determine changes in welding conditions and act under the equipment with appropriated action to do adjustments". In this process, sensors are used to detect problems and the controller performs the necessary changes in welding parameters, in real time, to produce sound welds. This type of welding is performed without intervention and supervision of the human.

Accord to the classification presented, the stick electrode welding is a manual process since the welder is responsible for all the activities, while GMAW is a semi-automatic process. This is so because the arc start and maintenance and wire feeding is made by the machine while the torch manipulation is made by the welder. When this manipulation is made by a mechanical device the processes is classified as mechanized.

Independent of the automation degree, its focus is costs reduction by reducing the number of people involved in the production and increasing productivity and quality of the final product by the rational control of the process parameters. An automatic equipment may be projected and programmed to perform a unique task (fix automation) or may be projected for multitasks, by programming, allowing to perform distinct tasks accord with the manufactured product (flexible automation) (Welding Handbook, 2001; Romano, 2002).

In manufacturing area, the term "automation" means that all the functions or steps of an operation are executed or controlled, in sequence, by mechanical and/or electronics devices replacing the human efforts, observation and decision.

The automation involves more than equipments or control by computer and may or may not include charge or discharge of components in an operation. The automation may be partial, with some functions or steps executed manually or may be total, where all the functions or steps are executed by the equipment, in a certain sequence without any adjustment by the operator (Romano, 2002).

To properly classify the automation level of a given process, the first step is to define the activities related to it. The responsible for the execution (execution agent), for the control (control agent) and for the sequence of activities (sequential agent) must be defined. Additionally, it is necessary to define which activities need to be treated as isolated and which must be included in the process operational cycle.

In the arc welding processes, the first activity to execute a weld is to specify which welding procedure must be used. This definition is also known as WPS - Welding Procedure Specification. Considering used material and the weld morphology wanted, the welding parameters related to the process may then be determined. The executor, controller and sequence specification agent of a WPS is the human. Even with a help of a computational program to choose the best parameters or the ones that will be used for the self adjustment of the machine, the final decision for the WPS is the human decision. Between the WPS elaboration and the welding beginning, a time interval for procedure preparation is necessary. Nowadays, this preparation occurs independent of the application method to be used and human direct interference are often necessary.

There is still a lack of a good system to automatically prepare the WPS and, from its preparation, immediately initiate the welding. Of course this system must be universal, since among the parameters to be chosen for the WPS there are the welding processes. It is, however, expected that such activities (WPS and welding start) be treated as isolated activities, because of the human involvement on them. The activities related to the welding cycle, the ones that will define the degree of automation of the process, must be the ones that allows the instantaneous sequence of the process.

The easiest way to relate such activities is to associate them to a welding process. It is unquestionable to say that welding with covered electrode (SMAW) is manual. To start the arc, the welder must approximate the tip of the electrode to the base metal, touch it and slowly and gently pull up such that the arc is established. After, he must translate and feed the electrode, such that the distance between the electrode tip and the weld pool is constant (arc length) until the end. To finish, he must smoothly pull the electrode, getting it apart from the base metal, extinguishing the arc. Spiral movements some times are also used to fill the crater. As observed, in this operation the execution and control agent is the human. All of the described tasks may, however, be made automatic, if needed, not necessarily for productivity improvement, but for security reasons, as hot tapping of tubes and underwater welding. Section 3 shows some results of the robotic SMAW.

According to the classification shown in Table 1, a typical semiautomatic process is the GMAW. To begin the welding, after some preparation, the welder must place the torch close to the base metal and after pushing the trigger, start the wire feeding. As the wire touches the base metal, the arc is established and the welder must translate the torch. As the welder translates it, the machine feeds the wire into the weld pool. To stop the welding, the welder needs only to release the trigger. The machine stops to feed the wire and the arc is extinguished. The arc opening and extinguishing are associated to the wire feeding by the machine (execution agent). However, who decides the feeding start and finishing moment is the human (controller and sequence agent), since he needs to push the trigger.

GMAW is the most suitable arc welding process to be carried by the robot. If the robot substitutes the human welder by translating the torch with a predefined trajectory (been classified as automatic welding according to Table 1), the improvement on repeatability is huge, as the robot will always make the same trajectory. However, to improve the quality, the trajectory needs to be programmed at least as good as the human does it. If the trajectory is not well programmed, the robotic weld will never be better than the human weld. Section 4 shows a case where the trajectory study was crucial to improve the welding quality for the robotic welding.

The execution of torch movement with a mechanical device with mechanical or electronic control (as the robot) is a necessary condition, but no sufficient, to a welding be automatic. In this case, the easiest way to differentiate a mechanized system from an automatic system is to base on the concept of automatic equipment. It is either an equipment designed and programmed to execute an unique task (fix automation) or a flexible equipment that, with reprogramming, allows the performing distinct tasks accord to the product to be manufactured (flexible automation) (Welding Handbook, 2001; Romano, 2002).

An industrial robot is an example of a flexible automatic system. Accord to RIA (Robotic Industries Association), "a robot is a reprogrammable manipulator, multifunctional, projected to move materials, parts, tools and specialized devices by programmed movements to perform many tasks" (Rivin, 1988). The development of this type of machine introduced an elevated degree of flexibility to the production environment.

The main condition for a welding equipment to be robotic is that it should be programmable. The most used industrial robots for welding are the anthropomorphic with six degrees of freedom. They are reprogrammable and multifunctional. This means that these robots may be used to weld different parts, needing only the reprogramming for the new part to be welded.

There are also robots designed for specific tasks. These robots are not multifunctional. A typical case is a robot used for welding, designed to execute an unique type of weld. An example is the robot developed to weld pipelines and will be presented in Section 5. This robot has its movements limited to rotate around the pipe while it stays stopped. Only pipes can be welded with this type of robot. This robot is then called "a dedicated robot".

Finally, to differentiate automatic system from mechanized system is a hard task. This is because the automation may be partial or total and there is not a 100% automatic yet. Regarding to welding systems what can be said is related to a flexible or dedicated (fix) system.

As general rule an automatic process is more productive than a mechanized process which is more than the manual. In welding, the gain in productivity many times is related to the reduction in time with reworking, close arc time and preparation to begin the welding cycle. On the other hand, also as general rule, the cost for implementation increases from manual welding to automatic welding. Allowing to say that one disadvantage of the automatic welding is its initial cost. Detailed studies of economical viability show that the benefits against costs to implement such systems are becoming satisfactory.

In general, if a welding process can be mechanized it can be automatized. The question is when a process should be mechanized and when it should be automatized. Additionally, if this automation needs or not a robot, i.e., it is a fix or a flexible automation.

Many factors must be considered to define the best execution method for a welding process, as type of process, part geometry, weld complexity, amount of welds and desired weld quality.

All these factors must be considered and also the advantages and disadvantages of each method. The more dependable way to define the appropriate method to produce a determined part is studying the economic viability. This should be done because, independent of the automation degree, what is seen is the reduction of manufacturing costs. Using automatic systems this can be reached by reducing the number of people involved in the welding, the increase in productivity and the increase in quality, through the use of more rational process parameters. Also, with automatic systems, the history of the welding and all the preparation also can be stored. This, together with the repeatability, allows the traceability of welded parts.

The following sections show some examples of welding automation in different levels for different applications.

3. Robotic shielded metal arc welding

One of the main problems with the shielded metal arc welding process is the bead weld quality, related to its microstructure homogeneity and its physical and dimensional aspect. These factors are directly related to the fact of such process to be, currently and predominantly manual and even the best welder is incapable to weld with absolute repeatability all the weld beads. This process mechanization already exists and increases the repeatability. However it has limits with bead geometry, which is determined by the mechanism assembly. In Figure 1 is shown a device which uses gravity to move the electrode holder (a) along a fixed trajectory (b) as the electrode (c) is melted.

There are many applications for the manual SMAW process but two of them are more specific and there is no other process that can be used. One application is underwater welding, as shown in Figure 2. For a long time many tries have been made to replace coated electrode in this type of welding, without success. It is easy, versatile and the chemical control of the weld metal is the most acceptable. Another application is hot tapping of tubes as shown in Figure 3. In this application, the welder has to weld a tap tube to the main line with inflammable fluids passing inside. As the main line cannot be emptied, this is a dangerous procedure to the welder, however it is the only way acceptable nowadays.

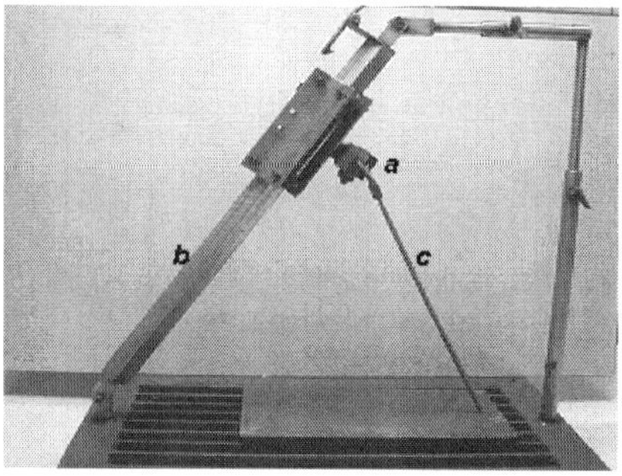

Fig. 1. Device used for gravity welding with covered electrode.

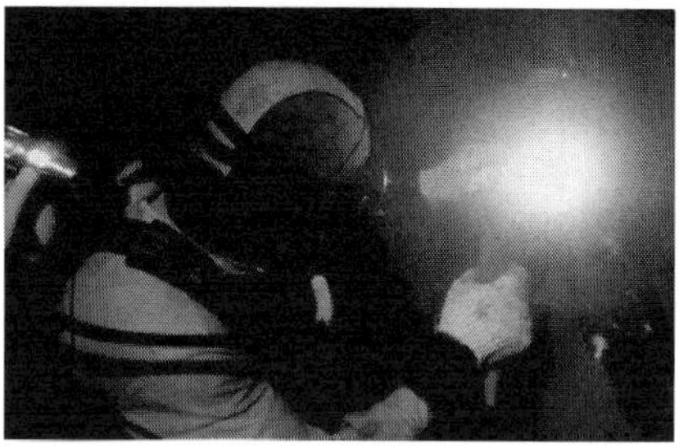

Fig. 2. Underwater welding with SMAW.

Fig. 3. Hot tapping with SMAW.

Aiming the improvement of the weld quality allied to the repeatability proportionated by the mechanization and the manual process flexibility, the process robotization appears as a solution. However, the robotization brings the problem that, depending on the electrode diameter and the weld current, the melting rate is not constant during all the electrode length. This is because the welding current crosses all the electrode length, causing its heating by Joule effect. This heating facilitates the melting of the electrode, which increases as the electrode is consumed. Thus, if the weld is made using a constant feed speed, it will obtain a bead with non homogeneous dimensional characteristics (Bracarense, 1994). Its morphology (width and reinforcement) increases as the material is deposited, since the melting rate, and consequently the material deposition rate increases as the weld is performed. Experimental results (Oliveira, 2000) had shown that, beyond of getting an irregular bead and without penetration, a constant feed speed can cause the electric arc extinction just after the beginning of the weld.

3.1 Trajectory generation

Due to this melting rate variation, this welding process cannot be programmed with the simple teaching of an initial and a final point to the electrode holder, as in this case it would be obtained a constant feed speed. Moreover, it is not possible to precisely calculate, before starting the welding, the melting rate behavior, as it depends on a number of process variables, as the electrode temperature, welding current, air flow etc.

So, to robotize the shielded metal arc welding, it is not sufficient to follow a predefined trajectory over the groove, as in the GMAW and FCAW processes, in which the wire feeding is automatic. In SMAW, it is necessary to make the feeding movement, in order to maintain constant the electric arc length. As the melting rate is not constant, the feeding speed has to be regulated during execution time.

The methodology presented by Lima II and Bracarense (2009) allows the Tool Center Point (TCP) movement programming in a similar way as in GMAW and FCAW, in a transparent way to the user. So, it is only necessary to program the weld bead geometry or trajectory over the groove without caring about the electrode melting.

The electrode is considered as a prismatic joint of the robot. Considering the joint length given by the electrode length, the TCP moves on the programmed trajectory and, at each sampling period, the new joint displacement is calculated and updated in the robot kinematics model. So, the diving movement of the electrode-holder is made independently of the welding movement.

Considering the initial and final electrode holder positions shown in Figure 4 and melting rate experimentally obtained by Batana and Bracarense (1998), Figure 5(a) shows the TCP and electrode holder trajectories during welding. The electrode tip moves along predetermined trajectory while the electrode holder makes the diving movement. In this case, as the electrode is parallel to the Z_0 axis, the electrode holder diving movement is made in this direction, as it moves in X_0 direction. The independence among the TCP advance movement and the electrode holder diving movement is easily stated. However, considering now a welding angle of 45º, these movements are not independent (Figure 5(b).

(a) (b)

Fig. 4. Initial (a) and final (b) robot positions during shielded metal arc welding for a 90º welding angle.

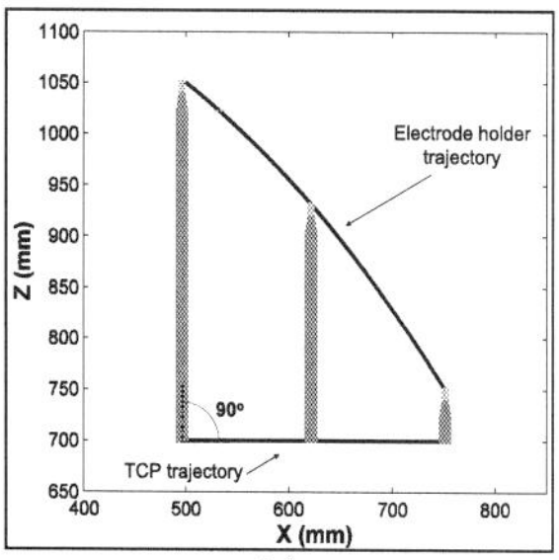

(a)

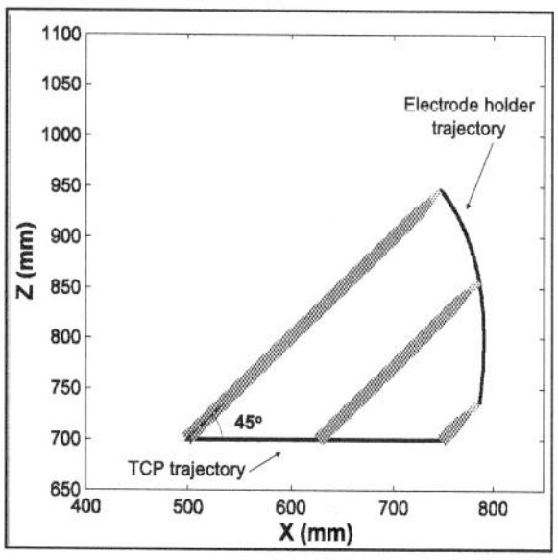

(b)

Fig. 5. Tool Center Point and electrode holder trajectories for welding angles of (a) 90° and (b) 45°.

This methodology can be extended to non linear trajectories, as in the orbital welding or welding for hot tapping in pipelines. The operator only has to program the welding trajectory in the same way as it is done in welding processes with continuous wire feeding. Figure 6(a) shows the programmed TCP trajectory on the tube and the electrode holder trajectory for 90° of welding angle and Figure 6(b) shows those trajectories for 45° of welding angle. More complex welding trajectories may be programmed by using a sequence of linear and circular movements as in other welding processes.

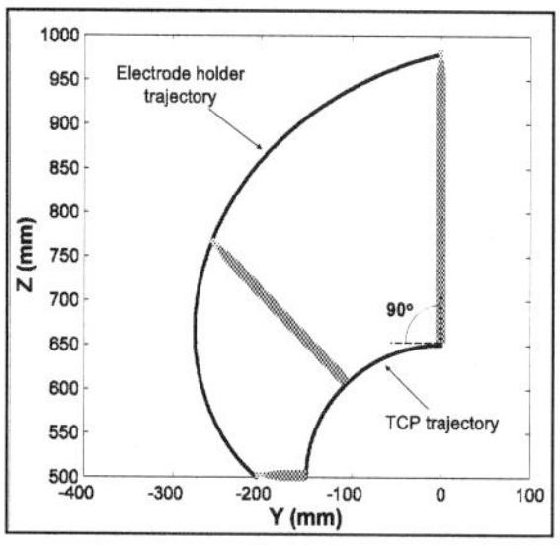

(a)

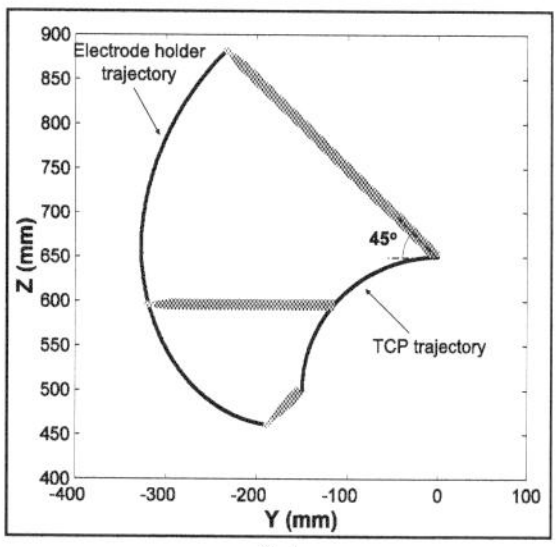

(b)

Fig. 6. Tool Center Point and electrode holder trajectories for 90° (a) and 45° (b) welding angles in orbital welding.

3.2 Electric arc length control

Previous works (Oliveira (2000); Batana & Bracarense (1998); Quinn et al. (1997)) seeking the robotization of the welding process with covered electrodes suggested the development of models for electrode melting rate considering current and temperature, to determine the speed of the electrode holder diving. Thus, making the diving movement at speeds equal to the melting rate, the arc length should remain constant throughout the welding. However, imperfections in the models, errors in current and in temperature measurements and other disturbances cause small differences between the value of the calculated melting rate and real melting rate. These differences, even if small, can cause great variation in the arc length, since it depends on the integral of the instantaneous difference. This shows that an "open loop control", as used by Oliveira (2000) is not suitable for the system.

The solution used here is to make a measurement of the arc length to determine the diving speed and use it in a "closed loop controller". In this case, a reference value for the arc length is given and the error is calculated as the difference between the reference and the actual arc length measured from the electric arc.

One solution for the problem of measuring the arc length would be to measure the voltage in electric arc (V_{arc}), since they are directly related. In the process a constant current power source is used. The problem is that it is not possible to directly measure the arc voltage, because, during welding, the electrode tip, near the melting front, is not accessible. It is possible, however, to measure the voltage supplied by the power source (V_{source}) through the entire electrical circuit, as shown in Figure 7, which includes the voltage drop in the cable, in the holder, in the base metal ($V_{c1}+V_{c2}$) and, mainly, along the extension, not yet melted, of the electrode, V_{electr}.

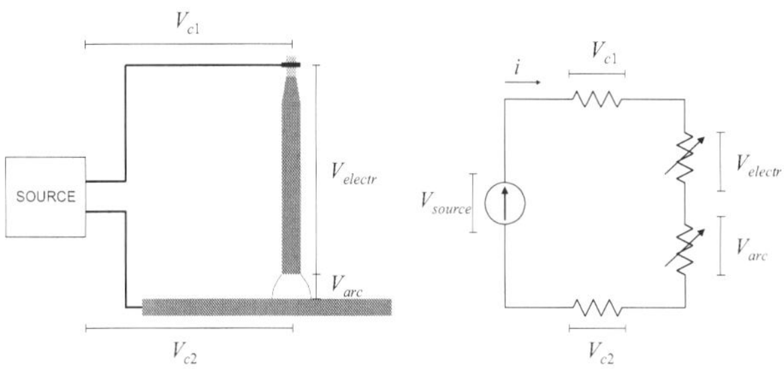

Fig. 7. Electrical circuit of covered electrode welding

It may be considered that the sum of the voltage drop in the cable, in the electrode holder and in the base metal ($V_{c1}+V_{c2}$) are constant along the welding since the welding current is kept constant by the power source. However, the voltage drop along the electrode that has not yet been melted, V_{electr}, is not constant, due to the reduction on its length and due to the increase of its electrical resistivity with temperature. Thus, even if the controller keeps the V_{source} constant through the control of the diving speed, it does not guarantee that V_{arc} is constant throughout the process, which does not guarantee, therefore, a constant arc length. In this study, a model of the electrode voltage drop, as a function of temperature to compensate for the effect of its variation was used.

The electrode voltage drop V_{electr}, may then be modeled as:

$$V_{electr} = \rho(T)\frac{l_{electr}(t)}{A}I\,, \tag{1}$$

where ρ(T) is the electrode electrical resistivity as a function of temperature, $l_{electr}(t)$ is the electrode length not yet melted, A is the area of the electrode wire and I is the welding current. As the electrical conductivity of the core wire is two orders of magnitude greater than the coating (Waszink & Piena, 1985), one can consider only the resistivity and cross sectional area of it.

As the electrical resistivity ρ of the core wire material varies with its temperature, it is important to know the temperature behavior along the electrode during the process. In Felizardo (2003) the authors conclude that the longitudinal temperature profile along the covered electrode is practically constant. Its heating is due to the Joule effect caused by the high electric current crossing the electrode. The conduction of the heat generated by the electric arc to the electrode is often slower than the fusion rate, which causes the temperature to be constant along the electrode length. Then, temperature can be measured during welding using thermocouples (Dantas et al., 2005) placed under the coating near the electrode holder.

3.3 Results

To validate the methodology, an anthropomorphic industrial robot, with 6 rotational degrees of freedom was used. This robot uses a controller that allows programming from simple, linear and circular join-to-joint movements to creation of complex programs, including changes of parameters at run time (KUKA, 2003). These characteristics make possible the implementation of the proposed methodology for trajectory generation and control of the electric arc length during welding. To perform data acquisition, a modular system I/O-SYSTEM 750 from WAGO® was used. This system communicates with the robot controller by a *DeviceNet* interface. For the tests, a constant current power source, capable to supplying currents up to 250A, and an open circuit voltage of 70V was used. A drill chuck was used as electrode holder (Dantas et al., 2005). The supply current is made through the jaw of the chuck, which is in turn electrically isolated from de holder by a part of nylon. To enable the arc initiation in the welding start point, it was used a composite specially developed to burn when submitted to electric current (Pessoa et al., 2003). When the composite is burned, the arc is established and the robot starts the movement. At the end point the current is interrupted by a fast movement of the electrode and the arc is terminated.
Using the robot routines to define tools, the Tool Center Point models with the complete electrode and with the melted electrode were defined (Figure 8).

Fig. 8. Complete electrode and melted electrode frames.

The proposed methodology allows welding with covered electrode of any length, diameter and type of coating, since it performs the closed loop control of the process. Thus, the proposed methodology was validated with rutile type covered electrodes (E6013) of 4mm in diameter, and with basic type covered electrodes (E7018) of 3.25 mm diameter. The welding current ranged between 150A to 180A as indicated by the manufacturer. Plates and tubes of carbon steel were used for linear and non-linear (circumferential) welding trajectories.

During the process, it was possible to observe that although the robot can keep the mean voltage constant, the arc length increases significantly at the end of the weld, as discussed above. To compensate this effect, the model of the electrode voltage drop in function of its length and temperature was used to correct the feedback signal used by the controller. For this, tests were made to obtain the curve of temperature versus time. Thermocouples type K were used for monitoring temperature during welding (Dantas et al., 2005).
Welding tests were then made using this compensation. The reference voltage (V_{ref}) was set to 21V. Figure 9 shows the voltage on the electrode (V_{electr}) as a function of time. Despite the voltage drop compensation in the electrode varies of only 0.5V, it was observed that the length of the arc remained constant throughout the execution of the weld, reinforcing the need for such compensation.

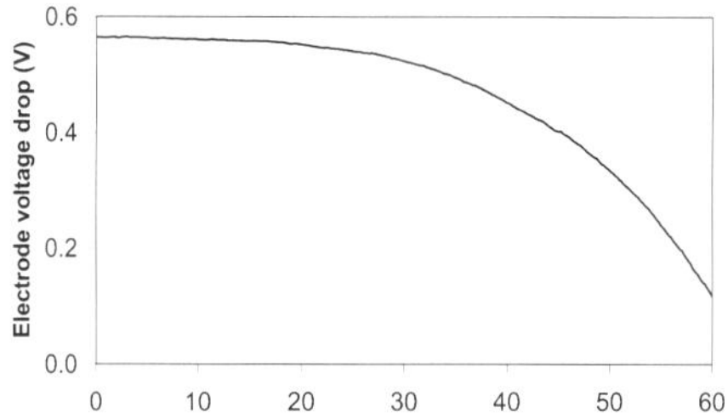

Fig. 9. Electrode voltage drop during welding.

To prove the repeatability achieved with the automation of the process, several beads on plate were performed using the E6013 electrodes with 4mm diameter, welding current of 175A, reference voltage of 21V and welding speed of 2.5 mm/s. Figure 10 shows the appearance of the welds. Despite the spatter problem it is possible to observe that all the welds are identical, demonstrating the repeatability obtained with the robotization of the process.

Fig. 10. Beads on plate performed by the robot using E6013 electrodes, demonstrating the repeatability of the process.

Aiming to demonstrate the flexibility of the used methodology with respect to the variety of electrodes, tests were made using E7018 electrodes of 3.25 mm in diameter. The best welds were obtained using current of 150A, speed of 2.5 mm/s and the reference voltage of 26.5 V. Figure 11 shows the appearance of welds.

Fig. 11. Welds made using E7018 electrodes demonstrating the flexibility and repeatability of the process.

As can be observed, the welds are more uniform and with less spatter than the ones obtained with E6013 electrodes. It is important to note that the E7018 electrodes, despite producing best quality welds, have greater difficulty in manual welding. In the experiments, however, these electrodes did not present any operational difficulties in relation to E6013 electrodes, but was necessary to conduct some additional experiments to adjust the reference voltage as the voltage of the electric arc varies considerably with the change of the electrode coating.

To demonstrate the generality of the developed methodology for the trajectories generation, an orbital welding on a steel tube with 14 inches diameter was conducted. The welding started in the flat position, going downward in vertical position with the electrode in an angle of 45°, pulling the weld bead. E7018 electrodes were used with a current of 130A, welding speed of 5.5 mm/s and reference voltage of 18V. Figure 12 shows robot positioned with the electrode at the arc opening and after its extinction.

(a) (b)

Fig. 12. Robot positioning (a) before arc opening and (b) after arc extinction.

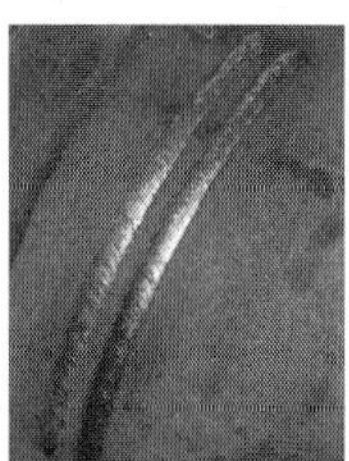

Fig. 13. Welds made on tube with E7018 electrodes.

Figure 13 shows the appearance of two welds made on the pipe with the same welding parameters, demonstrating the repeatability of the process.
The results show that is possible to automate an intrinsic manual process, bringing reliability and repeatability to it. Also it can be applied when the task is dangerous to be performed by the human welder.

4. Robotic GMAW

Before deciding for the automatization of a process using welding robots, various factors such as definition of the goals to be reached (production volume increase or quality improvement), necessity of improvement in the adjustment between the parts, among many factors must be verified (Bracarense et al., 2002).
This section shows the cooperation between University and Industry in the welding of scaffolds used in civil construction. The company wanted to use robots to improve the production, but was in doubt about the weld beads quality and the economic viability. The production line of scaffolds used manual welding and did not control the welding sequence nor the deposition rates. The University was then contacted to study the viability of using a robot to carry through these operations.

4.1 Scaffold welding study

Among many scaffold types manufactured by the company, the tubular scaffold was the one studied. These scaffolds are manufactured in three different models, with 1,0m by 1,0m, 1,0m by 1,5m and 1,0m by 2,0m, as shown in Figure 14.

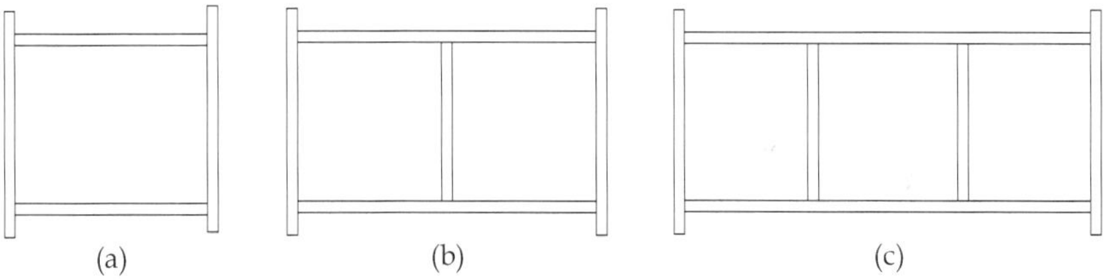

Fig. 14. Scaffold models manufactured by the company: 1,0m x 1,0m (a), 1,0m x 1,5m (b) and 1,0m x 2,0m (c).

In the manual process, before the welding, the scaffold joints are arc spot welded using Shielded Metal Arc Welding. Two operators work in this procedure: while one places the tubes on a jig, the other spot welds the joints in other jig. A great variation in the arc spot welding times is observed. For an average of 39,6s for arc spot welding of a complete scaffold, a standard deviation of 11,1s was obtained (Pereira & Bracarense, 2002).
Initially some problems, such as differences in tubes lengths (Figure 15a) and cut finishing (Figure 15b), beyond lack of parallelism in its extremities (Figure 15c) have been stated. These problems would compromise the robotic welding, since, although the manual welder perceives such differences and compensates them during the welding, the robot is not capable to make it, as its movements are based on a previous programming. To make possible using the robot, some modifications had been carried through in the cutting process in order to minimize such problems.

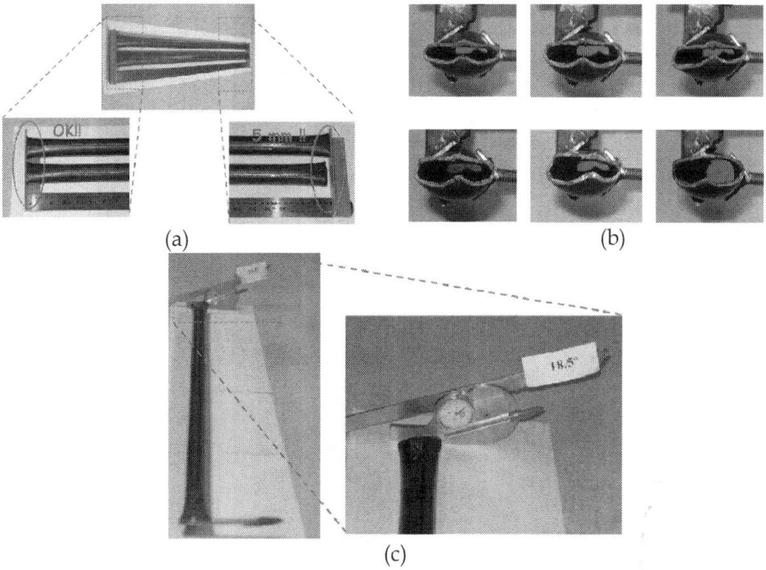

Fig. 15. Problems in the tubes preparation: difference in length (a), difference in the extremity sections (b) and lack of parallelism (c).

Aiming to define the size of the robot to be specified, simulations had been done using commercial software (Figure 16). The scaffold of 1,0m x 2,0m was considered in this simulation, because its bigger dimensions among the others to be produced. A MOTOMAN SK6 robot was considered the model since it was the one to be used in the laboratory.

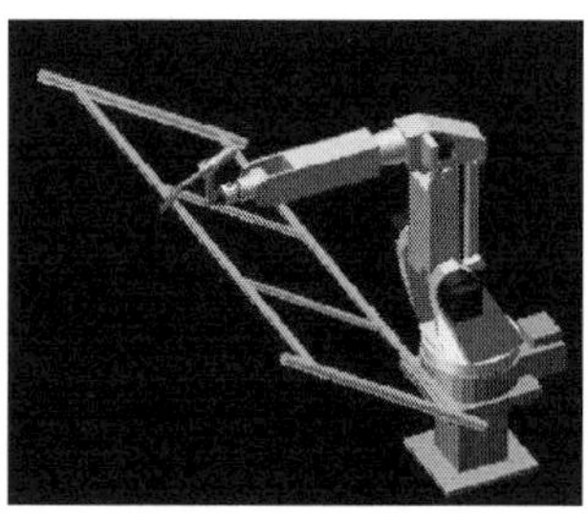

Fig. 16. Computer simulation of scaffold welding process.

The use of a simulation software allowed, beyond verifying if all the joints to be welded would be inside of the workspace of the robot, to verify if it would be possible to locate the tool with desired orientation in all the points to be welded, that is, if all the points would be inside of the robot dexterous workspace (Craig, 1989).

Then some welds had been carried through in the laboratory at the University within the objective to study the best welding parameters to be used (Figure 17).

Fig. 17. Tests at University laboratory for verification of the welding parameters.

Through tensile test, it was verified that the welds made using the robot are stronger than those manually welded (Pereira and Bracarense, 2002). Additionally, it was verified that the rupture occurs far from the HAZ (Heat Affected Zone), confirming the higher quality of the welds made by the robot. It was also stated that there was not any visible modification in the weld bead after the tests. Accepting the viability of using robots for the scaffold welding, the company decided to acquire a robotic welding cell.

4.2 Robotic cell conception and development

The Company acquired a KUKA KR16 robot, similar to MOTOMAN SK6, but with a wider workspace. The layout of the cell, projected by the University, consists of three jigs located around the robot (Figure 18). In this cell the operators can mount or remove two scaffold while the robot welds another in the third jig.

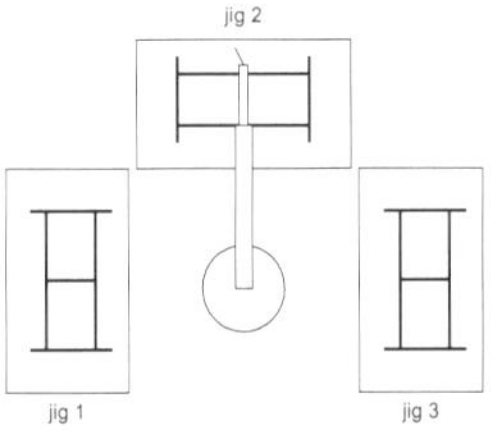

Fig. 18. Robotic cell conception.

The construction and installation of the robotic cell was supervised by the University and carried through the Company, which has resources to produce doors, gratings, tables and jigs (Figure 19).

Fig. 19. Installed robotic cell.

As it was conceived, with three jigs, the robotic cell allows getting a work cycle of practically 100%. Considering that the robotic arc spot welding process is faster than the manual scaffold assembly, it is possible to the robot to weld two scaffolds while two operators remove the welded scaffold and assembly new ones in the two other jigs.

4.3 Arc spot welding program development

Considering the problems observed in the tubes preparation, it was opted to initiate programming the arc spot welding of the scaffold to posterior manual welding by the operators. An operator would be trained on the robot programming and would follow the development of the arc spot welding program. This operator would be also responsible for determining changes in the cutting process, guiding the other employees to adapt it to the robot.

As commented before, the arc spot welding was originally made manually using SMAW. With the robot, the arc spot welding would be made with GMAW (Gas Metal Arc Welding), being, therefore, unnecessary to remove the slag after arc spotting, before welding of the joints.

The 1,0m x 1,5m scaffold has 6 joints to be welded. The complete arc spot welding consists on 12 spots with approximately 10mm, being 2 spots on each joint, as shown in Figure 20.

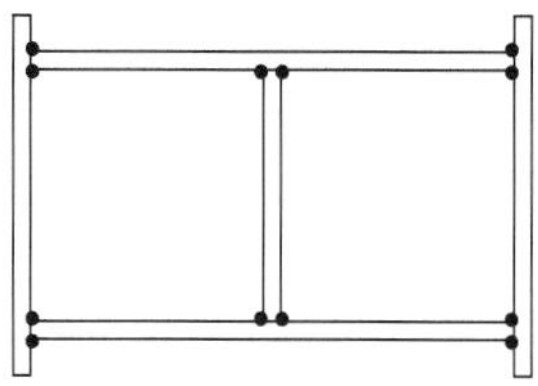

Fig. 20. Points to be arc spot welded in 1,0 x 1,5m scaffold.

To program each spot, it was identified two positions (one oscillation) for the robot program (Figure 21). The electric arc is opened in position 1, moves to position 2 and extinguish the electric arc.

Fig. 21. Positions used to program each arc spot welding: position 1 (a) and position 2 (b).

However, due to differences in the tubes lengths and lack of parallelism in the cuts, still present in its preparation, it was not possible to obtain a good repeatability in the spot welding. In some points occurred lack of fusion, being necessary manual rework.

It was opted then to program the arc spot welding using 3 oscillations with the torch moving twice to each position: after opening the electric arc, the robot would move from position 1 to position 2, back to position 1 and, finally, back to position 2, extinguishing the

electric arc. This way, it was possible to almost get a utilization of 100% in welds, although the lack of repeatability in the preparation of the tubes.
With this procedure, the arc spot welding program with 12 welds lasts on average 55 seconds. Considering three operators doing assembly of scaffolds on the jigs, the robot was capable to arc spot weld 520 scaffold per day.

4.4 Program optimization

This program, although efficient, was not productive, as the human operators are capable to arc spot weld the same number of scaffolds in less time. It was then performed a study to optimize the robot programming aiming to reduce the time cycle, without compromise the arc spot quality.
Initially, it was opted to reduce the number of torch movements on each spot, from three to two oscillations. To avoid problems caused by differences on the tubes, it was necessary to adjust the weld parameters, increasing the voltage, keeping the current constant, to increase the bead width without increasing the heat input.
Some tests were done to verify the arc spots quality, resulting in satisfactory joints (Figure 22). Even using only two torch oscillations, it was obtained almost 100% of good welds.

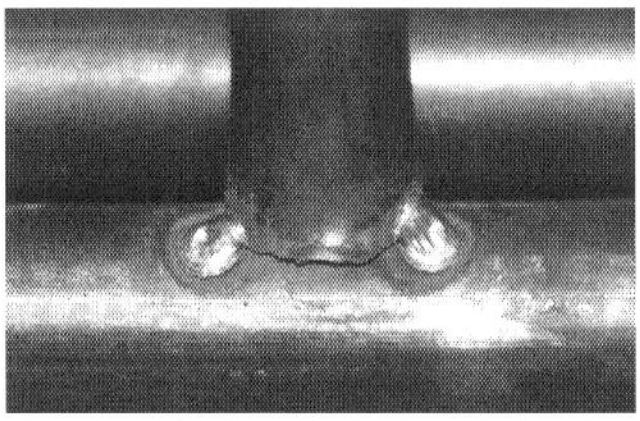

Fig. 22. Arc spot welded joint.

The average cycle time was reduced to 47 seconds, resulting in a daily production of 610 scaffolds.
Considering that this number was still low, it was opted to develop a new configuration of arc spot welds. In this case, instead of using two spots on each joint, the program was changed to make two spots in the inner joints and only one spot in the outer ones, as shown in Figure 23. This way, the number of arc spots to be make in each joint decreased from 12 to 8 spots.

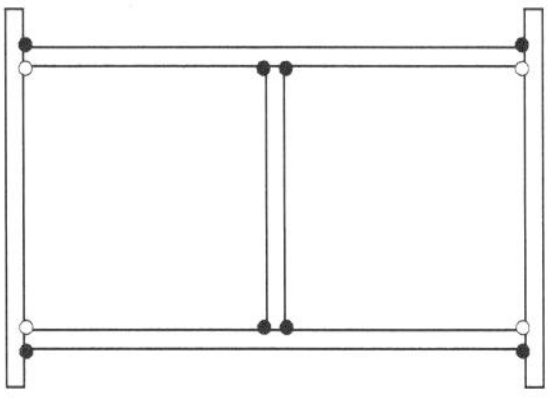

Fig. 23 Arc spot welds configuration using only one spot in the outer joints.

Figure 24 shows a spot in one of the outer joints.

Fig. 24. Outer joint spot.

The average cycle time was reduced to 31 second, resulting in a daily production of 920 scaffolds, over the company expectation that was of 900 scaffolds per day. However, those scaffolds presented a low flexural stiffness resistance, which could cause deformations during the transport and posterior welding.

It was then used a new spots configuration, in which the inner joints have only one weld spot, while the outer ones have one or two, alternatively, as shown in Figure 25. The average cycle time was not changed as the number of spots is also 8, maintaining a daily production of 920 scaffolds.

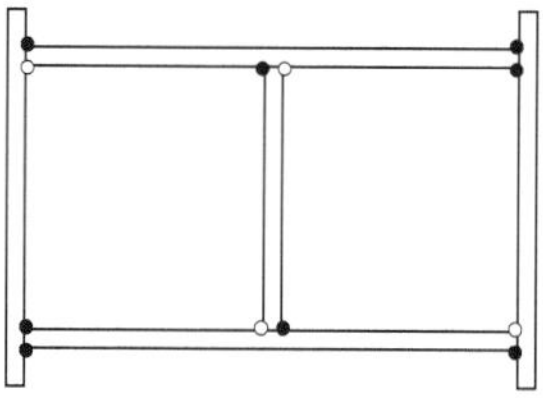

Fig. 25. Arc spot welds configuration to increase flexural stuffiness resistance using only 8 spots.

Figure 26 shows an arc spot weld made in one inner joint. It was observed a significant increase in the flexural stiffness resistance. However, it was still lesser than using 12 arc spot welds. It was then used a new configuration, as shown in Figure 27. This new configuration, in theory, would increase the resistance as it locates the isolated spots in points with less flexor moment.

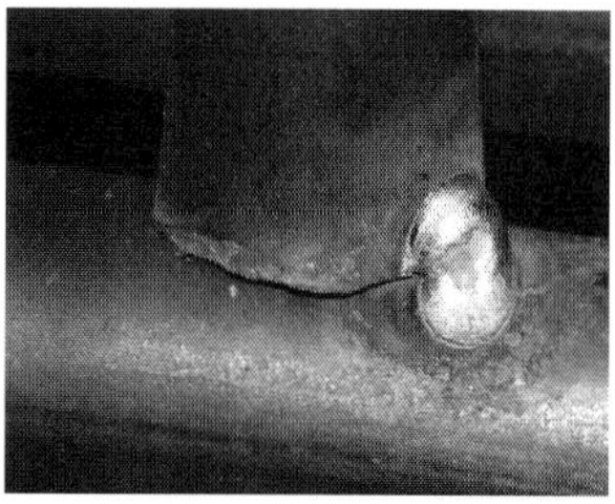

Fig. 26. Weld spot in one inner joint.

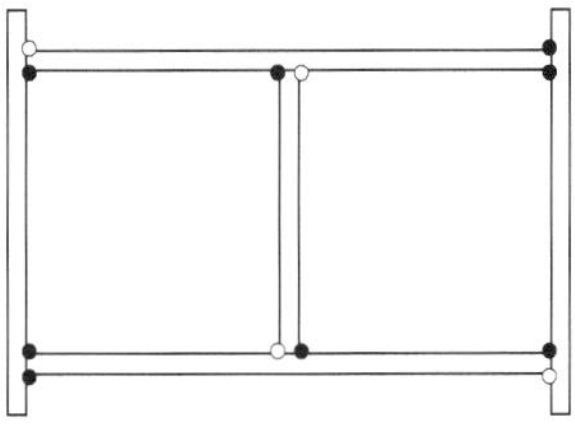

Fig. 27. Alternative configuration of arc spots to increase flexural stiffness resistance using 8 spots.

As foreseen the new configuration produces scaffolds with a greater stiffness resistance, if compared to the previous configuration.
The next step was to program the complete welds; however, the repeatability of the tubes preparation was not adequate to the robotic welding process. Some tests showed that it is possible to weld tubes with joints with almost 3mm of gap by changing the welding parameters in order to obtain wider beads (Fig. 28a); however, if the gap is greater than this value, it is not possible to obtain adequate beads (Fig. 28b).

(a) (b)

Fig. 28. Weld obtained changing welding parameters for small gaps (a) and great gaps (b).

This work shows that a robotic system is not always able to solve practical problems, as its programs just repeat the previously programmed trajectory and parameters. To solve this problem it would be necessary an adaptive control system to measure the gap and change welding parameter for each joint.

5. Development of a robot for orbital welding

Manufacturing of oil and gas lines is made through the union of metallic pipes, which length, in average, changes from 12 to 14 meters, in order to produce lengthier ducts. Figure 29 shows pictures obtained in a pipe work performed in Brazil. Figure 30 shows the welding procedure. The process used to weld these pipes is called circular or orbital welding. As can be seen in the figure, in Brazil the welding of pipes is all manually carried through with the GTAW process (Gas Tungsten Arc Welding) and coated electrode - SMAW (Shielded Metal Arc Welding).
The manual welding is not just ergonomically improper to the human been because the pipes are welded in loco and near to the floor but also it does not guarantee the desired productivity and repeatability. The great challenge then was the development of a robot for the orbital welding of pipes aiming to better comply the work with requirements. This process has as operational characteristic the fact that each welding bead is composed by 4

different welding positions. The positions are the plain position, the over-head position, the ascending vertical position and the vertical descendant position. In each one of them the optimal welding parameters are different and a robot must to self adjust to them.

Fig. 29. Preparation of a pipeline assembly. (a) pipes positioning and (b) root pass.

Fig. 30. Pipeline manual welding procedure.

As commented before, the definition of a robot is a "reprogrammable multifunctional manipulator designed to move materials, parts, tools or specialized devices through variable programmed motions for the performance of a variety of tasks". From this definition, it can be said that the devices for the orbital welding shown in literature up to now (Blackman and Dorling 1999; Yapp and Blackman 2004) are not robots, because they do not allow the programming of trajectories or parameters of welding. For these reasons, it is said that the currently process of orbital welding is mechanized, not robotic.

Weld pipes *in loco* using anthropomorphous industrial robots would be possible, however impracticable, due to the great weight which would have to be dislocated to each new pipeline bead. The device developed and that will be presented here, in such a way, makes possible the programming of trajectories, so as programming the welding parameters for the orbital welding of pipes. Therefore, such device can be called a robot due to its capability of being completely programmable and automatically carrying through all welding activities: opening and closing the electric arc, moving the welding torch (controlling the welding speed, the torch angle and stick-out) and controlling the welding current and the electric arc voltage. However, incapable of being completely multi-functional (it could not be used for a generic task, being limited to movements around of the pipe), such mechanism can be defined as a "robot designed to special tasks".

The development of the welding started with some tests been performed by a qualified welder using GMAW and FCAW processes in order to obtain optimal orbital welding parameters (Soragi, 2004). Beads on pipe were made in the four welding positions - flat, overhead, vertical up and vertical down. For every sample produced, metallographic tests were performed to determine the bead quality and the best welding parameters. An anthropomorphous industrial robot was also used to check the repeatability and weld quality in the four positions.

From the obtained results, best parameter tables were generated indicating the optimal parameters (voltage, current, welding speed, torch angle and stick-out) for each welding position mainly with tubular wire. In the intersection or where the welding changes from one position to other the parameters were interpolated in a small interval to avoid discontinuities and heterogeneity of the bead.

It was observed, however, that the optimal parameters for the descendant vertical welding substantially differ from those at vertical up, flat and overhead. Choosing to weld all around the pipe would introduce then unnecessary difficulties in the regions where the parameters must change from one position to other. It was opted then to perform welding only in the following sequence: overhead, vertical up and flat. Thus, the robot must weld one side of the pipe, extinguish the arc, go down to the other side and perform the other bead in the same sequence. This, of course, has the inconvenience of having electric arc extinguished. However, it allowed using short cables to connect the robot to the controller and to the welding machine.

In order to change parameters during the welding process, it was necessary to know the position of the robot in relation to the flat position. This positioning can be provided by a sensor that informs the inclination where the robotic system is at every moment (the inclinometer).

The robot was projected and constructed with 4 degrees of freedom: movement around the pipe, torch angle, stick out and torch lateral motion. Figure 31 shows these 4 degrees. As the robot has to weld pipes near the floor, it needs to be compact. Many versions were studied. Figure 32 shows the first and the 6th version.

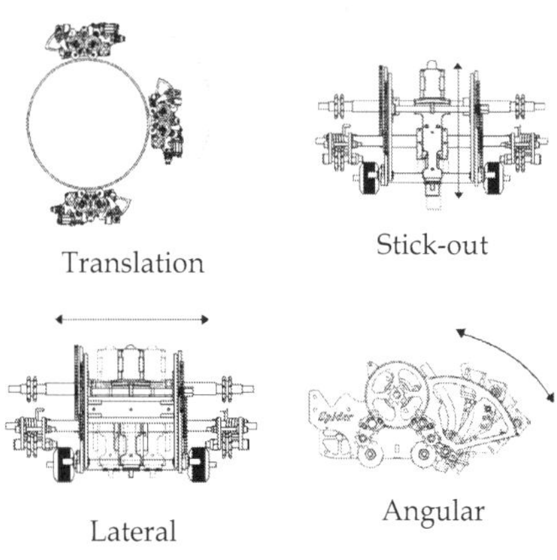

Fig. 31. Degrees of freedom of the robot.

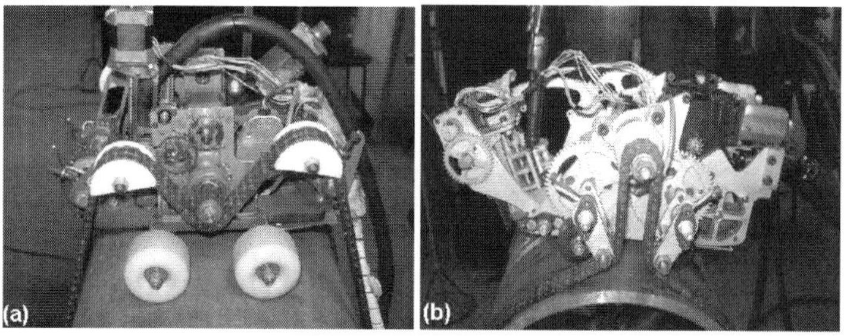

Fig. 32. Versions of the robot (a) the first and (b) the 6th.

In order to drive the movement around the pipe and to control its speed, a DC motor was selected, driven by PWM (Pulse Width Modulation). For the stick out, inclination and lateral motion it have been selected step motors which although its reduced dimensions, provide high torque. Moreover, for these movements, position control must be precise, what makes the step motors the perfect choice. The robot controller is implemented in a PC in which digital output and input boards were added in order to make possible to drive and control the robot axles, as well as the welding machine. During the program execution, the controller generates set point values to the speed of the first axle and position of the three following axles. The values of welding speed, the torch angle and stick out are informed through the parameters look-up table. Thus, for each position of the robot around the pipe (which is read from the inclinometer sensor), it is possible to generate the set points with the optimal values for such parameters.

Knowing the reference values, the controller implements the speed control of the movement around the pipe. The speed measuring is performed by means of an encoder located in the axle of the driving motor. Using the encoder pulses frequency, the real speed of the robot is determined with precision. When some error between the reference and the real speed exists, the driving voltage of the driving motor is modified so that the error heads to zero. After calculating the new driving voltage, an analogical signal is generated through a D/A board and sent to the PWM which amplifies the signal power and drives the DC motor.

In the case of this robot, the controller must be as robust as possible, as many factors have influence in the system dynamics: the traction in the chains, the robot position (flat, overhead, descendant vertical and ascendant vertical), the pipe diameter etc. On the other hand, in the positioning control of the step motors are used drivers that feed the coils in the right order, so as to put them into motion according to the signal sent by the PC.

The conventional welding source was modified in order to have two independent wire feeders allowing simultaneous use of two robots. Originally, the weld font had a potentiometer to adjust the welding voltage. Each one of the feeders has a potentiometer to adjust the welding current (wire feeding speed). Both potentiometers were manual. So, the operator would have to regulate voltage and current before starting the welding.

To the robotic process, however, it is needed that the welding parameters (current and voltage) be regulated by the robot itself. Thus, an electronic board was developed to work as the interface between the robot controller and the welding machine. The values of current and voltage to be used are determined by means of the parameters look-up table, in accordance to the robot position around the pipe. The digital values for regulation of the welding font are determined by means of a calibration curve from the welding power source.

Figure 33 shows a weld bead made by the robot. It can be observed the homogeneity of the bead, even where the welding position is modified. This is achieved by the gradual variation of the parameters of the table during the welding.

Fig. 33. Weld bead.

The orbital welding process robotization brings enhancement in the final product quality, considerable increase of the repeatability, reduction of rework and reduction of the weld execution time. At the very least, the robot is capable to reproduce the work (the weld bead) of the best human welder, through the use of the same parameters contained in a reference table. Moreover, it is possible to optimize such parameters, in order to increase the quality and to reduce the weld execution time through the welding speed increase.

The use of the robot in the welding with GMAW and FCAW revealed to be extremely viable. It was shown that the bead aspect did not suffer great variations from a welding position to another one, if a gradual change of the parameters (voltage, current, welding speed, torch angle and stick-out) is executed. In pipes with larger diameters, it is still possible to use two robots simultaneously, decreasing even more the closed arc time, which consequently increases the work factor.

6. Conclusions

This chapter discussed the many levels of automation of the arc welding processes, from the manual process (with no automation) to the adaptive control. To implement the automation of a process and to decide which level should be implemented, some aspects need to be studied as financial viability and number and variability of welds. If it is an extremely variable process, it should be considered no automation at all, as the setup and programming would take more time than the welding itself. On the other side, if it is a repetitive process with an adequate preparation of the parts to be welded, a robotic system with a preprogrammed task would guarantee repeatability and productivity to the process.

7. References

AWS A3.0:2001, (2001), *Committee on Definitions, Standard Welding Terms and Definitions*, American Welding Society (AWS), Miami, USA, 2001, pp. 11-67.

Batana, J. R. & Bracarense, A. Q. (1998). Monitoração e controle do processo de soldagem com eletrodo revestido visando a robotização. *XXIV Encontro Nacional de Tecnologia da Soldagem*. ABS.

Bracarense, A. Q. (1994). *Shielded Metal Arc Welding Electrode Heating Control By Flux Ingredients Substitution*. PhD Thesis, Colorado School of Mines. Golden, Colorado.

Bracarense, A. Q., Bastos Filho, T. F., Felizardo, I., Rogana, W. G. (2002). Soldagem Robotizada. In: Romano, V. (Org.). *Robótica Industrial. Aplicação na Indústria de Manufatura e de Processos*. São Paulo: Edgard Blücher. cap. 8, p. 139-155.

Cary, H.B., (1994), *Modern Welding Technology*, 3º Edition, Englewood Cliffs, New Jersey: Regents/Prentice Hall.

Craig, J. J. (1989) *Introduction to Robotics Mechanics and Control, 2nd ed.* Addison-Wesley Publishing Company.

Dantas, C. M.; Lima II, E. J. & Bracarense, A. Q. (2005). Development of an instrumented gripper for robotic shielded metal arc welding. *Proceedings of the 18th International Congress of Mechanical Engineering*. ABM.

Felizardo, I. (2003). *Estudo Experimental e Numérico do Aquecimento de Eletrodos Revestidos Durante a Soldagem*. Doctorate thesis. Minas Gerais, Brazil, Universidade Federal de Minas Gerais.

KUKA Roboter GmbH (2003). *KUKA System Software (KSS) Expert Programming*.

Lima II, E. J. & Bracarense, A. Q. (2009). Trajectory generation in robotic shielded metal arc welding during execution time. *Industrial Robot Journal*, 36 (1): p.19-26.

Oliveira, H. B. (2000). *Estudo para implementação de um Sistema para Monitoração e Controle na Soldagem Robotizada com Eletrodo Revestido*. Dissertação de Mestrado, Universidade Federal de Minas Gerais. Departamento de Engenharia Mecânica.

Pessoa, E. C. P.; Bracarense, A. Q.; Liu, S. & Guerrero, F. P. (2003). Estudo comparativo do desempenho de eletrodos revestidos E6013, E7024 e E7018 em soldagem subaquática em água doce do aço A36 a profundidades de 50 e 100 metros. *Anais do 2ºCOBEF*. ABCM.

Pereira, A. G. & Bracarense, A. Q. (2002). Soldagem robotizada de andaimes tubulares. In: *II Congresso Nacional de Engenharia Mecânica*, 2002, João Pessoa, Pb, Brasil. CONEM, ABCM.

Quinn, T. P.; Bracarense, A. Q., & Liu, S. (1997). A melting rate temperature distribution model for shielded metal arc welding electrodes. *Welding Journal*, 76(12):p.532s-538s.

Rivin, E. (1988), *Mechanical Design of Robots*, McGraw-Hill Inc., First Edition, New York, 1988.

Romano, V. F. (2002), *Robótica Industrial: Aplicação na Indústria de Manufatura e de Processos*, 1ª Edição, Edgard Blücher Ltda, Manet, São Paulo, SP, 2002, pp. 256.

Waszink, J. H. & Piena, M. J. (1985). Thermal process in covered electrode. *Welding Journal*, 64(2):p.37s-48s.

Welding Handbook, (2001), *Welding Science & Technology*, Volume 1, 9º Edition, American Welding Society (AWS), Miami, USA, 2001, pp 452-482.

4

WHI Formula as a New Criterion in Automatic Pipeline GMAW Process

Alireza Doodman Tipi and Fatemeh Sahraei
Kermanshah University of Technology, Pardis, Kermanshah, Iran

1. Introduction

Pipeline welding is one of the most significant applications of GMAW process. Automatic welding for pipelines has been developed from early 1970's. In these systems the welding robot moves around the two pipe's seam and welds the pipes by arc welding machine. Depending on the pipe thickness, weld process is repeated in several passes while the seam is filled of weld mass. The automatic pipeline welding systems has been recently paid more attention [1, 2].

In order to achieve sufficient performance in the process, the input parameters must be chosen correctly [3]. Welding parameter designing is a complicated step in the GMAW process, because of the large number of parameters and complexity of dynamic behavior. This complexity is particularly intensified in automatic pipeline systems, because of the complex seam geometry, wide range of the angle variations and strict quality requirements [1].

The most important input parameters in the automatic pipeline GMAW process are: welding current, arc voltage, travel speed, wire feeding speed, Contact Tube to Workpiece Distance (CTWD), welding position (angle), gas type, pipe type/thickness and seam geometry [4, 5]. The output parameters of the process are usually defined as either mechanical properties or weld bead geometry [6]. Weld bead geometry method considers the relationships between the input parameters and weld bead dimensions (penetration, width, reinforcement height, and width to penetration ratio and dilution [3, 7, 8].

Appropriate melting of the seam walls is certainly one of the most important conditions to achieve a proper dimension in fusion zone. A fusion zone with a sufficient width is necessary to prevent from some defects like lack of Fusion (LOF) [9, 10]. Having a direct contact between the arc and seam walls and receiving enough energy to the walls led to suitable wall melting and appropriate fusion zone [11, 12].

Some criteria such as heat input are related to the total energy which is given to the weld region without considering the amount of energy required to melt the wire. Principal parameters to calculate the heat input value are: welding current, arc voltage, travel speed and welding efficiency [11, 13].

During the welding process, part of the arc energy is spent to melt the wire [12]. Seam geometry also plays an important role in the amount of arc energy that directly reaches the walls. However a more general formula is not yet introduced.

In this paper WHI introduced as a new criterion, which is related to the arc energy that directly reaches the walls considering the both required energy to melt the wire and seam geometry. This criterion has the capability to be used for designing the welding parameters and for welding analysis applications. Section 2 contains theoretical parts to achieve WHI criterion, which includes wire melting energy (section 2.1), WHI formula (section 2.2), and wall geometry calculations (section 2.3). In section 3 two experimental tests are performed using the fabricated automatic system [15] in order to validate the obtained results from the presented formula.

Value (unit)	*Symbols*	*Nomenclature*
(mm^2)	A	*Molten area for wire (front view)*
(mm)	A_{osc}	*Torch oscillation amplitude*
500 (J/kg.°C)	c_{st}	*Specific heat for steel*
7800 (kg/m^3)	d_{st}	*Steel density*
(J/mm)	E_i	*Heat input*
7.7 (J/mm^3)	E_{st}	*Energy density for steel melting*
(J/mm)	E_w	*Wire melting energy*
(J/mm)	E_{wall}	*Wall energy*
(J/mm^2)	E_{wd}	*WHI*
2.48×10^5 (J/kg)	F_{st}	*Heat of fusion for steel*
(V)	V	*Arc voltage*
(A)	I	*Welding current*
--	η	*Arc radiation lost coefficient*
(m/s)	w_s	*Wire feed rate*
(m/s)	T_s	*Travel speed*
(mm)	r	*Wire radius*
(mm)	l	*Wall cross length with molten metal and arc (front view)*
(mm)	l_a	*Arc length*
(mm)	l_h	*Seam floor length (front view)*
(mm)	l_v	*Side wall length (front view)*
27 (°C)	T_0	*Heat of environment*
1510 (°C)	T_{mst}	*Melting point for steel*
(mm)	R	*radius of the seam shape*
(mm^3/mm)	V_w	*Wire volume per travel speed*
(mm)	W_a	*Arc width*
(deg)	a	*Wall angle*
(deg)	δ	*Arc angle*

Table 1. Variables and material properties.

2. WHI theory

In this section, firstly the required energy for melting the wire is calculated and the wall length of the groove face in contact with the arc is computed as well, remaining arc energy on the seam walls is named WHI.

2.1 Wire melting energy

Weld metal area (cross section, in front of view) is a function of wire radius, wire feeding speed and travel speed. The deposition rate (w_s/T_s) is usually considered to be fixed for designing of welding parameters. Therefore the molten metal cross section (A) is counted as an assumed parameter in design procedure.

$$A = \pi r^2 \frac{w_s}{T_s} \tag{1}$$

The amount of heat input over the length unit of travel axis is computed considering the radiation energy [11].

$$E_i = \eta \frac{VI}{T_s} \tag{2}$$

The mass of 1mm^3 steel is equal to $7.8 \times 10^{-9} kg$. Therefore melting of 1mm^3 steel (with 300°K primary temperature) needs approximately 7.7J energy according to eqn. (3).

$$E_{st} = 10^{-9} d_{st} c_{st} \left(T_{mst} - T_0 \right) + 10^{-9} d_{st} F_{st} \tag{3}$$

The volume of the molten wire poured down inside the seam along l_t mm of the travel axis is obtained using eqn. (4)

$$V_w = A l_t \tag{4}$$

The value of the required Energy for melting the wire poured inside the seam (versus travel axis unit (J/mm)) can be computed as below:

$$E_w = E_{st} V_w \tag{5}$$

2.2 WHI formula

Arc energy is spent to melt both the filler wire and the walls (eqn. 6). Therefore the remaining energy which directly contacts and melts the walls is named "wall energy" (E_{wall}).

$$E_i = E_w + E_{wall} \tag{6}$$

Energy density with respect to the seam wall length is computable through dividing the energy of the wall by the seam wall length (l) (front view). Therefore WHI can be defined as the below equation.

$$E_{wd} = \frac{E_{wall}}{l} \tag{7}$$

So WHI can be shown by welding parameters as below:

$$E_{wd} = \frac{\eta \frac{VI}{T_s} - 7.7\pi r^2 \frac{w_s}{T_s}}{l} \tag{8}$$

2.3 Wall geometry

The front view of the arc and bevel for the welding of the first pass in a U-type bevel can be seen in fig.1.

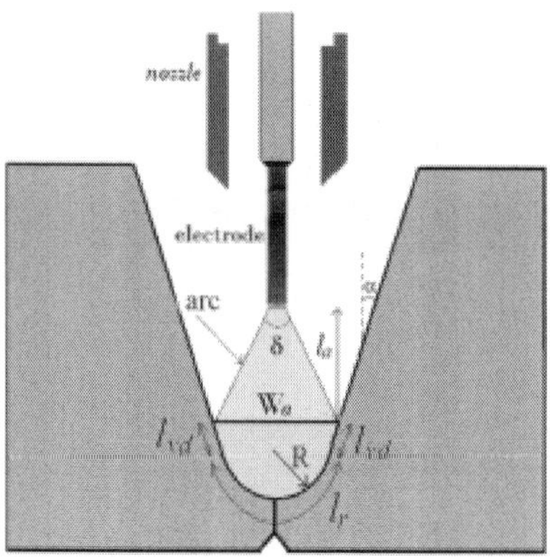

Fig. 1. A schematic view of the arc, melting wire area and seam walls

The arc width (Fig. 1) is calculated using eqn. (10) [14].

$$W_a = 2l_a \tan(\frac{\delta}{2}) \tag{9}$$

The wall length involving with the arc edges is calculated by eqn. (10-13).

$$\frac{W_a - 2R}{2} = l_{vd} \sin(\alpha) \tag{10}$$

$$l_{vd} = \frac{W_a - 2R}{2\sin(\alpha)} \tag{11}$$

$$l_r = \pi R \tag{12}$$

$$l = 2l_{vd} + l_r \tag{13}$$

For the other passes (except for the root pass), front view of the arc and seam (for U-type seam) is like Fig. 2.

Seam walls length (front view) involving with the arc edges is computable using eqn. (14-15).

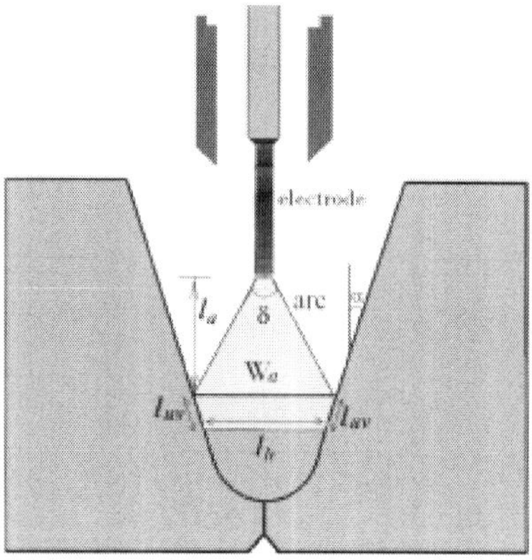

Fig. 2. A schematic view for melting area in the second pass

$$l_{vu} = \frac{W_a - l_h}{2\sin(\alpha)} \tag{14}$$

$$l = 2l_{vu} + l_h \tag{15}$$

If there is torch oscillation amplitude (see Fig. 3) the effective arc width can be computable by eqn. (16) [15].

$$W_a = A_{osc} + 2l_a \tan(\frac{\delta}{2}) \tag{16}$$

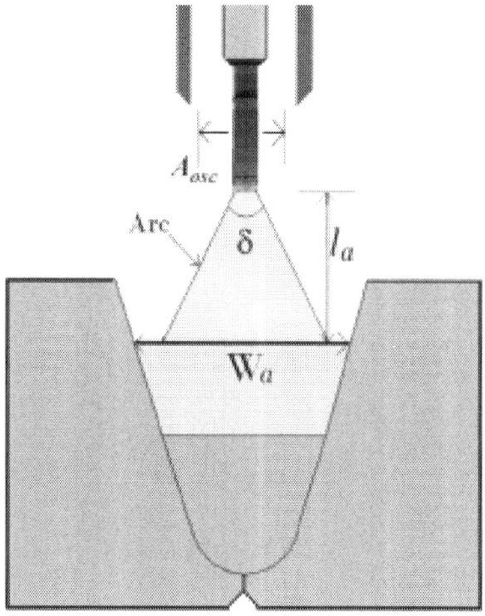

Fig. 3. Seam and arc with nozzle oscillation amplitude

If the effective arc width is too much compared to the melting wire area, some of the wall energy will be lost over the seam without any involving with the molten metal. Furthermore if the center of the oscillation and the seam centerline are not identical (see Fig. 4(left)), the fusion area in the both sides, will not be same.
A schematic view and a real test result are shown in Fig. 4. In this figure the center of the oscillation and the seam centerline do not coincide, additionally, the oscillation amplitude is too much, hence Fig. 4(right) has been resulted in the experiment.

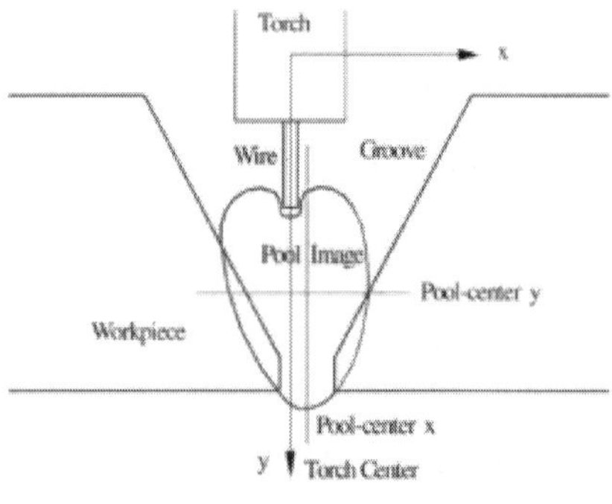

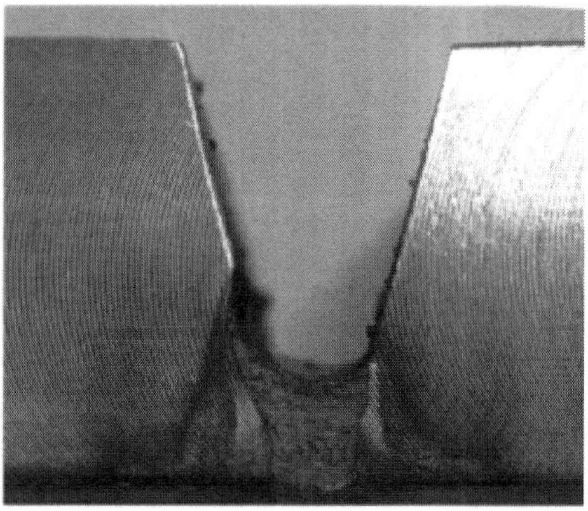

Fig. 4. Unsymmetrical and extra amplitude of arc width compared to molten wire area, schematic view (left), real test result (right)

3. Experimental results

3.1 Setup

The automatic pipeline welding system used in the experiments [15] has been shown in Fig. 5. The welding progresses downward semi-circularly from top (0°) to the bottom (180°) of the pipe on each side. The solid wire was ER70S-6(SG3), having diameter of 1 mm, the shielding gas is the mixture of Argon and CO_2 by 82/18 proportion.

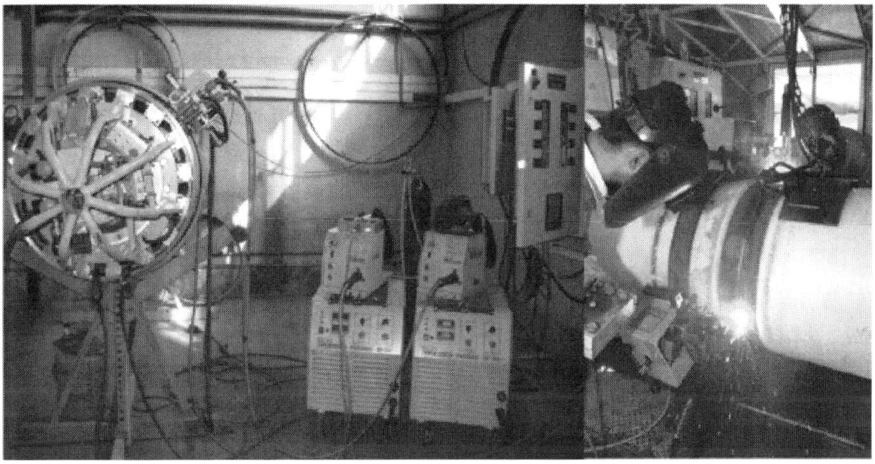

Fig. 5. Automatic pipe line welding system in the experiments (made by Novin Sazan Co.)

The pipe material is API 5L x65 HSLA steel with the thickness of 20.6 mm and the outside diameter of 32 inches. A U-type joint design is used according to fig6. Seam area is about 130mm^2, that is filled with several weld passes.

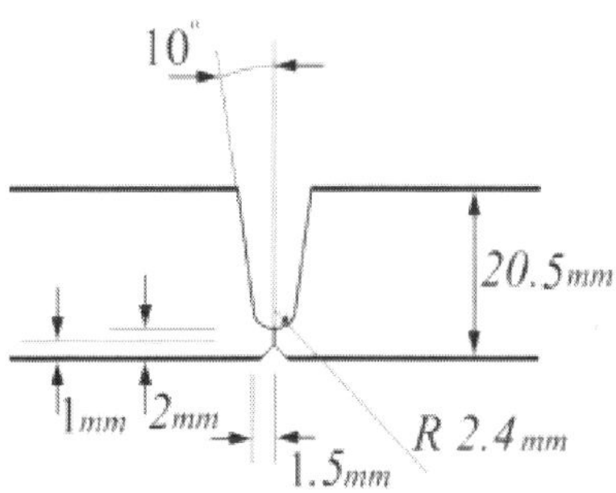

Fig. 6. Joint configuration in the experiments

3.2 Experiments

Ex. 1: WHI value has been chosen equal to 32.3 J/mm^2,the other parameters have been computed as the first row in Table 2 (only root pass parameters are shown). These parameters implemented on the system. Longitudinal cross section of the weld metal (front view) is shown in Fig. 7 (left), moreover root pass reinforcement (from inside the pipe) is shown in Fig. 8 (left).

	V(V)	I(A)	W_s(mm/s)	T_s(mm/s)	A(mm^2)	L(mm)	E_i(J/mm)	WHI(J/mm^2)
Ex. 1	24.4	276	251.7	14	15.1	8.84	393	32.3
Ex. 2	22.6	255	166.7	14	9.3	7.54	337	35.2

Table 2. Welding parameters for two experimental tests

The molten base metal area is about 66mm². This area is highlighted in Fig. 9 (a). this area has been calculated by computer and image processing algorithms using MATLAB.
Base metal molten area over to total molten area (summing of base metal and melting wire area) is defined as relative molten area, is about 34% for this example (*Ex. 1*).

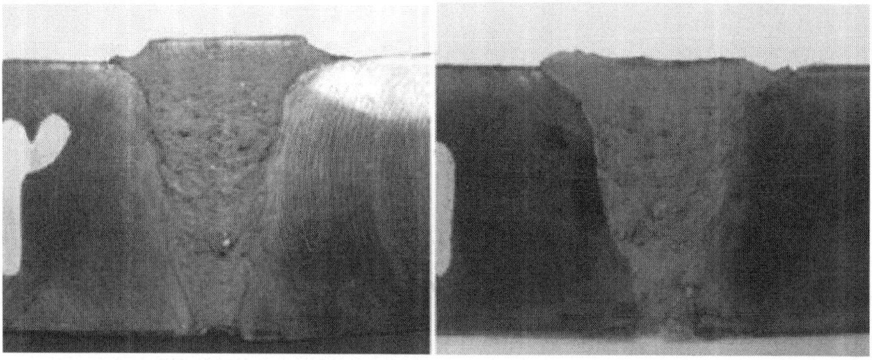

Fig. 7. Front view of the weld sections, 32.3J/mm² WHI and 393 J/mm heat input according to *Ex. 1*parameters (left); 35.2J/mm² WHI and 337 J/mm heat input with *Ex. 2* parameters (right)

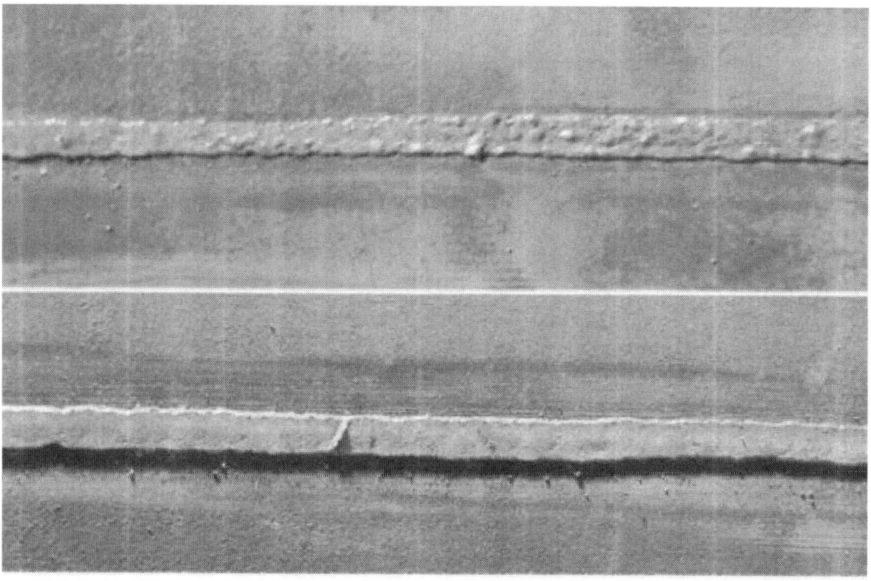

Fig. 8. Back view of welds from inside the pipe; for *Ex.1* (up) and for Ex. 2 (down), *Ex. 2* has more penetration than *Ex. 1* related to the more WHI

Ex. 2: WHI value has been selected equal to 35.2 J/mm^2, and Parameters are calculated according to the WHI value. Welding parameters are shown in the second row of the Table 2 (only for root pass). Comparing parameters of *Ex. 1* and *Ex. 2* indicates an important point: WHI increases but total heat input decreases because of the decreasing of the wire feeding rate (in *Ex. 2* than *Ex. 1*). Cross section and back view (from inside the pipe) is shown in Fig. 7(right) and Fig. 8(right) respectively. Molten base metal area (walls molten area) is estimated to be about 105 mm^2, which is shown in Fig. 9(b). Relative molten area is also estimated to be about 45%. Because of the more WHI (despite decreasing the heat input), fusion zone and penetration has been increased.

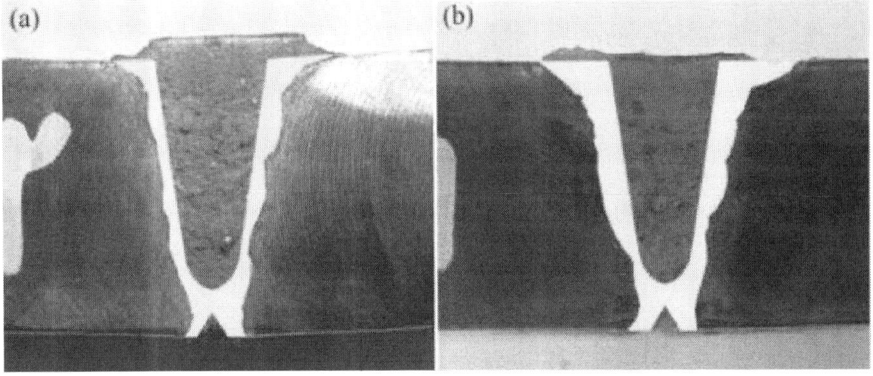

Fig. 9. Fusion zones of metal base in *Ex. 1* (32.3J/mm^2 WHI) (a), and for *Ex. 2* (35.2J/mm^2) (b)

4. Conclusion

In this paper WHI was introduced as a new criterion for designing of the welding parameters and welding analysis. This criterion calculates some of the heat input that is directly given to the walls from the arc. This formula considers the effects of the both required wire melting energy and seam geometry on the input energy. It was shown that WHI has a more correlation with fusion zone area compared to the heat input formula. In the other view the obtained results can be extended to the other welding processes and other applications. However WHI has a good feature in welding parameters designing to achieve some appropriate welding properties like the fusion zone.

5. References

[1] Lopes AGT (2006) *Arc-Based Sensing in Narrow Groove Pipe Welding*. Ph.D. Thesis, Sch. Ind. Manuf. Sci., Cranfield U.

[2] Blackman SA, Dorling DV (2000) Advanced welding processes for transmission pipelines. *In: 3rd Int. Conf., Pipeline Technol. Proc.*

[3] Murugan N, Parmar RS (1994) Effects of MIG process parameters on the geometry of the bead in the automatic surfacing of stainless steel, *J Mater Process Technol* 41: 381-98

[4] Thomsen JS (2004) *Advanced control methods for optimization of arc welding*, Ph.D. Thesis, Department of Control Engineering, Aalborg University, Denmark, June
[5] Connor LP (1991) *Welding handbook-welding processes*. 8th edi. American Welding Society
[6] Benyounis KY, Olabi AG (2008) Optimization of different welding processes using statistical and numerical approaches-A reference guide, *Adv Eng Soft* 39: 483-496
[7] Raveendra J, Parmar RS (1987) Mathematical models to predict weld bead geometry for flux cored arc welding. *Journal of Metal Constructions* 19: 31-35.
[8] Kim IS, Son JS, Kim IG, Kim OS (2003) A study on relationship between process variable and bead penetration for robotic CO2 arc welding. *Journal of Materials Processing Technology* 136: 139-145.
[9] Mendez PF, Eagar TW (2003) Penetration and Defect Formation in High-Current arc welding, Weld J, 82(10)296s-306s
[10] Okui N, Ketron D, Bordelon F, Hirata Y, Clark G (2007) A Methodology for Prediction of Fusion Zone Shape, weld J, 35s-43s
[11] Lancaster JF (1986) *The physics of welding*. Pergamon Pub., 2nd edi.
[12] Lin ML, Eagar TW (1985) Influence of arc pressure on weld pool geometry, Weld J, 64 163s-169s
[13] Lancaster JF (1993) *Metallurgy of welding*. Chapman & Hall pub., 5th edi.
[14] Guoxiang XU, Chuansong WU (2007) Numerical analysis of weld pool geometry in globular-transfer gas metal arc welding, Front. Mater. Sci. China, 1(1): 24–29
[15] Doodman AR, Mortazavi SA (2008) A new adaptive method (AF-PID) presentation with implementation in the automatic welding robot, *IEEE/ASME Int. Conf. Mechat. Emb. Sys. Appl., (MESA08).*

5

Sensors for Quality Control in Welding

Sadek C. Absi Alfaro
The Brasilia University, UnB
Brasil

1. Introduction

The welding process is used by many manufacture companies and due to this wide application many studies have been carried out in order to improve the quality and to reduce the cost of welded components. Part of the overheads is employed in final inspection, which begins with visual inspection, followed by destructive and non-destructive testing techniques. In addition to cost raise, final inspection is conducted when the part is finished only. When a defect occurs during welding, it can be reflected in the physical phenomena involved: magnetic field, electric field, temperature, sound pressure, radiation emission and others. Thus, if a sensor monitor one of these phenomena, it is possible to build a system to monitor the weld quality.

For the automation and control of complex manufacturing systems, a great deal of progress came up in the last decade, with respect to precision and on-line documentation (bases for the quality control). With the advent of electrically driven mechanical manipulators and later the whole, relatively new, multidisciplinary mechatronic engineering, the need of information acquisition has increased. The acquisition is, in many cases, distributed through the system, with strong interaction between the robot and its environment. The design objective is to attain a flexible and lean production. The requirement of real time processing of data from multisensor systems with robustness, in industrial environment, shows the need for new concepts on system integration.

A Multisensor system represents neither the utilization of many sensors with the same physical nature nor many independent measurement systems, but mainly sensor fusion, the extraction of global information coming from the interrelation data given by each sensor. Some examples are the estimation of the slope of any surface using two or three individual sensors, the simultaneous acquisition of the parameters of the automatic welding process MIG/MAG ("Metal Inert Gas/ Metal Active Gas") or the direct observation of the welding pool related to the control of current, voltage, wire speed and torch welding speed.

Technology advancements seek to meet the demands for quality and performance through product improvements and cost reductions. An important area of research is the optimization of applications related to welding and the resultant cost reduction. The use of non-destructive tests and defect repair are slow processes. To avoid this, online monitoring and control of the welding process can favor the correction and reduction of many defects before the solidification of the melted/fused metal, reducing the production time and cost.

With continuing advancements in digital and sensor technology, new methods with relatively high accuracy and quick response time for identification of perturbations during the welding process have become possible. Arc position, part placement variations, surface

contaminations and joint penetration are key variables that must be controlled to insure satisfactory weld production (Chen et al., 1990).
The techniques related to welding process optimization are based on experimental methodologies. These techniques are strongly related to experimental tests and seek to establish relations between the welding parameters and welding bead geometry. The introduction of close or adaptive control to welding processes must be done by monitoring a variable or set of variables which can identify a process disturbance. For each practical implementation of an adaptive system to a welding process one should identify the "envelope" or the set of monitoring variables. These variables must be used as a reference value in the process control, making the system control start with a parameter adjustment (welding current, voltage, etc.) to guarantee bead characteristics close to desirable values. The welding parameters vary in accordance to base material, type of chosen process, plate dimensions and welding bead geometry, so the adjustment of the reference value of a monitored variable will depend on the establishment of a set of optimized parameters which provide a welding bead with desirable specifications.
Researches related to adaptive systems for welding seek the improvement of welding bead geometry with direct (if based on monitoring sensors) or indirect monitoring techniques. The indirect monitoring systems are the more used, looking to link elements such as welding pool vibrations, superficial temperature distribution and acoustic emissions to size, geometry or welding pool depth (Kerr et al., 1999). The most used approaches in welding control are infrared monitoring, acoustic monitoring, welding pool vibrations and welding pool depression monitoring (Luo et al., 2000).
Aiming to optimize human analysis during the defect identification process, many researches were conducted to develop alternative techniques for automatic identification of defects considering different classes of signals such as plasma spectrum (Mirapeix et al., 2006), ultrasonic (Fortunko, 1980), computer vision (Liu et al. 1988), etc.
Three levels of "on-line" quality control have been adopted by the industry (see Fig. 1). In the first level, it should be able to automatically detect "on-line" bad welding joint production. In the second level, it should be able to search and to identify the fault and which are the reasons for the fault occurrence (changes in welding process induced by disturbances in shielding gas delivery, changes in wire feed rate and welding geometry, etc). In the third level, it should be able to correct welding parameters during the welding process to assure proper weld quality (Grad et al., 2004). The conventional parameters are usually used to detect and to identify defects. Moreover, the non-conventional parameters, at the present, are not used enough to evaluate the welding quality. They are some non-contact methods for welding monitoring process as acoustical sensing (Drouet, 1982; Mansoor, 1999; Yaowen, 2000; Tam, 2005; Poopat, 2006; Cayo, 2007, 2008, 2009), spectroscopy emission (Lacroix, 1999; Alfaro, 2006; Mirapeix, 2007), infrared emission (Nagarajan, 1992; Wikle, 1999; Fan, 2003) and sensoring combination (Alfaro et al. 2006).

2. Case studies

2.1 Spectroscopy

The science responsible for the study of the radiation emission is called spectroscopy. The physical phenomena consist on a photon emission in a determined wavelength or frequency after the absorption of some energy. Atoms, ions and molecules can emit photons in different wavelengths, but a wavelength is related only to one atom or ion or molecule. This can be compared to a fingerprint. Thus, with this property it is possible to know what chemical element, ion or molecule is found at the reading area.

It is possible to improve a non-destructive and on-line weld defects monitoring system through the radiation emitted by the plasma present in the electric arc. Some spectral lines involved in the welding process are chosen and their intensity is measured by a spectrometer sensor. One objective is to evaluate whether the spectrometer is capable of sensing disturbances in the electric arc. Another goal is to determine change detection techniques able to point those disturbances.

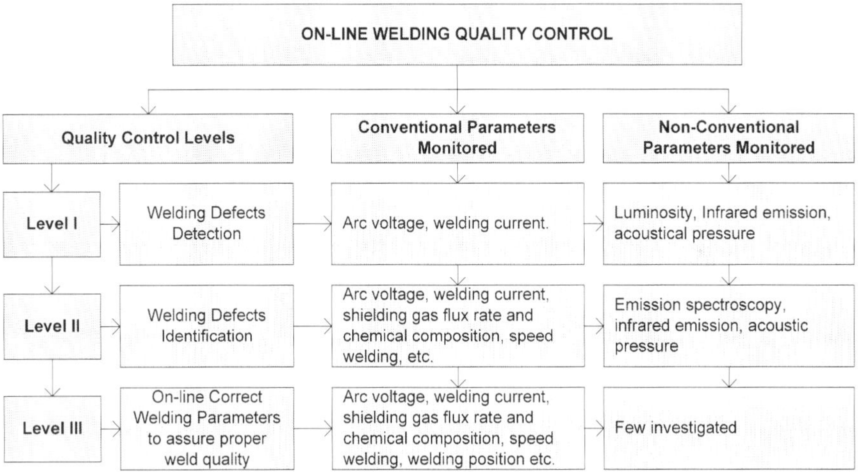

Fig. 1. On-line Welding Quality Control Levels.

Two analyses can be made with this information: qualitative and quantitative. In a qualitative approach, one is concerned in what elements are found on the plasma. And as a quantitative study, the objective is to evaluate some information extracted by the spectral taken. Therefore, a spectrometer could be applied as a sensor in a manufacturing process, such as welding, to detect the presence of some chosen elements or substances, like Iron, Cooper, water, grease; or to monitor significant changes of the energy emitted by some elements.

For example, in a stable GTAW the spectrum of the electric arc is stable as well. The amount of shielding gas, vaporized and melted steel, and other elements found at the electric arc are quite constant; therefore, if reflects on a stable spectrum. If a quantity of any element changes it will reflect on higher or lower emission energy. If different elements are introduced on the process, it will raise the energy of those elements.

An ordinary factor applied as a quantitative evaluation is the calculation of the plasma Electronic Temperature. Another that can be applied is the intensity of radiation emitted by some spectral lines. The Electronic Temperature can be calculated with different techniques, one is the relative intensity of spectral lines, of the transition from the level m to r of one line and from j to i of the other line, given by Equation 1 (Marotta, 1994)

$$T_e = \frac{E_m - E_j}{K_B \cdot \ln\left(\frac{E_m \cdot I_{ji} \cdot A_{mr} \cdot g_m \cdot \lambda_{ji}}{E_j \cdot I_{mr} \cdot A_{ji} \cdot g_j \cdot \lambda_{mr}}\right)} \quad (1)$$

Where: E is the energy level, KB is the Boltzmann constant, I is the spectral line intensity, A is the transition probability, g is the statistical weight and λ is the wavelength. These values can be found at the (NIST, 2010), except for the intensity, given by the sensor.

2.1.1 Change detection

The key idea behind change detection techniques is given by its name. It is to evaluate a signal and if there is an appreciable change in its behavior (frequency, magnitude or abrupt peaks), the system must be capable of detecting it. These perturbations can be defects on the welding process and a schematic diagram is given in Figure 2 (Gustafsson, 2000).

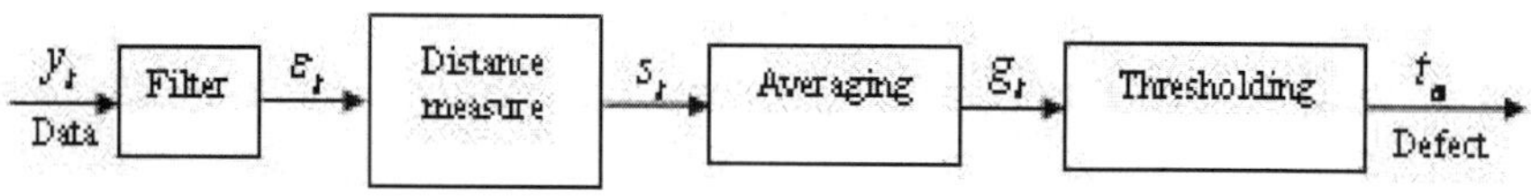

Fig. 2. Change detection diagram flux.

The blocks are explained separately. However, it is important to present the applied model first. The model proposed for the spectrometer reading is showed in Equation 2.

$$y_t = \theta_t + v_t \quad (2)$$

The signal given by the sensor (yt) is the radiation emitted by the plasma (θt) added by a noise (vt). The noise is a random variable with normal distribution with zero mean and variance R. Firstly, there is the filter block. It estimates the radiation intensity (θt) found in the reading model, Equation 2, and calculates the residual (εt). The next block calculates the distance. It is the difference between the sensor reading and the estimation. It is based on the residuals or it can be the value itself. A statistic test (gt) based on the distance is given in the third block. Finally, gt is compared to a threshold (h) to decide if there is a disturbance in the plasma. If the value is lower than the reference, it is assumed that the welding process is normal. Although, if the statistic test value is greater than the threshold, it is possible that a defect had occurred.

There are many change detection algorithms. It will present three of them. One widely used, Cusum LS Filter (Gustafsson, 2000), other developed by (Appel, 1983) and another here proposed, which applies steps from different algorithms. The Cusum LS Filter, Equation 3, presents a Least Square filter to estimate the radiation intensity, $\hat{\theta}_t$. The distance, s_t, is given by the residuals, ε_t, and the statistic test, g_t, is given by a cumulative sum. The factor σ is subtracted at each time instant *t* to avoid false alarms. Its value is chosen by the designer. And finally, the comparison of the statistic test to a threshold *h*. If its value is greater than *h*, an alarm is set and its instant, t_a, is recorded, the statistics tests are reset and t_0 becomes t. The input parameters are *h* and *σ*. With the values of the alarm instants, it is possible to indicate the defects position, once the weld speed is constant.

$$\textit{Filter} \qquad \hat{\theta}_t = \frac{1}{t-t_0} \cdot \sum_{k=t_0+1}^{t} y_k \qquad \text{Distance } \textit{measure} \qquad \varepsilon_t = y_t - \hat{\theta}_{t-1} \tag{3}$$

$$\textit{Averaging} \qquad g_t^{(1)} = \max\left(g_{t-1}^{(1)} + s_t^{(1)} - \sigma, 0\right), \quad g_t^{(2)} = \max\left(g_{t-1}^{(2)} + s_t^{(2)} - \sigma, 0\right)$$

$$\textit{Thresholding} \qquad \textit{if} \;\; g_t^{(1)} > h \;\; \textit{or} \;\; g_t^{(2)} > h \;\; \begin{cases} \textit{alarm}: \; t = t_a \\ g_t^{(1)} = g_t^{(2)} = 0 \\ t_0 = t \end{cases}$$

The other algorithms are based on sliding windows. There are two different proposals. A scheme can be seen in Figure 3. One proposal, first scheme, presents the idea to compare two models (two filters). Both models present the notation of Equation 2. The slow filter, that estimates M1, uses data from a very large sliding window. The fast filter estimates M2 by a small window. Then, two estimates, $\hat{\theta}_1$ and $\hat{\theta}_2$ with variances $\hat{R}_1$ and $\hat{R}_2$, are obtained. If there is no abrupt change in the data, these estimates will be consistent. Otherwise, an alarm is set.

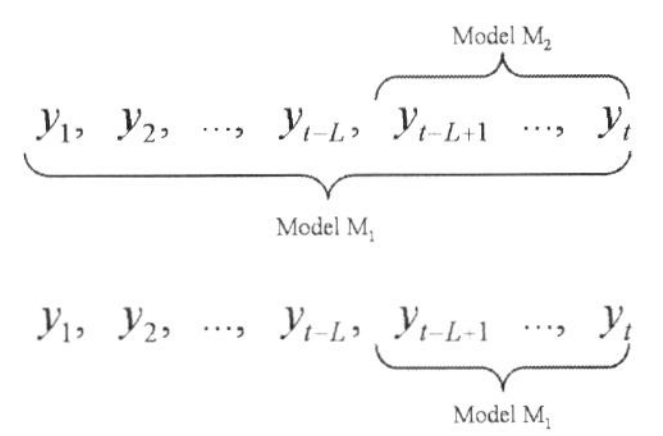

Fig. 3. Schemes for sliding windows.

An algorithm that adopts this scheme is the Brandt's GLR, proposed by (Appel, 1983). The algorithm is given by Equation 4. The variable $\hat{\theta}$ is the estimation of θ, L is the size of the sliding window, $\hat{R}$ is variance estimation and ta is the alarm time instant. Input parameters are L, h and, if the estimations do not converge, R_1 and R_2.

The other proposal, given by the second sliding window scheme, presents two Kalman Filters, Equation 5. One filter estimates the data in a sliding window, Model M_1 and the other estimates the past data at time t-1, Model M_2. The factor K is the filter gain and the Ps is the covariance matrix. For the distance measurement it was chosen the Brandt algorithm, Equation 4. The statistic test and thresholding are based in the CUSUM LS Filter, Equation 3. The algorithm parameters input are h, σ and R.

$$\text{Filter - Model M1} \qquad \qquad \text{Filter - Model M2}$$

$$\hat{\theta}_t^{(1)} = \frac{1}{t-t_0} \cdot \sum_{k=t_0+1}^{t} y_k \qquad \hat{\theta}_t^{(2)} = \frac{1}{t-L} \cdot \sum_{k=L+1}^{t} y_k \tag{4}$$

$$\hat{R}_t^{(1)} = \frac{1}{t-t_0} \cdot \sum_{k=t_0+1}^{t} \left(y_k - \hat{\theta}_t^{(1)}\right)^2 \qquad \hat{R}_t^{(2)} = \frac{1}{t-L} \cdot \sum_{k=L+1}^{t} \left(y_k - \hat{\theta}_t^{(2)}\right)^2$$

Distance Measure

$$\varepsilon_t^{(1)} = y_t - \hat{\theta}_{t-1}^{(1)} \qquad s_t = \log\left(\frac{\hat{R}_t^{(1)}}{\hat{R}_t^{(2)}}\right) + \frac{\left(\varepsilon_t^{(1)}\right)^2}{\hat{R}_t^{(1)}} - \frac{\left(\varepsilon_t^{(2)}\right)^2}{\hat{R}_t^{(2)}}$$
$$\varepsilon_t^{(2)} = y_t - \hat{\theta}_{t-1}^{(2)}$$

Averaging

$$g_t = g_{t-1} + s_t$$

Thresholding

$$if \quad g_t > h \begin{cases} alarm: \ t = t_a, \ t_0 = t \\ g_t = 0 \end{cases}$$

Window Model M1 (5)

$$\varepsilon_t^{(1)-} = y_{t-L} - \hat{\theta}_{t-1}^{(1)} \qquad \varepsilon_t^{(1)} = y_t - \hat{\theta}_t^{(1)-}$$
$$K_t^{(1)-} = \frac{P_{t-1}^{(1)}}{P_{t-1}^{(1)} - R} \qquad K_t^{(1)} = \frac{P_t^{(1)-}}{P_t^{(1)-} + R}$$
$$\hat{\theta}_t^{(1)-} = \hat{\theta}_{t-1}^{(1)} + K_t^{(1)-} \cdot \varepsilon_t^{(1)-} \qquad \hat{\theta}_t^{(1)} = \hat{\theta}_t^{(1)-} + K_t^{(1)} \cdot \varepsilon_t^{(1)}$$
$$P_t^{(1)-} = \left(1 - K_t^{(1)-}\right) \cdot P_{t-1}^{(1)} + Q_1 \qquad P_t^{(1)} = \left(1 - K_t^{(1)}\right) \cdot P_t^{(1)-} + Q_1$$

Model M2

$$\varepsilon_t^{(2)} = y_t - \hat{\theta}_{t-1}^{(2)}$$
$$K_t^{(2)} = \frac{P_t^{(2)}}{P_t^{(2)} + R}$$
$$\hat{\theta}_t^{(2)} = \hat{\theta}_{t-1}^{(2)} + K_t^{(2)} \cdot \varepsilon_t^{(2)}$$
$$P_t^{(2)} = \left(1 - K_t^{(2)}\right) \cdot P_t^{(2)} + Q_2$$

Distance measure

$$s_t = \frac{\left|\varepsilon_t^{(1)} - \varepsilon_t^{(2)}\right|}{\sigma} = \frac{\left|\hat{\theta}_t^{(1)} - \hat{\theta}_t^{(2)}\right|}{\sqrt{P_t^{(1)} + P_t^{(2)}}}$$

Thresholding

$$if \quad s_t > h, \ alarm: \ t = t_a$$

Figure 4 shows the results of the experiment based on the variation of the shielding gas flow rate for the CUSUM LS Filter algorithm, the Figure 5 shows for the proposed algorithm and Figure 6 for Brand Algorithm. The welding parameters chosen were the industrial standard ones. The spectral line chosen was Argon 460.9 nm. The algorithms were capable of detecting fine changes. The ellipses indicate where initially the rate had changed.

The results suggest that a spectrometer can be applied as a sensor for detecting disturbances in the electric arc during welding. These disturbances can be related to weld defects. The radiation emission was analyzed instead of the electronic temperature, once the interest was only in signal changes, not in its absolute value. Bearing that in mind, the computational effort is lower. The change detection technique can be applied to point out disturbances in the sensor signal. The algorithms chosen to analyze the signal with the selected spectral lines presented satisfactory performances; the best being obtained using the CUSUM LS Filter was for gas flow variation. Other parameters and different elements (or ion) wavelengths can be applied. More than one spectral line can be monitored with different algorithm input parameters to ensure disturbance detection. The selection of the spectral line to be monitored depends on the welding process, weld parameters, weld material and defects or disturbances to be monitored. Using this system, only the regions indicating defects will have to be inspected and reworked, therefore, shorter working hours and lesser consumables will be requiring, thus reducing production costs.

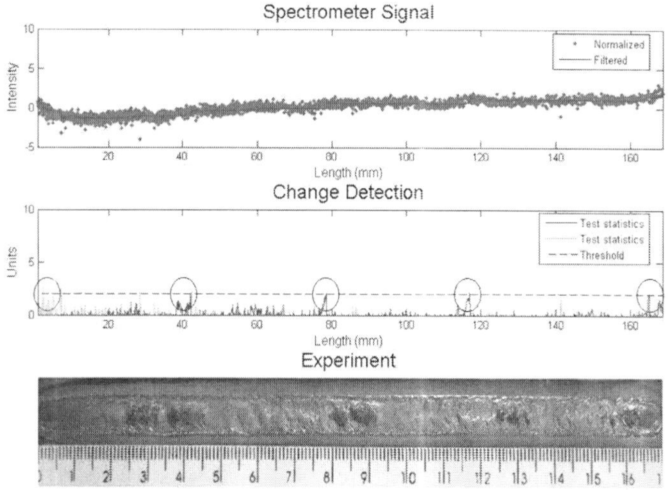

Fig. 4. Gas flow variation - CUSUM LS Filter.

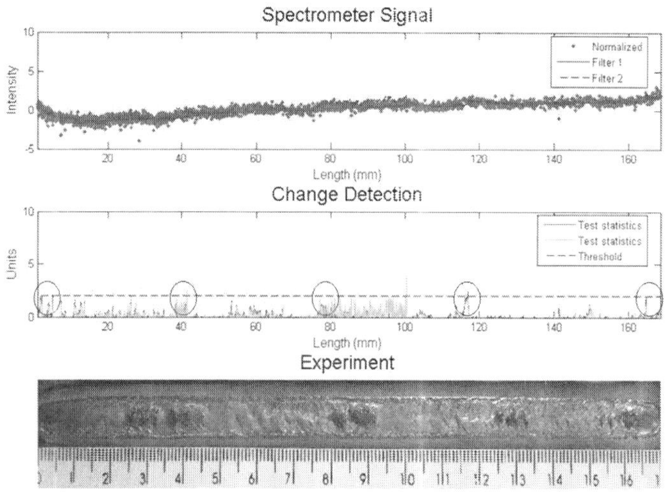

Fig. 5. Gas flow variation - proposed algorithm.

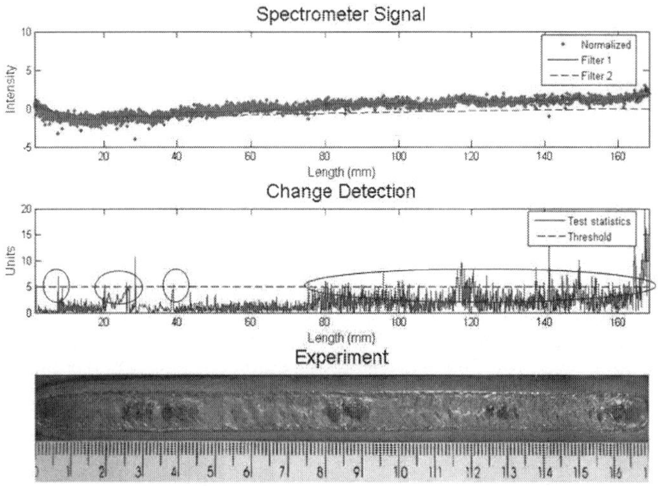

Fig. 6. Gas flow variation - Brandt algorithm.

2.2 Infrared monitoring

During the welding process, the high temperature associated with the arc and appropriate thermo physical properties such as thermal diffusivity cause strong spatial temperature gradients in the region of the weld pool. Convection in the weld pool, the shape of the weld pool and the heat transfer in both the solid and liquid metal determine the temperature distributions in the plate and on the surface. For an ideal weld with stable conditions, these surface temperatures should present repeatable and regular patterns. Perturbations in welding penetration should be clearly identifiable from variations in the surface temperature distribution (Nagarajan et al, 1989).

Infrared emissions indicate the heat content of the weld. For example, deeper penetration tends to correlate with increased heat input (caused by higher current or slower weld speed). Greater heat input results in higher temperatures and increased infrared emissions (Sanders et al., 1998). The temperature may be monitored by a pyrometer, but depend on the kind of sensor is using, due to the slow response time of the system and the presence of an intense thermal signal from the welding focused area (saturation problems). According to (Sanders et al., 1998), a better indicator is the infrared energy emitted by the weld, including both the contributions from the weld pool and plasma.

It is necessary to carry out the temperature measurement with a sensor that doesn't introduce defects during the welding process. It is for this reason that non-contact temperature sensors are more suitable. An infrared thermometer measures temperature by detecting the infrared energy emitted by all materials which are at temperatures above absolute zero, (0 Kelvin). Arc welding is intrinsically a thermal processing method. To this end, infrared sensing is a natural choice for weld process monitoring. Infrared sensing is a non-contact measurement of the emissions in the infrared portion of the electromagnetic spectrum.

The infrared monitoring techniques for weld pool are: area scanning and point monitoring. Area scanning provides a bidimensional view of the surface temperature distribution profile, making possible a complete analysis of the heat transfer process during welding (Nagarajan, 1989; Chen, 1990). Considering that we are dealing with bidimensional images, the application of area scanning demands a better computational structure (hardware and software), requiring a longer processing time (Venkatraman et al., 2006). On the other hand, the point monitoring technique demands little computational structure, requiring a shorter processing time, which makes it more appropriate for controlling in real time (Chin, 1999; Wikle, 2001). A recent study presented the adaptive control of welding through the infrared monitoring of the weld pool using point sensors (Araújo, 2004). The most basic design consists of a lens to focus the infrared (IR) energy on to a detector, which converts the energy to an electrical signal. This configuration facilitates temperature measurement from a distance without contact with the object to be measured (Merchant, 2008).
To make a correct measurement with this class of sensors, it is necessary to focalize the area that is going to be measured; this is possible by knowing the focal distance of the lens. Figure 7 shows a focal distance for one infrared sensor. In this case, a focal distance has a length of 600 mm and a radius of 4 mm (waist radius).

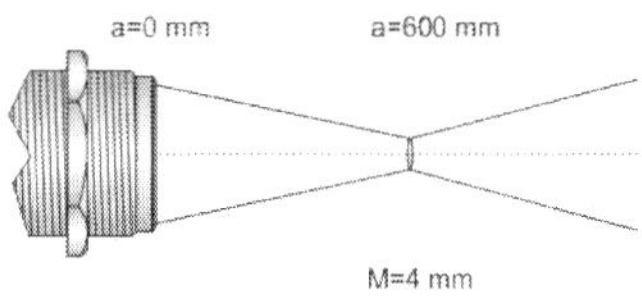

Fig. 7. TL-S-25 Infrared Sensor focus[1].

2.2.1 Failure detection

This study compares two algorithms for defect detection. The first one used is the conventional Kalman filter together with the Mahalanobis distance calculus to evaluate the presence of failures. In the second, the linear regression Kalman filter-LRKF and the generalized likelihood ratio test (Appel et al., 1983) are used to determine the distance between the autoregressive model and the signal read.

2.2.2 Change detection

This is a statistical technique that can detect abrupt changes in signals. Since welding is a stochastic process (Alfaro, 2006), some properties and algorithms can be applied. It consists basically on the flux of Figure 2.
Under certain model assumptions, adaptive filters take the measured signals and transform them to a sequence of residuals that results in a white before the change occurs (Gustafsson, 2000). If there is no change in the system and the model is correct, then the residuals are a sequence of independent variables with zero mean and known variance. When a change occurs, it can reflect on some variation in the mean, variance or both values that makes the residuals greater. The main point is to establish how great is this value to assume that a change had occurred. The statistical test decides whether the deviation is significant or not. The evaluation is usually made on four situations, change in the mean, change in the

[1] Calex Electronics Ltd

variance, change in correlation and change in signal correlation. In this work the evaluation was made on the mean and it is based on the residuals.
The stopping rule is based on the distance measurement. Many change detection algorithms make a decision between two hypotheses:

$$H_0 : E(s_t) = 0,$$
$$H_1 : E(s_t) > 0 \tag{6}$$

This rule is achieved by the value calculated by the low-pass filter s_t and compares to a threshold. If the value is greater, an alarm is set.

2.2.3 Kalman filter

A simple description of the infrared signal behavior as a discrete temporal series can be made in terms of an autoregressive model (AR) of order m. The present sample value is represented by the linear combination of m past samples incremented by a parameter of uncertainty. For a temperature registry of a component z[t], According to (Pollock, 1999) the model AR of order m is given by Equation (7):

$$Z[t] = \sum_{i=1}^{m} a_i z[t-i] + \varepsilon[t] \tag{7}$$

where ai = {a1, . . . , am} are the coefficients of model AR and ε[t] is the noise component to represent the inaccuracy of the signal reading during welding. It is supposed that the sequence ε[1 : t] = {ε [1], ..., ε [t]} is independent and identically distributed (i.i.d) Gaussian with mean E{ε [n]} = 0, variance E{(ε[n])2} = σ2.
From the observation of different statistic characteristics in the noise residues and the presence of defect in a model AR of order m, it is possible to establish a recursive estimation system using a stochastic filtration technique to observe and track the temperature interval in which the gaussianity of the sequence is preserved. One of these tools is the Kalman filter. The state vector is given by Equation (8) (Jazwinski, 1970):

$$x[k] = A[k]x[k-1] + w[k] \tag{8}$$

where x[k] is the state vector of dimension n, A[k] is a square matrix of state transition, w[k] is a sequence of dimension n of Gaussian white noise of null mean. The observation model is given by Equation (9):

$$z[k] = H[k]x[k] + v[k] \tag{9}$$

in which z[k] is the observation vector of dimension m, H[k] is the measuring matrix and v[k] represents Gaussian white noise of null mean. It is supposed that the w and v processes are non-correlated and also:

$$E\{w[k]w[i]^T\} = \begin{cases} Q[k], & if \quad k = i \\ 0, & if \quad k \neq i \end{cases}$$
$$E\{v[k]v[i]^T\} = \begin{cases} R[k], & if \quad k = i \\ 0, & if \quad k \neq i \end{cases} \tag{10}$$

In this system, the initial state $x[0]$ is a random Gaussian variable of mean $\hat{x}[0]$ and matrix of covariance P[0]. $x[0]$ is supposedly non-correlated to the w and v processes. The basic problem of the Kalman filter is to obtain an estimation $\hat{x}[k|k]$ of $x[k]$ from the measurement {z[1], z[2], . . . , z[k]}, in order to minimize a metric of mean square error. This metric is given by the trace of the *a posteriori* error covariance matrix as presented in Equation (11):

$$P[k|k] = E\left\{(x[k]-\hat{x}[k|k])(x[k]-\hat{x}[k|k])^T\right\} \quad (11)$$

Fortunately, this estimation problem presents a recursive solution. This solution is given in two stages. First there is a prediction stage (between observation Equation (12, 13)), in which:

$$\hat{x}[k|k-1] = A[k]\hat{x}[k-1|k-1] \quad (12)$$

$$P[k|k-1] = A[k]P[k-1|k-1]A[k]^T + Q[k] \quad (13)$$

Then, there is the correction stage in which the actual observation is used to correct the prediction $\hat{x}[x|k-1]$:

$$\hat{x}[k|k] = \hat{x}[k|k-1] + H[k](Z[k] - H[k]\hat{x}[k|k-1]) \quad (14)$$

$$P[k|k] = P[k|k-1] - K[k]H[k]P[k|k-1] \quad (15)$$

in which:

$$K[k] = P[k|k-1]H[k]^T\left[H[k]P[k|k-1]H[k]^T + R[k]\right]^{-1} \quad (16)$$

is named Kalman gain.

The main idea concerning the defect identification is related to the use of a statistic test that, jointly with stochastic filtration, verifies if the infrared samples properties are related to the estimation of the model AR given by the Kalman filter. If the test fails, it is supposed that the actual sample correlates with the presence of defect.

The comparison between the infrared signal sample and the recursive estimation consists in the Chi-square probabilistic hypothesis through the Mahalanobis distance (Duda, 2001). Such a distance is a natural measurement that indicates, in a probabilistic sense, how much of the registry of the actual sample is compatible with the estimated infrared signal model, estimated by the Kalman filter.

Figures 8 and 9 show an experiment in which the defects were introduced through the presence of water during the welding process. Figure 8 shows an analysis done with the generalized likelihood ratio test, and Figure 9 shows an analysis done according to the Mahalanobis distance. In Figure 8 we observe four clearly detected defects, it is observed that the last two defects remained on the limit of the Stopping Rule and two more anomalies were detected around 108 and 110 mm. In just one of them (108 mm) a variation in the form of a bead is observed.

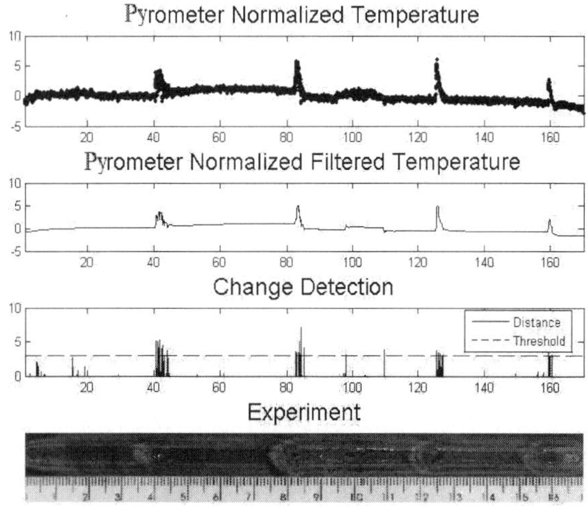

Fig. 8. Plate with water defects and change detection analysis.

Figure 9 shows an analysis according to the Mahalanobis distance. It is observed that the distance to the region where there is no presence of defects (constant temperature) is located below the threshold proposed by the statistic test. During the presence of the defect, the residue between the real sample Z[k] and the sample estimated by the AR coefficient increases at such a rate that the distance Z[k] surpasses the established threshold where the defect presence is verified. We should also note that this test could not detect the anomalies presented around 105 mm.

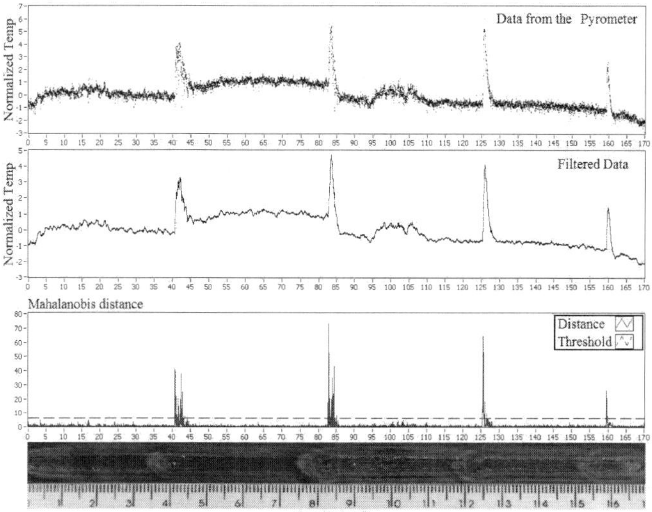

Fig. 9. Current Plate with water defects and Mahalanobis distance.

Infrared weld pool monitoring in the GTAW process provides information about penetration depth. It also shows that infrared signal variations in DC are related to weld penetration depth, while AC portions of the output can be correlated with surface irregularities.
Together with a change detection algorithm, the system monitors the residual of the regression algorithm, looking for changes in the mean. The proposed method maintains a regression model where residuals are filtered by a Kalman filter. A Mahalanobis distance algorithm monitors significant changes in the output of the Kalman filter. The Kalman filter has a good performance in detecting real changes from noisy data. The simplicity of the proposed algorithm permits its implementation in systems for monitoring, detection and localization of events in real time.

2.3 Acoustic sensing

By monitoring arc voltage and welding current allowing the detection of the arc perturbations during the welding process and depending of its profile, these perturbations can be interpreted as defects on the welded joints. The on-line detection and localization of the defects reduces the severity and time consuming of the quality control tests. Most of the commercial equipment for arc voltage and welding current monitoring use sensors based in voltage divisors and Hall Effect and they are installed directly on the welding process. The sensors connected directly on the welding process present two considerable disadvantages: The stability on the welding process, due to its high sensibility, can be interfered by the sensors with electrical connections altering the electrical arc impedance and generates undesirable arc instabilities and the electromagnetic arc interferences alters considerably the measure makes by the sensors. The electrical arc generates physical phenomena like luminosity, infrared radiation electromagnetic fields and sound pressure. It is known that the specialized welders use an acoustic and visual combination for the monitoring and control of the welding process (Kralj et al., 1968). In the end 70 years' measurement of electrical arc voltage was successfully carried out by acoustical methods (Drouet, 1979, 1982). The welding arc sound represents the behavior of the electrical parameters of the welding arc, consequently this fact make possible monitoring the stability of the welding process through the sound. The arc sound of the GMA welding in the short circuit transfer mode can represent the extinction and ignition sequence of the arc voltage and therefore it opens the possibility to detect acoustically perturbations in the welding arc (Cayo, 2008). The main advantage of the sound monitoring system lies on the fact that there is no need to have electrical connections to the welding process since the sound is transmitted from the welding arc to acoustic sensor through the air. This fact make eases the installation of the sensor and reduces the possibility to alter the electric parameters of the welding process and reduces the influence of the electromagnetic field on the acoustic sensor. It can be found in literature some acoustical monitoring systems for GMA welding process, but it not yet are used in the industry (Mansoor, 1999; Grad, 2004; Poopat, 2006; Cayo, 2007). In the present work was developed a weld defect detection technique based on the welding stability evaluation through sound produced by welding electric arc.

2.3.1 Welding electrical arc and acoustical signals

The relationship between sound pressure, sound pressure level and spectrum frequency profile behavior with the arc voltage and welding current have been studied. The sound pressure is a longitudinal mechanical wave, produced by the difference of pressure in a medium that can be solid, liquid or gaseous; in this work the transport medium is the air. The metallic transference in the welding process produces changes in the air volume on the electric arc environment. This change produces pressure variations that are airborne transported and sensing by the microphone. The sound pressure from electric arc is a consequence of amplitude modulation of arc voltage and welding current (Drouet, 1979, 1982). This relation is expressed by the equation (17).

The sound pressure level - SPL also called as equivalent continuous sound pressure level, is a comparative measurement with the microphone sensitivity. It is defined as twenty times the logarithm in base ten of the ratio of a toot-mean - square sound pressure during a time interval to the reference sound pressure. The equation (18) expresses the sensibility function.

$$S_a(t) = \frac{d(k.V(t).I(t))}{dt} \tag{17}$$

Where, *Sa (t)* is the sound signal (V), *V (t)* the arc voltage (V), *I(t)* welding current (A) and *K* the geometric factor.

$$SPL = 20.Log\left[\sqrt{\frac{1}{\Delta t}\int_t^{t+\Delta t} P^2(\xi)d\xi}\Big/ p_o\right] \tag{18}$$

The relation between the microphone pressure response and its sensibility is given by the equation (19) and therefore the SPL in function of the sound pressure is given by the equation (20). Relating the equation (17) and (20) results the equation (21) that expresses the SPL in function of the arc voltage and welding current.

$$P(\xi) = \frac{S(\xi)}{50E-3} \tag{19}$$

$$SPL = 20.Log\left[\sqrt{\frac{1}{\Delta t}\int_t^{t+\Delta t}\left(\frac{S(\xi)}{50E-3}\right)^2 d\xi}\Big/ p_o\right] \tag{20}$$

$$SPL = 20.Log\left[20\sqrt{\frac{1}{\Delta t}\int_t^{t+\Delta t}\left(\frac{d(k*V(\xi)*I(\xi))}{d\xi}\right)^2 d\xi}\Big/ p_o\right] \tag{21}$$

In which SPL is the sound pressure level, V the arc voltage, I the arc current, K the geometrical factor, Po the reference sound pressure (20 uPa), ξ is a dummy variable of time integration over the mean time interval, t the start time of the measurement, Δt the averaging time interval, S the sound signal.

For the spectrum frequency profile analysis it has been used the continuous Fourier transform; it is a linear transformation that converts the acoustic pressure signal from time domain to frequency domain.

This transformation is made using the Discrete Fourier Transform - DFT and it is expressed by the Eq. 22.

$$S(k) = \frac{1}{N}\sum_{n=0}^{N-1} s(n)e^{-j2\pi kn/N} \tag{22}$$

The octave frequency fractions analysis allows to evaluate the frequency strips behavior instead of any frequency. A frequency octave is defined as an interval among two frequencies where one of them is the double of the other. The octave band limits are calculated by Eq. 23, to 25. After obtaining the acoustic pressure spectra S(k), is obtained the octave frequency strips G(n) from Eq. (26).

$$f_{C_{n-1}} = 2^m f_{C_n} \tag{23}$$

$$f_{L_n} = \frac{f_{C_n}}{2^{m/2}} \tag{24}$$

$$f_{U_n} = 2^{m/2} f_{C_n} \tag{25}$$

$$G(n) = \sqrt{\frac{1}{(f_{U_n} - f_{L_n})} \sum_{f(k)=f_{L_n}}^{f_{U_n}} [S(f)]^2} \tag{26}$$

Where m is the octave band fraction, f_{Cn} central frequency of the n band, f_{Ln} the inferior limit of the n band and f_{Un} the superior limit of the n band.

2.3.2 Quality control in the GMA welding process

The quality control study in the welding processes is a main subject of many researchers. The evaluation task of weld quality is not trivial, even for the experienced inspector. This is particularly true when it comes to specifying in quantitative terms what attributes of the weld affect its quality and in what extent. Different types of discontinuities have been categorized for this purpose, such as cracks, porosity, undercuts, microfissures, etc. (Cook, 1997). Generally, good quality GMA welds are uniform and contain little or no artifacts on the bead surface. Furthermore, the bead width is relatively uniform along the length of the bead (Cook, 1995). To reaches the standard weld quality is fundamental maintain continuity on the welding stability and this happens when the mass and heat flow of the end consumable electrode until fusion pool through arc maintains uniformity in the transference; possible discontinuities and/or upheavals in the transference could originate weld disturbances. The stability of the short circuit gas metal arc welding process is directly related to weld pool oscillations. Optimal process stability corresponds to maximum short-circuit rate, minimum standard deviation of the short-circuit rate, a minimum mass transferred per short circuit and minimum spatter loss, (Cook, 1997; Adolfsson, 1999; Wu, 2007). In the present work, the welding stability was evaluated using the sound pressure through the acoustic ignition frequency (AIF) and sound pressure level (SPL) signatures.

The metallic transference on the short circuit mode in GMA welding (GMAW-S[2]) is characterized by a sequence cycles of ignition and extinction arcs. The Figure 10 - a show the behavior of arc welding current and voltage signals as well as the welding arc sound and

[2] Gas Metal Arc Welding in short circuit transfer mode

the resultant signal using the equation (1). The sequence cycles of the welding metallic transference is replicated in the electrical measured signal as well as in the welding arc sound. The arc ignition produces a great acoustical peak and the arc extinction produces a small amplitude acoustical peak. Although, it is possible to observed a delay 'Δt' on the similarity signal of calculated and measured arc sound (see Fig. 10 - b). Some studies in psychoacoustic have determined that while the delay of welding arc sound signal does not exceed 400 ms, the sound will be a good indicator of welding process behavior (Tam, 2005). In our case the Δt delay measured was approximately 0,6 ms, this value undertakes the reliability to monitoring the welding process behavior.

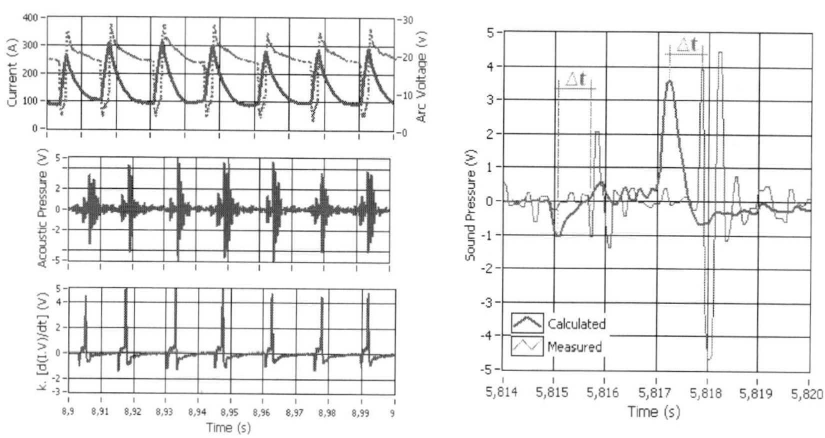

Fig. 10. (a) Welding Signals (b) Sound pressure calculated and measured.

The welder uses his experience and ability to learn and know acoustical signatures from quality welds. The figure 11 shows the measured and calculated SPL (using the equation 5) with its respectively sound pressure signal. It can be observed that the shielding gas flux is sensed by the sound signal, but as the SPL calculated is only function of welding current and arc voltage, this sound signal is not taking into the account. It can be obtained others information from the SPL sound pressure like the arc welding ignition and extinction average frequency, the average period of the transferences cycles and its standard deviations.

The measurement of ignition and extinction average frequency from welding arc is a method for evaluate welding stability (Adolfsson, 1999). As it was explained before, the arc sound pressure follows the arc ignition and extinction sequence (Fig. 10 - b). The acoustic amplitude pulses produced by the arc ignitions are greater than acoustic amplitude pulse produced by the extinctions (short circuits). These acoustical impulses sequence occurs together with chaotic transients and noise oscillations and in order to reduces it and to obtain only the ignition and extinction average frequency, the envelope sound signal was extracted from acoustic sound signal (see Fig. 12).

The envelope sound signal was obtained using a quadratic demodulator. Squaring the signal effectively demodulates the input by using itself as the carrier wave. This means that half the energy of the signal is pushed up to higher frequencies and half is shifted towards

DC. The envelope can then be extracted by keeping all the DC low-frequency energy and eliminating the high-frequency energy. However, the statistical filter called "kalman filter" was used due to the sound pressure have a stochastic behavior and were low-pass filters is needed with an elevated order, this order produce a pronounced delay and deformation in the envelopment signal. This statistical filter instead of letting pass low frequencies, it follows the statistical tendency of the squared signal obtaining in the envelopment sound pressure (Cayo, 2008). In Figure 12, the 150 ms moving window data was extracted from the sound pressure signal. From envelop sound pressure signal is calculated the arc ignitions for each moving window data. An ignition takes place whenever the envelop sound pressure signal surpasses the ignition threshold established (k = 0,2).

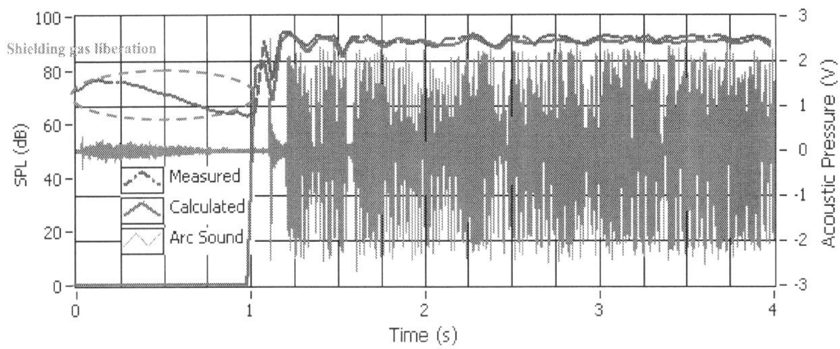

Fig. 11. Measured and Calculated SPL.

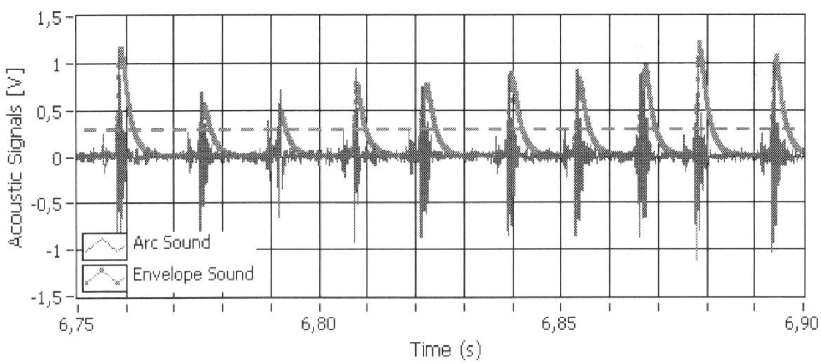

Fig. 12. Acoustical Pressure and its envelope Signals.

2.3.3 Weld bead quality profile identification

Previous to the interferences detection, many weld experiments for finding the optimal set of the welding parameters was carried out. The satisfactory parameters selection allows reaching the maximum stability in welding, but to reach that was necessary chooses the adequate arc voltage.

The figure 13 – a, illustrates the relationship between the short circuit frequency calculated by the arc voltage and arc sound and in figure 13-b is showed the respective standard deviation. The average short circuit frequency obtained from arc voltage and the arc sound pressure show a similar result. Moreover, in both short circuit frequency and standard deviation have some differences between calculating methods, but this differences are minimal and the acoustical method to calculate these statistical parameter can be considered as reliable. To choose the arc voltage value that generated the best quality bead (continuity and uniformity on the bead ruggedness) following related research (Adolfsson et al., 1999) that concluded that the best quality is reached basically when the short circuit rate is maximum and its standard deviation is minimal was discover some unexpected results. The arc voltage that generates the maximum short circuit rate (17,5 V) is not the weld with the better quality inside the weld run experiments (Fig. 12). Considering the arc voltage that generates the minimal standard deviation from short circuit rate obtained from arc voltage (23,0 V) is not the weld with the better quality and considering the minimal standard deviation obtained from arc sound also is not the weld with the better quality. The visual inspection of the bead set shows that the weld bead run with the voltage range between 19,0 V and 20,5 V can be considered as the best quality weld bead inside this set.

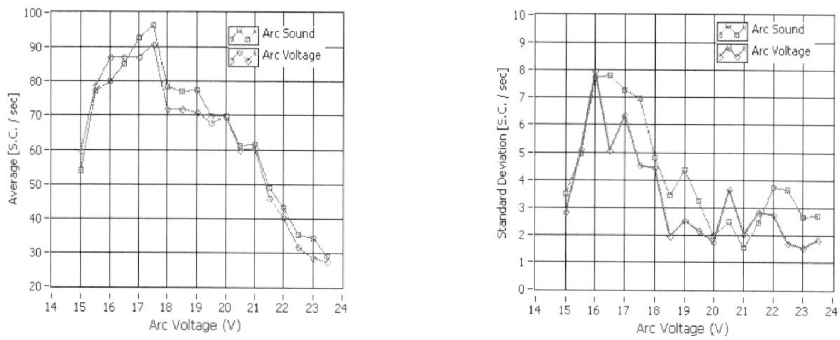

a) Short Circuits Frequency b) S. C. F. Standard Deviation

Fig. 13. Short Circuits vs Arc Voltage.

In order to chosen the arc voltage value that generates the best weld bead quality was carried out a second statistical analyze. The Figures 14 – a, illustrate the relationship between the transference cycle period average obtained by the arc voltage and arc sound at the same different arc voltage values analyzed previously. In this case the results obtained by the electrical and acoustical methods show a narrow similarity with differences of milliseconds. The arc voltage that generates the least transference cycle period average was also 17,5 and comparing with the weld bead quality shows not the best. The transference cycle period uniformity is measured by its standard deviation as show in the figure 14 – b. This standard deviation distribution has more uniformity than the short circuit frequency standard deviation and the arc voltage that generates the minimal value into standard deviation distribution (20,0 V) also generates the weld bead with more geometrical ruggedness and uniformity and can be considered as the best quality weld bead. Consequently, the remaining weld run experiment for interferences detection will carry out using the welding parameters showed in table 1 with arc voltage adjusted to 20,0 V.

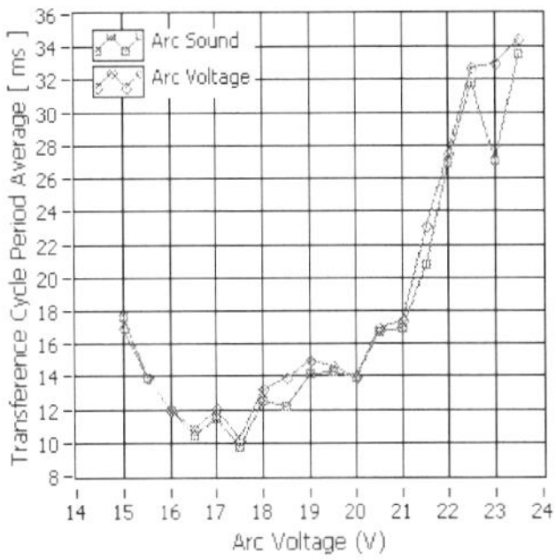

a) Short Circuits Frequency

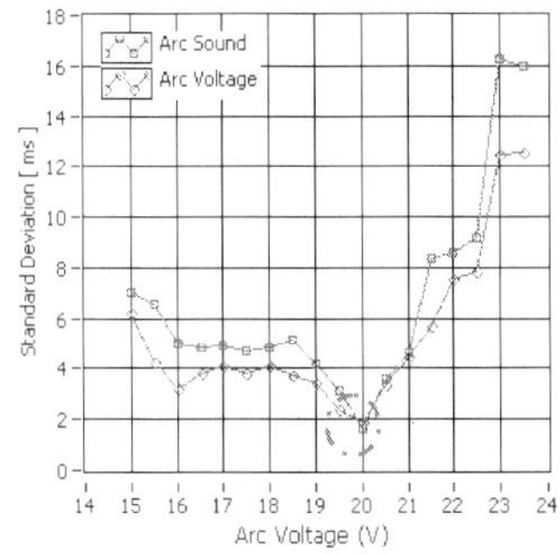

b) S. C. F. Standard Deviation

Fig. 14. Transference cycle Period .vs. Arc Voltage.

2.3.4 Acoustical profiles to interferences detection

The general approach used to evaluate GMA welds was examining the variation of weld profiles, sampled across the weld bead in a sequence of locations. A high quality weld would generally yield small variation in the weld profiles, while low quality weld profiles would vary substantially, as irregularities and various discontinuities are encountered in the distinct profile scans (Cook, 1997). The initial weld profiles tested were the arc sound ignitions frequency, the average sound pressure level and the power spectral density using the continuous and octave fraction frequency domain. The figure 15 - a, shows the AIF and 15 - b, the SPL weld profiles signal behavior. These profiles were tested on welding runs with and without presences of disturbances. Both profile signals were determined using a moving windows signal applied on the arc sound pressure signal. The moving window was fixed in 150 ms considering that the data sample rate was 20 kHz.

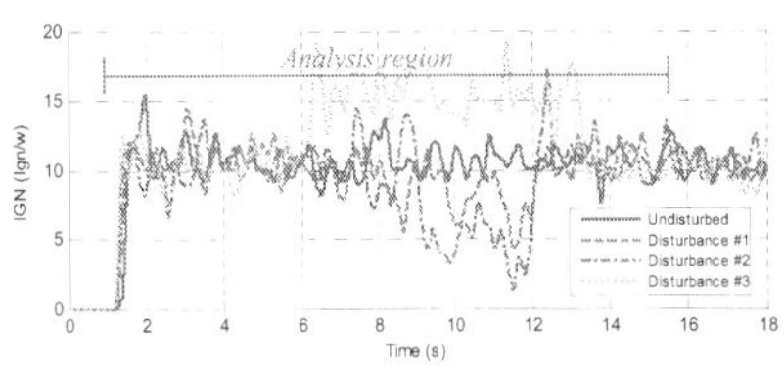

a) Ignition Frequency - AIF

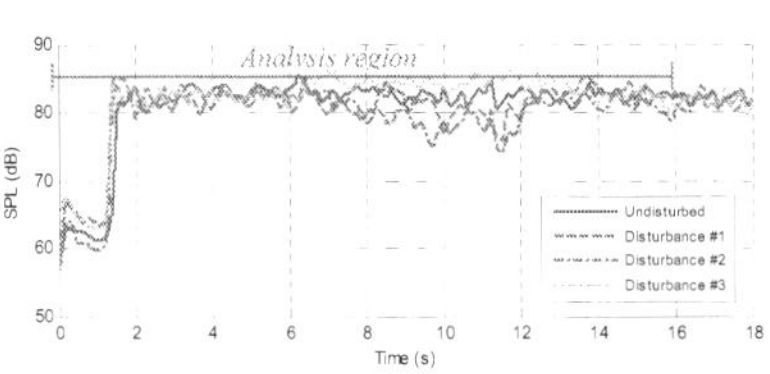

b) Sound Level Pressure - SPL

Fig. 15. Time Profiles to welding arc sound.

Signature analysis of the short-circuiting frequency by using time - frequency analysis method was applied to welding arc sound. On this signal there are two regions (undisturbed - UR and disturbed - DR regions). On these regions a spectral analysis was made at continuous and an octave fractions frequency. When the welding entered on the interfered region the spectra frequency varies its magnitude on the overall frequency. The spectra frequency magnitude varies on the two analyzed regions; there are approximately dominant frequency components at 2 kHz.

The octave fraction spectra also have greater amplitude inside this band frequency, but it does not show frequency bands signatures that vary pronouncedly to disturbed sound signals. As there are not signature bands in fraction octave analysis, the continuous frequency versus time behavior was analyzed. The figure 10 shows the spectrogram for welding runs with induced interferences on the plate. The amplitude variations on the sound spectra imply that there are signature variations on the time domain. The welding arc sound have many chaotic transients between impulses, these fluctuation have a stochastic nature. Fourier analysis is very effective in problems dealing with frequency location. However, there are severe problems with trying to analyze transient signals using classical Fourier methods (Walker, 1997). This is the principal reason for not distinguishing clear signatures frequencies on the arc sound spectra that can identify disturbances. The signatures on the time domain describe in figure 15 shows pronounced signatures when the welding enters to interfered regions.

2.3.5 Interferences monitoring and detection

The disturbances detection was made using a limit control based on the third standard deviation method. This control shows optimal results in the disturbances detection of both, electric arc voltage and welding current monitoring signals (Cook, 1995, 1997). As already explain, the signature signals analysis was made on the 150 ms moving window. The equation (27), (28) and (29) represents the average, standard deviation and disturbances control limits for each moving window data respectively.

$$\overline{x_i} = \frac{1}{n}\sum_{j=1}^{n} x_i = \frac{1}{n}(x_1 + \ldots + x_n) \qquad (27)$$

$$S_i = \sqrt{\frac{1}{n}\sum_{j=1}^{n}(x_j - \overline{x_i})^2} \qquad (28)$$

$$Control\_Limits = \overline{P_N} \pm 3 \times S_P \qquad (29)$$

Where:

$\overline{x_i}$ Average parameter of the *i* analysis moving windows (150 ms)

x_i data *i* from the moving window

n Data component number from analysis moving window

S_i The standard deviation for the *i* analysis moving window data,

x_j The j component data from analysis moving window data,

$\overline{P_N}$ Average established for each parameter,

S_P The standard deviation established for each acoustical parameter without

In order to validate the disturbances detection method and base on the acoustics GMAW-S signal, a total of forty welding runs were carried out. The average short circuit numbers per seconds obtained from arc voltage and the arc average ignition numbers per seconds obtained from sound pressure show a similar result as shown in figure 16 - a. These results were obtained in the first weld experiments group without induced disturbances. The minimal standard deviation 16 – b confirms that the arc sound pressure can well represent the behavior of the GMAW-S metallic transference.

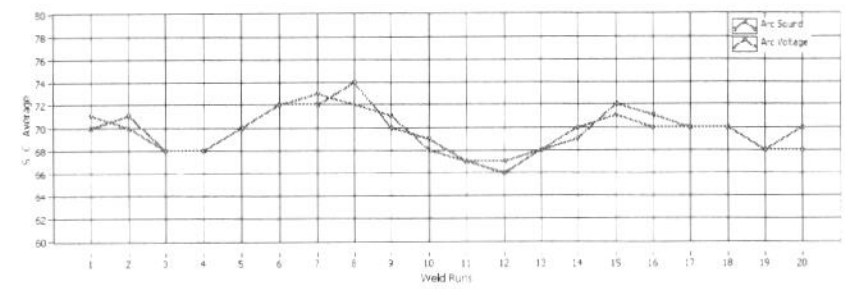

a) S. C. Average

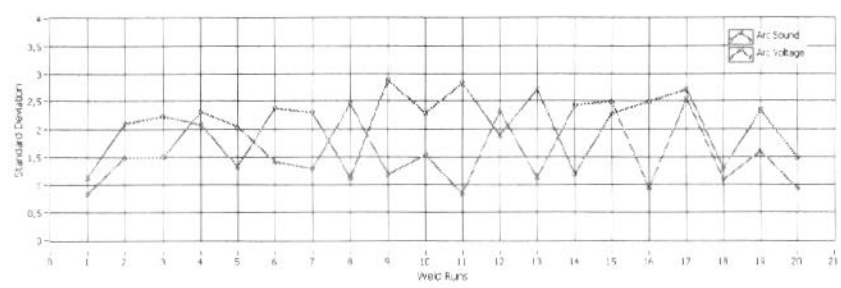

b) S. C. Standard Deviation

Fig. 16. Short circuits average and standard deviation.

The initial time profiles have temporarily instabilities [Figures 15 (a) and (b)]. In order to avoid these initial instabilities influence in the quality control evaluation, the analysis region is established from second 2 to second 18. From the figure 17 and 18, (a), (b) and (c) respectively, are showed the acoustical parameters behavior and it can also be observed a baseline and two threshold levels, one superior and another inferior. These established limits are three times the standard deviation on the average of each parameter in stable conditions welds (without the presence of disturbances). When the parameters are within these two threshold limits are no apparent disturbances in the welds. Therefore, when the parameters exceed the established threshold limits implies in having detected some disturbance that possibly could originate some weld defect. Figure 17 (a) and (b) show the AIF and the SPL behaviors, respectively obtained from the acoustical of arc without induced disturbances. In the Figure 17 (c) is showed the aspect of the welding bead and even appearing oscillations on the signal, it does not exceed the established threshold limits. These oscillations appear due to stochastic behavior of the acoustical pressure emitted by the electric arc of the welding process and do not necessarily represent disturbances presence.

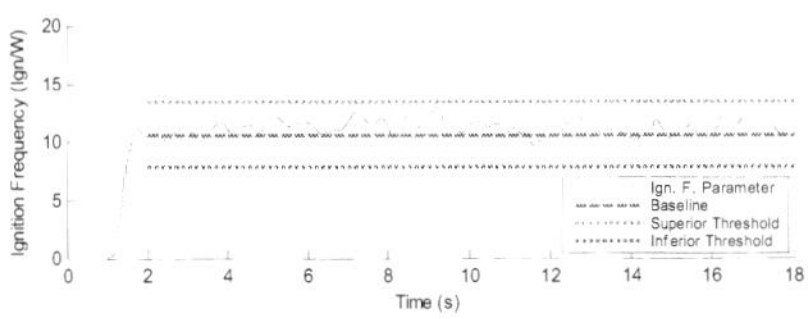

a) Ignition Frequency

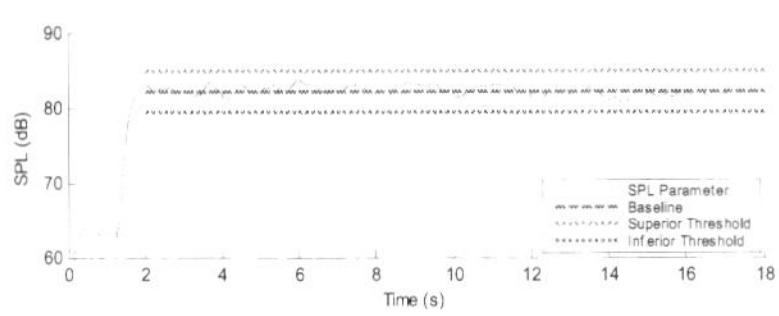

b) Sound Pressure Level

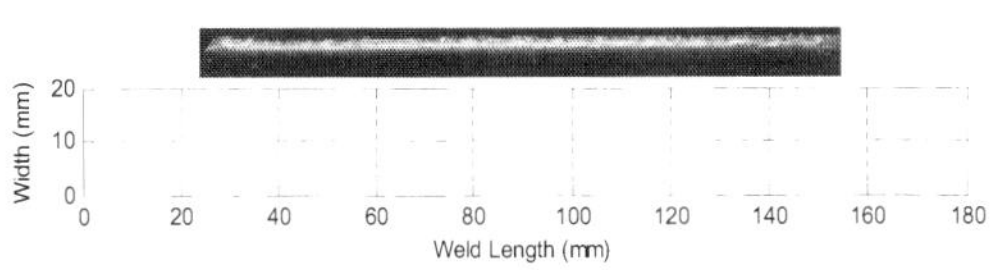

c) Weld Bead without Disturbance

Fig. 17. Parameters without Disturbance.

The Figure 18 (a) and (b) show the AIF and the average SPL per moving window respectively. Both graphics were obtained from the acoustics of a welding arc with induced disturbance originated by the arc length variation. The Figure 18 (c) shows the visual aspect of the weld bead. The instabilities only occurs when the weld bead pass through the beginning, end and the holes of the added plate, respectively. In Figure 18, (a) and (b) can also be observed that the ignitions frequency and the SPL do not present oscillations that exceed the established threshold limits, before and after the weld bead pass throughout the interference region. When the weld passage throughout the interference region, abrupt changes of signal level are produced in each parameter. These instabilities exceed the level control previously established. When the CTWD length varies, the arc length varies too, these variations produce instabilities in the arc ignition. It can also be observed that the parameters return to inside threshold limits even without leaving the disturbance region due to the arc reaches a new level of stability.

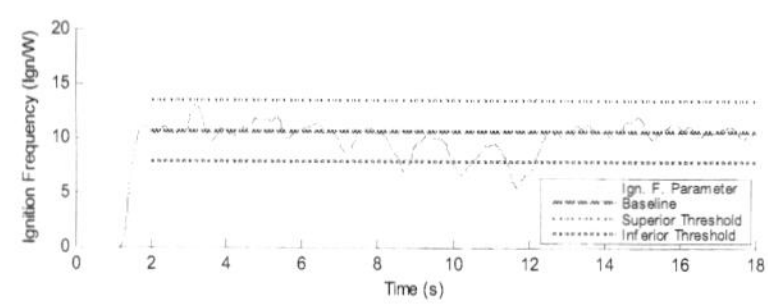

a) Ignition Frequency

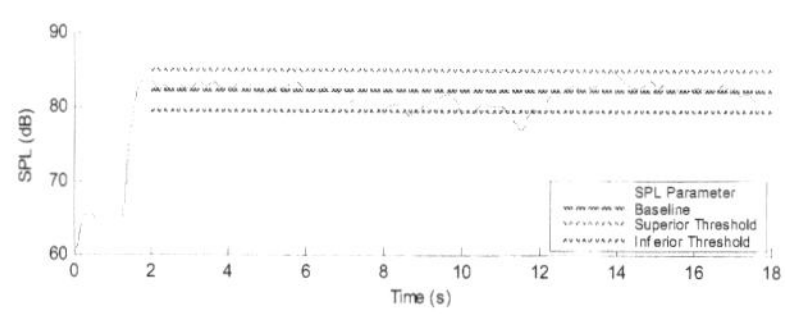

b) Sound Pressure Level

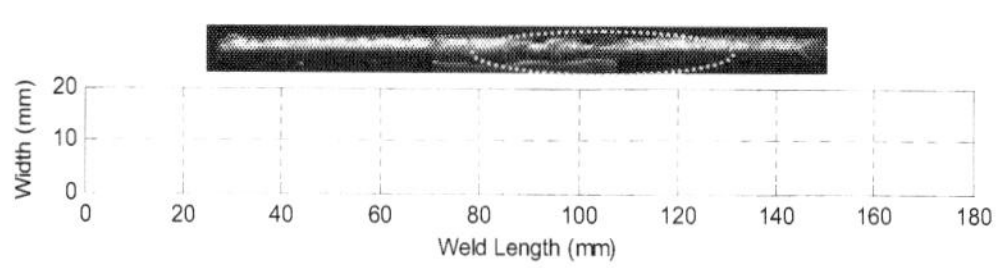

c) Weld Bead with Disturbance 1

Fig. 18. Parameters with disturbance.

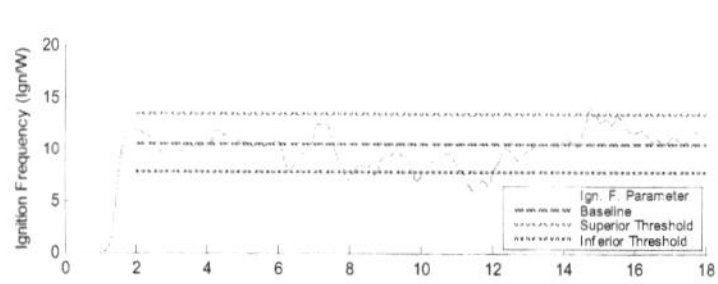

Fig. 19. a) Ignition Frequency

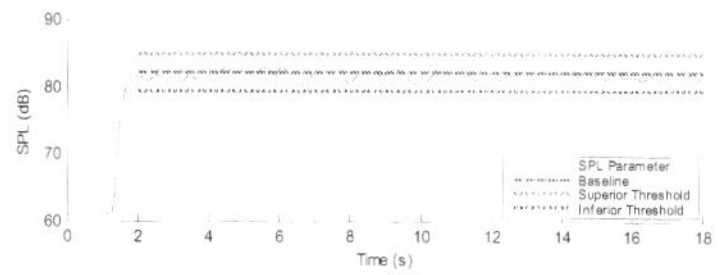

b) Sound Pressure Level

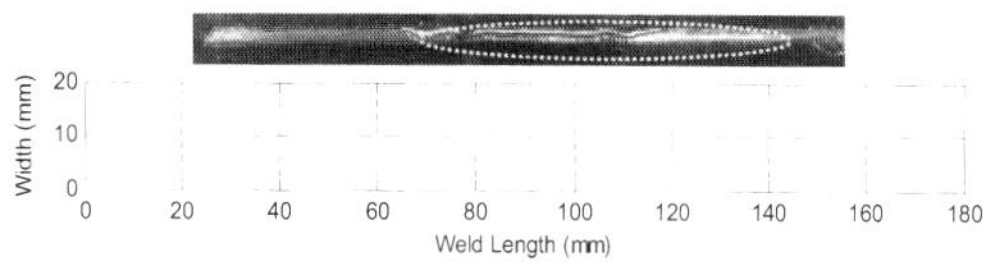

c) Weld Bead with Disturbance 2

Fig. 19. Parameters with Disturbance 2.

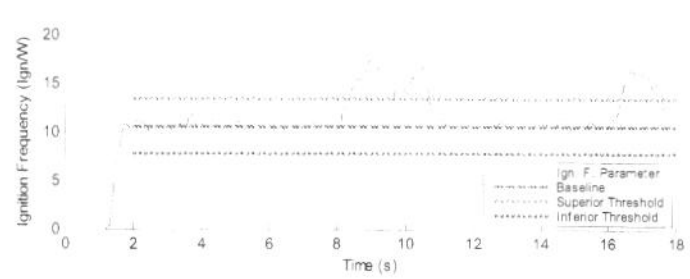

a) Ignition Frequency

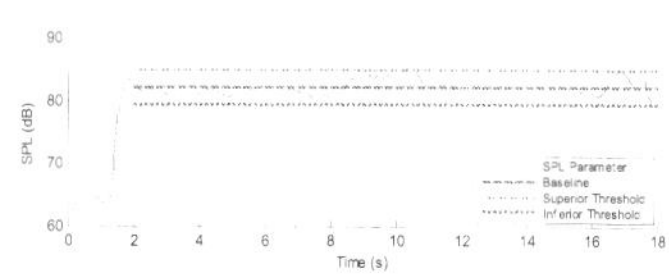

b) Sound Pressure Level

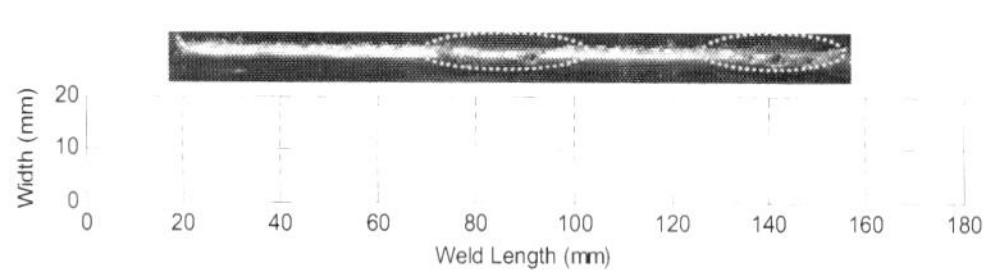

c) Weld Bead with Disturbance 3

Fig. 20. Parameters with Disturbance 3.

In the Figure 19 (a) and (b) is showed the acoustical stability parameters behavior for a weld bead with a disturbance induced due to grease presence in the welding trajectory (see Fig. 19-c). When the welding run pass the disturbance region, instabilities in the arc ignitions was observed and unexpected upheavals in the metallic transference cycles occurs forming weld bead deposit interruptions in all welding trajectory. Initially, when the weld run reach the grease do not take place oscillations in the ignition frequency due to the grease borders are evaporate caused by the thermal welding cycles. In the AIF parameter can be observed that the interference is noticeable as a chaotic decrement, however this behavior overcame slightly the control limits (see Fig. 19-a). The average SPL parameter has little abnormal variations when the weld run pass on the interference. The induced interference generates structural discontinuities in weld bead but only in AIF parameter is most evident that the SPL parameter. The Figure 20 (a) and (b) show the AIF and SPL parameter behavior in welding experiment without shield gas. In this case the induced perturbations were localized in two regions on the weld bead. These interferences have led to porosities formation (see Fig. 20-c) and higher spatter level. The AIF parameter has incremented suddenly and this behavior is showed in both instabilities and exceeds the control limit established. The AIF were incremented because the absence of the shield gas causes contamination in the arc welding environment originating the incomplete metallic transference, increasing the short circuit and ignitions rate. This increment noticeable in the AIF is also noticeable in the SPL, nevertheless these variations do not exceed the threshold limit control.

3. Conclusions

Signals of the arc voltage and sound pressure were tested for monitoring the welding process. As both signal have the same behavior, it can be concluded that the sound pressure can be used for welding monitoring. From sound pressure were calculated two parameters: AIF and SPL. The monitoring of the GMAW-S process by digital analysis of acoustical welding parameters enable to detect disturbances related to phenomena that take place in welding arc and can influence its stability. This fact shows that the arc acoustics is a non-intrusive potential tool that could be used for the weld quality evaluation.

4. References

Adolfsson, S.; Bahrami, A.; Bolmsjö, G.; Claesson, I. On-line quality monitoring in short-circuit gas metal arc welding Weld. J. 1999, 3, 59s–73s.

Alfaro, S., D. Mendonça, M. Matos, "Emission Spectrometry evaluation in arc welding monitoring system", Journal of Materials Processing Technology, No. 179, 2006, Pages 219-234.

Alfaro, S., P. Drews, "Intelligent Systems for Welding Process Automation", Journal of the Brazilian Society of Mechanical Sciences, Vol. 28, No. 1, 2006, Pages 25-29.

Alfaro, S.C.A.; Carvalho, G.C.; Da Cunha, F.R. A statistical approach for monitoring stochastic welding processes. J. Mater Process Technol. 2006, 175, 4–14.

Appel, U.; Brandt, A.V. Adaptive sequential segmentation of piecewise stationary time series. Inform. Science 1983, 29, 27–56.

Araújo, C.F.B. Estudo da Monitoração por Infravermelho como Indicador de Penetração em Soldas Obtidas no Processo TIG. Masters Dissertation, University of Brasilia: Brasilia, Brasil, 2004.

Cayo, E., S. Alfaro, "Welding Quality Measurement Based On Acoustic Sensing", COBEM-2007, Brasília. 19th International Congress of Mechanical Engineering. São Paulo : ABCM, 2007. v. 1. Pages 2200

Cayo, E. H.; Alfaro, S.C.A. Medición de la Calidad en Soldadura Basado en Sensoreamiento Acústico. In 8th Congreso Iberoamericano de Ingeniería Mecánica: Cuzco, PE, October 2007, pp.1068-1079.

Cayo, E., "Monitoring, Detection and Localization System for Welding Defects based on the Acoustic Pressure of the GMAW-S Process Electric Arc". Master Dissertation in Mechatronics Systems, FT, The University of Brasilia, DF, Brazil, 2008, Pages 108.

Cayo, E. H. Monitoring, Detection and Localization System for Welding Defects based on the Acoustic Pressure Electric Arc of the GMAW-S Process Master Dissertation in Mechatronics Systems, FT University of Brasilia, DF, Brazil, 2008.

Cayo, E., S. Alfaro, "GMAW process stability evaluation through acoustic emission by time and frequency domain analysis", AMME' 2009, June 2009, Gliwice, Poland, 2009, p. 157-164.

Chen, W.; Chin, B.A. Monitoring Joint Penetration Using Infrared Sensing Techniques. Weld J. 1990, 69, 181s–185s.

Chin, B.A.; Zee, R.H.; Wikle, H.C. A Sensing System for Weld Process Control. J. Mater. Process. Technol. 1999, 89–90, 254–259.

Cook, G.E.; Barnett, R.J.; Andersen, K.; Springfield, J.F.; Strauss, A.M. Automated Visual Inspection and Interpretation System for Weld Quality Evaluation, Thirtieth IAS Annual Meeting, IAS '95, Conference Record of the 1995 IEEE: Orlando, FL, USA, October 1995; pp.1809 -1816.

Cook, G.E.; Maxwell, J.E.; Barnett, R.J.; Strauss, A.M. Statistical Process Control Application to Weld Process, IEEE Transactions on Industry Applications 1997, 33, No. 2, 454s-463s.

Drouet, M.; Nadeau, F. Pressure waves due to Arcing Faults in a Substation, IEEE Transactions on Power Apparatus and Systems 1979, 5, 98s.

Drouet, M.; Nadeau, F. Acoustic measurement of the arc voltage applicable to arc welding and arc furnaces Phys J. 1982, 15, 268s-269s.

Duda, R.O.; Hart, P.E.; Stork, D.G. Pattern Classification; 2nd ed. John Wiley and Sons, Inc.: New York, NY, USA, 2001.

Fan, H., N. Ravala, H. Wikle III, B. Chin, "Low-cost infrared sensing system for monitoring the welding". Journal of Materials Processing Technology, no. 140, 2003, Pages 668-675.

Fortunko, C.M. Ultrasonic Detection and Sizing of Two-Dimensional Weld Defects in the Long-Wavelength Limit. Ultrason. Symp. 1980, 862–867.

Grad, L., J. Grum, I. Polajnar, J. Slabe, "Feasibility study of acoustic signals for on-line monitoring in short circuit gas metal arc welding", International Journal of Machine Tools and Manufacture Volume 44, Issue 5, April 2004, Pages 555-561.

Gustafsson, F. Adaptative Filtering and Change Detection; John Wiley & Sons: New York, NY, USA, 2000.

Jazwinski, A.H. Stochastic Processes and Filtering Theory; 1st ed.; Academic Press: New York, NY, USA, 1970.

Kerr, H.W.; Hellina, M.C.; Huissoon, J.P. Identifying Welding Pool Dynamics for GMA fillet welds. Scien. Tech. Weld. Join. 1999, 4, 15–20.

Kralj, V. Biocybernetic investigations of hand movements of human operator in hand welding. Int. Ins. Weld. 1968, Doc. 212-140-68,

Lacroix, D., C. Boudot, G. Jeandel, "Spectroscopy Studies of GTA Welding Plasmas. Temperature Calculation and Dilution Measurement". Euro Physics Journal, AP 8, 1999, Pages 61-69.

Liu, Y.; Li, X.H.; Ren, D.H.; Ye, S.H.; Wang, B.G.; Sun, J. Computer vision application for weld defect detection and evaluation, Automated Optical Inspection for Industry. Theory Technol. Appl. II. 1998, 3558, 354–357.

Luo, H.K.; Lawrence, F.M.K.; Mohanamurthy, P.H.; Devanathan, R.; Chen, X.Q.; Chan, S.P. Vision Based GTA Weld Pool Sensing and Control Using Neurofuzzy Logic; SIMTech Technical Report AT/00/011/AMP; Automated Material Processing Group: Singapore Institute of Manufacturing Technology, Singapore, 2000, 1–7.

Mansoor, A., J. Huissoon, "Acoustic Identification of the GMAW Process", 9th Intl. Conf. on Computer Technology in Welding, Detroit, USA, 1999, Pages 312-323.

Marotta, A. (1994), "Determination of axial thermal plasma temperatures without Abel inversion", Journal of Physics D. Applied Physics, 27, 268-272.

Merchant, J. Infrared Temperature Measurement Theory and Applications, Mikron Instruments Company Inc, Application Notes, 2008.

Mirapeix, J.; Cobo, A.; Conde, O.M.; Jaúregui, C.; López-Higuera, J. M. Real-time arc welding defect detection technique by means of plasma spectrum optical analysis. NDT&E Int. 2006, 39, 356–360.

Mirapeix, J., A. Cobo, D. González, J. López-Higuera, "Plasma spectroscopy analysis technique based on optimization algorithms and spectral synthesis for arc-welding quality assurance". Optics Express, Vol. 5, no. 4, 2007, Pages 1884-1889.

Nagarajan, S.W.; Chen, H.B.; Chin, A. Infrared Sensing for Adaptive Arc Welding. Weld J. 1989, 68, 462s–466s.

Nagarajan, S., P. Banerjee, W. Chen, B. Chin, "Control of the Process Using Infrared Sensors", IEEE Transaction on Robotics and Automation, Vol. 8, no. 1, 1992, Pages 86-93.

NIST, National Institute of Standards and Technology (accessed in may 2010) http://physics.nist.gov/PhysRefData/ASD/lines_form.html

Pollock, D.S.G. A Handbook of Time Series Analysis, Signal Processing and Dynamics; 1st ed.; Academic Press: New York, NY, USA, 1999.

Poopat, B., E. Warinsiriruk, "Acoustic signal analysis for classification of transfer mode in GMAW by noncontact sensing technique", Journal of Science and Technology, University of Technology Thonburi, Thungkru, Bangmod, 2006, Bangkok, Thailand, Vol. 28, Issue 4, Pages 829-840.

Sanders, P.G.; Leong, K.H.; Keske, J.S.; Kornecki, G. Real-time Monitoring of Laser Beam Welding using Infrared Weld Emissions. J. Laser Appl. 1998, 10, 205–211.

Tam, J., "Methods of Characterizing Gas-Metal Arc Welding Acoustics for Process Automation", Master Dissertation in Mechanical Engineering, University of Waterloo, Canada. 2005, Pages 120.

Tam, J.; Huissoon, J. Developing Psycho-Acoustic Experiments in Gas Metal Arc Welding International Conference on Mechatronics & Automation: Niagara Falls, CA, July 2005, pp.1112-1117.

Tam, J. Methods of Characterizing Gas-Metal Arc Welding Acoustics for Process Automation Master Dissertation in Mechanical Engineering, University of Waterloo, CA, 2005.

Venkatraman, B.; Menaka, M.;.Vasudevan M.; Baldev R. Thermography for Online Detection of Incomplete Penetration and Penetration Depth Estimation, In Proceedings of Asia-Pacific Conference on NDT, Auckland, New Zealand, November 5th–10th, 2006.

Walker, J. S.. Fourier Analysis and Wavelet Analysis, Proceeding of the AMS 1997, 44, No. 6, 658s-670s.

Wikle, H. III, R, Zee, B. Chin, "A Sensing System for Weld Process Control", Journal of Materials Processing Technology, no. 89-90, 1999, Pages 254-259.

Wikle III, H.C. Kottilingam. S.; Zee, R.H.; Chin, B.A. Infrared Sensing Techniques for Penetration Depth Control of the Submerged Arc Welding Process. J. Mater. Process. Technol. 2001, 113, 228–233.

Wu, C.S.; Gao, J.Q.; Hu, J.K. Real-Time Sensing and Monitoring in Robotic Gas Metal Arc Welding, Phys J. 2007, 18, 303s-310s.

Yaowen, W., Z. Pendsheng, "Plasma-arc welding Sound Signature for on-line Quality Control", Materials Science & Engineering Department, Taiyuan University of Technology, Taiyuan 030024, China, 2000, Pages 164-167.

Part 3

Weldability of Metals and Alloys

6

Weldability of Iron Based Powder Metal Alloys Using Pulsed GTAW Process

Edmilson Otoni Correa
Universidade Federal de Itajuba
Brazil

1. Introduction

In the last decades, powder metal (PM) iron-based alloys have been extensively used as structural parts in mechanical components due to their good balance between ductility and tensile strength, low cost, high performance, flexibility of manufacturing, good magnetic properties and corrosion resistance. Consequently, such PM components have emerged as an effective alternative for replacing machined parts, castings and forgings in many engineering applications. However, continued efforts are demanded for obtaining optimum combination of properties to withstand various service conditions.

When replacing forged, cast or machined parts, weldability is one of the important requisites expected of P/M parts in actual service conditions like in structural and automobile applications. This is due to the fact of many of these parts occasionally need to be joined to one similar part or dissimilar material as integrated components. The welding of dissimilar metal is generally more challenging than that of similar metals because of considerably difference in the physical, thermal, electrical, mechanical and metallurgical properties of the parts to be joined. In order to take full advantage of the dissimilar metals involved, it is necessary to produce high quality joints between them.

The welding of powder metal parts is different from welding of cast, rolled and forged parts due to the presence of porosities in their microstructure. The nature of the porosity is controlled by several processing variables such as green density, sintering, etc. In particular, the fraction, size, distribution and morphology of the porosity have a profound impact on mechanical behaviour, especially in components under welding conditions.

Fusion welding methods have been successfully used to join powder metal parts and are more related to the welding of medium and high density powder metal parts. Welding process such as gas tungsten arc welding (GTAW) and gas metal arc welding (GMAW) have been often cited as feasible possibilities to join PM structural parts. However, very little experimental information about welding parameters used, more adequate filler metal, etc., is available on the application of these welding process to join PM components.

This chapter will present a review of the main characteristics of the PM parts which differs them of the materials fully dense materials as it pertains to joining and the weldability of powder metal iron alloys using the pulsed gas tungsten welding process with filler metal.

2. Benefits of the powder metallurgy

The variety of materials, complexity of components and advances in manufacturing process make powder metallurgy (PM) an established process for the production of structural parts in mechanical components.

In conventional PM process (German, 2005; Jenkins & Wood, 1997; Lenel, 1980; Thummler & Oberacker, 1993; Schatt & Wieters, 1997) the part is made by four basic production steps including mixing elemental or alloy powder, compacting, debinding and sintering. The compacting step may be subdivided into three steps, namely powder filling, powder compacting and green compact rejecting. Normal concerns for powder filling are powder particle flow, filling height, powder particle packing and powder particle segregation.

During the compacting step, some phenomena including particle deformation, cold welding at points of contact and interlocking between particles occur. Concerns for powder compacting are friction between powder particles and between powder particles and die walls and density distribution of the compact. According to author (Middle, 1981), due to these friction effects during the compacting, the density distribution of the PM part after sintering may be very uneven and may produce a irregular shrink and, consequently, the nucleation and propagation of cracks in the thermal affected zone of the material subjected to fusion welding.

During the rejecting the compact from the die (spring back), the compact volume expansion to release stored residual stress is the most important factor. If the spring back occurs too fast with high magnitude, it will cause undesirable deterioration of the compact.

Debinding and sintering, in general, are carried out in the same furnace. Parameters that influence both processes, including temperature, time and furnace atmosphere have to be optimized. The optimum conditions for debinding and sintering of a certain type powder metal part in the specified furnace are important for obtaining a good metallurgical bond between the powder particles and, consequently, a PM part with good mechanical resistance (German, 2005; Jenkins & Wood, 1997; Lenel, 1980; Thummler & Oberacker, 1993; Schatt & Wieters, 1997).

The PM process typically uses more than 95% of the starting raw material in the finished part and only minor machining is required. Because of this, PM process is an energy and material conserving process as well as cost effective in producing simple or complex parts at very close final dimensions in production rates which can range from a few hundred to several thousand parts per hour. PM process also may be sized for closer dimensional control for both, higher density or strength (Metal Powder Industry Federation [MPIF], 2004).

The versatility of PM is applied in numerous industries, including automotive, aerospace, electrical and electronic equipments, agricultural equipments and power tools. PM parts design serve these industries in a wide range of engineering applications which fall into two main groups: In one group are parts of difficult-to-fabricate materials by other manufacturing process such as tungsten and molybdenum, porous bearing, magnetic parts, etc. Another group consists of PM ferrous components increasingly attractive in replace machined parts, castings and forgings. In this group there are more material systems and requires to meet the requirements of more demanding applications.

The benefits of the PM process may be summarized as follows:

- Eliminates or minimizes the machining
- Eliminates or minimizes scrap losses

- Maintain close dimensional tolerances
- Permits a wide variety of alloy systems
- Produces good surface finishes
- Provides materials which may be heat-treated for increased strength or increased wear resistance
- Facilitates manufacture of complex or unique shapes which would be impractical or impossible with other metalworking processes.
- Suited to moderate-to high volume component production requirements
- Offers long-term performance reliability in critical applications
- Cost effective
- Provides controlled porosity for self-lubrification or filtration

3. Weldability of ferrous PM materials

3.1 Influence of the porosity

The most prominent microstructural feature of a PM component is its porosity, which affects virtually all its physical properties and, consequently, its weldability. The nature of the porosity is controlled by several processing variables such as green density, sintering temperature and time, alloying additions, and particle size and type of the initial powders. In particular, the fraction, size, distribution and morphology of the porosity have a profound impact on mechanical behaviour, especially in components under welding conditions (Chawla & Deng, 2005; Sudhakar et al., 2000).

Firstly, the pores act as thermal insulators which slow the transfer of heat, affecting considerably the thermal conductivity of the PM material to be joined. As the change in heat transfer naturally affects the welding parameters, the welder needs constantly to adjust them to assure the good quality of the weldment. Also, since the amount of porosity reduces the thermal conductivity, the cooling rate of the material also slows, reducing the hardening tendency (Hamill, 1993; Kurt et al., 2004; Kumar et al., 2007).

The thermal expansion is another important physical characteristic which is influenced by the porosity. Potential changes in the porosity volume fraction during welding, due to smaller particle melting or filler metal infiltration, can result in excessive shrinkage or growth. As a consequence, subsequent cracking can occur in the heat affected zone (HAZ) or in the fusion zone (ZF) of the PM base metal.

Porosity can also cause erratic fluctuations in welding performance as well as other welding defects because of entrapped oxides or impurities within the structure. These oxides and impurities may be originated from lubricant residues and quench oils.

3.2 PM welding process

The selection of the welding process more suitable to join PM parts should be made taking an account the requirements desired such as strength, environmental factors, appearance and the porosity volume. According to the literature (Hamill, 1993; Jayabharat et al., 2007), fusion welding processes are used successfully to join ferrous powder metal parts with high density (> 7.0 g/cm^3) once these high density PM parts typically have the same weldability as forged, rolled or cast materials. Indeed, research of Hinrichs et al showed that it is possible to obtain good quality dissimilar weldments of low and medium carbon PM steels joined with forged steels using the most common fusion welding processes. The researchers also showed that in the case of the welding of PM high-strength low-alloy (HSLA) steels, procedures such as pre-heating and hydrogen control should be adopted to guarantee the success of the joining.

Low and intermediate density PM parts (< 7.0 g/cm³) should be joined using welding processes which minimize the volume of molten weld metal such as resistance projection welding, friction welding and brazing.

The reason is that the low fracture resistance of these PM materials, caused by the small number of bonding between the particles, does not allow that these absorb the residual stress produced by the high densification that occurs in the heat affected zone and shrinkage of the weld metal, resulting in subsequent cracking in or near the weld interface. Additionally, when choosing the brazing process to join the low or intermediate density PM parts, a special attention should be paid to the capillary force o f the pores once the porosity near the joint wicks the copper brazing filler metal into the pores, leaving an insufficient amount of filler metal to establish the satisfactory weld strength. To overcome this problem, the PM parts must be copper infiltrated before brazing (Hamill, 1993; Jayabharat et al., 2007).

4. Pulsed GTAW process

The Gas Tungsten Arc Welding (GTAW) is a fusion welding process, where arc is produced between non-consumable tungsten electrode and base metal. This process provides suitable results in many situations because of its ability to control the welding parameters (heat input, travel speed, feed rate and type of filler metal) during the welding and subsequent weld metal and HAZ hardness. The higher control of the heat input in this process compared with other fusion welding processes may be mainly attributed to the fact of the welding arc does not suffer direct interference of the metal transference during welding.

Pulsed GTAW involves cycling of the welding current from a high level to a low level at a selected regular frequency. Thus, pulsing the current introduces additional operational parameters, which include peak current, base current, peak pulse time and base pulse time (Pawan et al., 2011). Figure 1 shows the representation for the pulsed current.

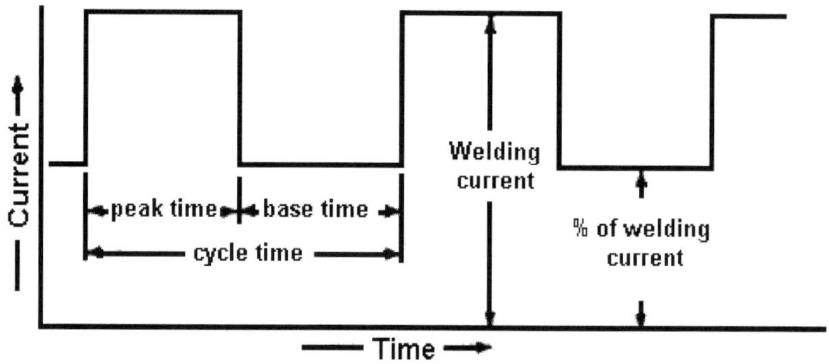

Fig. 1. Representation of the pulsed current.

Pulsed GTAW is frequently used for welding of several materials as heat input can be precisely controlled. Also, this process is strongly characterized by the bead geometry control, which plays an important role in determining the mechanical properties of the weldment (Juang & Tarng, 2002).
In contrast to continuous current welding, in the pulse mode the heat energy required to melt the base material is supplied only during peak current pulses for brief intervals of time allowing the heat to dissipate into the base material. With this, it is possible to achieve the maximum penetration without excessive heat build-up (Juang & Tarng, 2002).
As a result of all mentioned above, it is possible to obtain weldments with a narrower HAZ as well as reduction of segregation of alloying elements, of residual stress and of the hot cracking sensitivity (D' Oliveira et al., 2006; Wang et al., 2006). Current pulsing also results in periodic variations of the arc forces and in an increase of the melt pool agitation leading to additional fluid flow, which lowers the temperature in front of the solidifying interface. This temperature fluctuations leads to the continual changes in the weld pool size and shape favouring the growth of new grains (Pawan et al., 2011).
As a consequence of this grain refinement in the fusion zone, an improvement of the mechanical properties, such as tensile and fatigue resistance is achieved. Therefore, through the use of pulsed parameters, it is possible working with high current peaks without increase the average heat input to the base material, which enables itself as a good choice for welding powder metals alloys.

4.1 Weldability study of iron based powder metal alloys

According to the literature (Hamill, 1993), a unique conventional GTAW application involving welding of PM parts was the replacement of a casting for two welded PM components together for use in a commercial truck differential. It was found that the welded PM components exhibited higher and more consistent strength values than a bolted gray iron casting along with providing a 35% cost savings compared with the previous method of manufacture. However, there was not enough information about the welding procedures (parameters, heat input, gas shield flow, etc) used to carry out the welding.
This topic therefore intends to provide to the readers experimental information about welding of iron based PM alloys using the pulsed GTAW.
The materials involved in this study were three different iron based powder metal alloys whose compositions and features are given in table 1. As mentioned before, these alloys are the largest and the most effective alternative PM parts group for replacing castings, forged and machined parts, mainly in the automobile industry.

P/M alloys	Chemical composition (wt-%)			Raw material		
	Fe	Ni	P	Powder	Particle size range (μm)	Apparent density (g/cm³)
Fe	100	-	-			
Fe-Ni	96.00	4.00	-	Sponge iron	60-150	2.8-3.1
				Carbonyl nickel	3	-
Fe-Ni-P	95.75	4.00	0.25	Pre-alloyed Fe-P (20wt-% P)	< 44	-

Table 1. Specification of the powder metal alloys.

In order to obtain the PM alloys, the powder metal were first mixed with lubrificant (zinc stearate) in a ball mill according to the chemical compositions of the alloys given in table 1 to produce a homogeneous mixture of ingredients. After that, the mixed powder of each alloy was compacted to about 90% relative density in a press (green compact) and then sintered in a pure hydrogen atmosphere according to the thermal cycle shown in Figure 2 to complete the metallurgical bonds between powder particles.

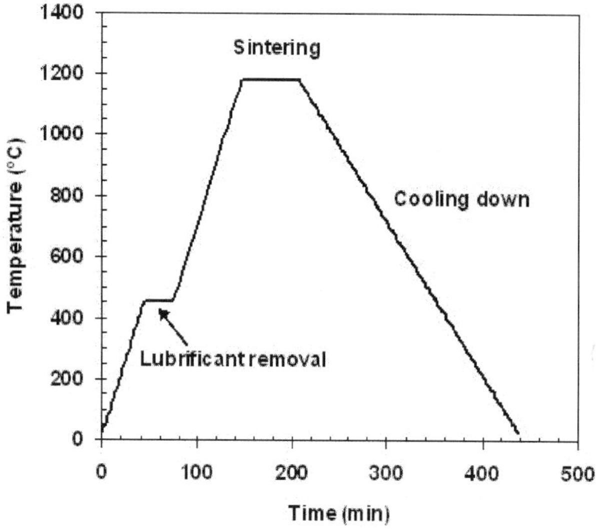

Fig. 2. Thermal cycle of sintering.

The powder metal samples (dimensions: 100 mm x 20mm x 7 mm) produced were welded in the butt joint, flat position with three different filler metals (AWS R 70s-6, AWS R 309L, AWS R Fe-Ni) using four passes weld by a manual pulsed GTAW process. The pulse welding parameters used were the same to the three different alloys and were chosen after preliminary tests that guaranteed an arc stability and lower heat input. A flow rate of argon (99.99% purity) of 7 l/min was used as a shielding gas. The travel speed was adjusted to give an adequate penetration and weld bead contour. These welding parameters are given in table 2.

	Parameters				
Alloy	Peak current (I_P) (Amps)	Base current (I_B) (Amps)	Peak time (T_P) (s)	Base time (T_B) (s)	Filler metal
Fe	140	80	0.40	0.40	AWS R 70S-6 (Ø 1,6 mm)
Fe-Ni	140	80	0.40	0.40	AWS R 70S-6 (Ø 1,6 mm)
Fe-Ni	140	80	0.40	0.40	AWS R FeNi (Ø 3,25 mm)
Fe-Ni-P	140	80	0.40	0.40	AWS R FeNi (Ø 3,25 mm)
Fe-Ni-P	140	80	0.40	0.40	AWS R 309L (Ø 2,4 mm)

·as: Argon, Shielding gas flow rate: 7 l/min, DCEN polarity, Backing shielding gas flow rate: 10 l/min

Table 2. Pulsed GTAW parameters.

After welding, the test samples were transverse sectioned, polished with Al_2O_3 and etched with 2% nital (HNO_3 + alcohol). Microstructural examination of the specimens was carried out using standard optical microscopy and scanning electronic microscopy (SEM). Vickers hardness values were taken across the transverse section using a 10 Kg load. The tensile test samples geometry is shown in Figure 3 and these were in accordance with ISO 2740 standard.

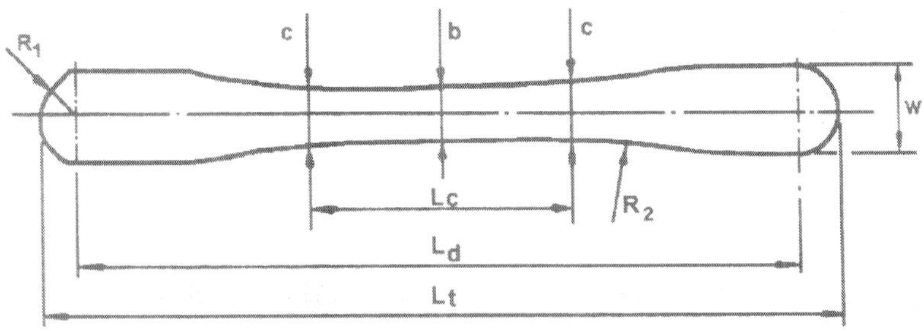

b	c	L_c	L_d	L_t	W	R_1	R_2
5.70 ± 0.02	b + 0.025	32	81.0 ±0.5	89.7 ±0.5	8.7 ± 0.2	4.35	25

Dimensions in (mm)

Fig. 3. Tensile test samples geometry.

4.1.1 Microstructural characterization of the weldments

a. Powder metal pure Fe and Fe-Ni alloy

As shown in Figures 4, 5 and 6, no difficulty concerning to the weldability of 7-mm thickness samples of powder metal pure Fe using filler metal of mild steel and Fe-Ni alloy using filler metal of mild steel and Fe-Ni alloy (60% Fe-40% Ni) was observed. Metallographic examinations showed the presence of small pores in the base metal of the alloys randomly distributed in a ferritic matrix. Meanwhile, no porosity and shrinkage cracks were observed in the weld metal and heat affected zone of the weldments. This may be mainly attributed to the high density after sintering (> 7.0 g/cm^3) and, in lesser extension, to the small size of the pores.

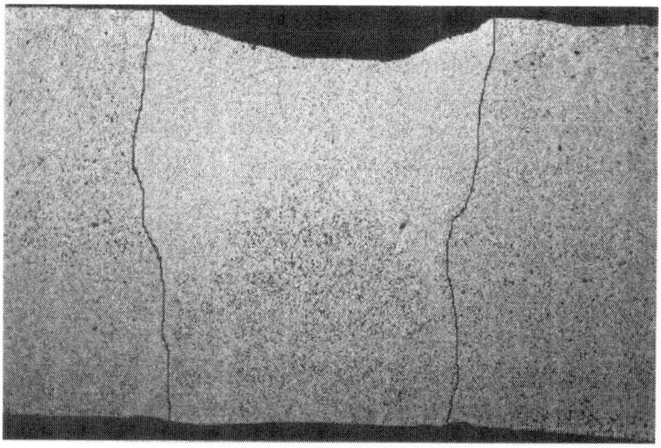

Fig. 4. Macrograph of weld of 7 mm powder metal pure Fe using filler metal of mild steel. The black lines outline the fusion zone. Etching: 2% nital. Magnification: 15x.

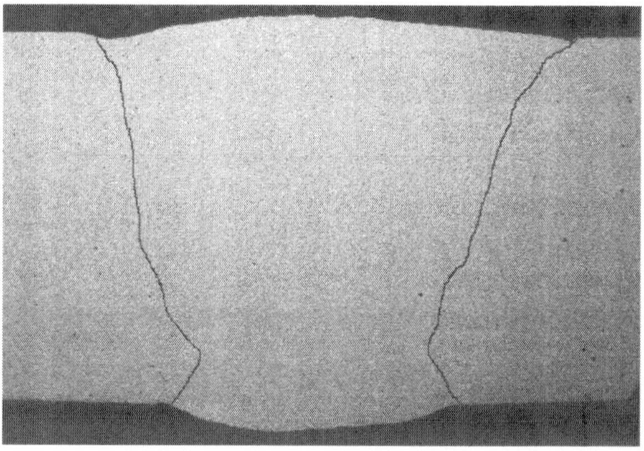

Fig. 5. Macrograph of weld of 7 mm powder metal Fe-Ni alloy using filler metal of mild steel. The black lines outline the fusion zone. Note the good mixing between the filler metal and base metal. Etching: 2% nital. Magnification: 15x.

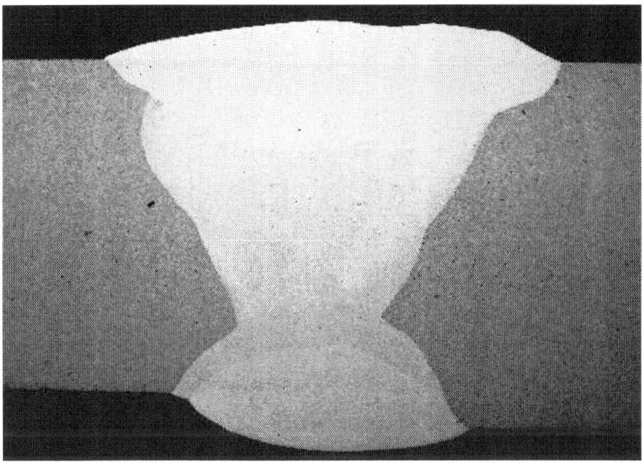

Fig. 6. Macrograph of weld of 7 mm powder metal Fe-Ni alloy using filler metal of Fe-Ni alloy (60%Fe-40% Ni). Etching: 2% nital. Magnification: 15x.

Figures 7, 8 and 9 showed that the powder metal pure Fe and Fe-Ni alloy did not evidence significant changes of hardness profile in the HAZ in comparison with base metal for the filler metals used in this study, which is an indicative of a good continuity of mechanical properties after welding.

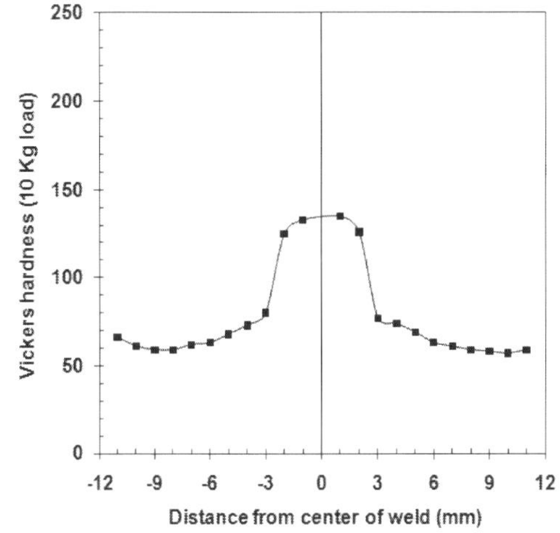

Fig. 7. Hardness profile through the powder metal pure Fe using filler metal of mild steel.

In general, phosphorus is intentionally added in powder metal iron based alloys to increase the densification of the iron powder once this element allows the formation of a transient liquid phase during sintering. Furthermore, phosphorus is known to improve the corrosion resistance and magnetic properties of the powder metal iron-based parts (ASM Handbook,

1999). However, phosphorus additions are not particularly attractive for fusion welding applications because its presence in the metal base composition is associated with the formation of the eutectic M_3P, which may promote solidification cracking in the fusion zone. Therefore, the amount of phosphorus added in the alloy must be rigorously controlled (Correa et al., 2008).

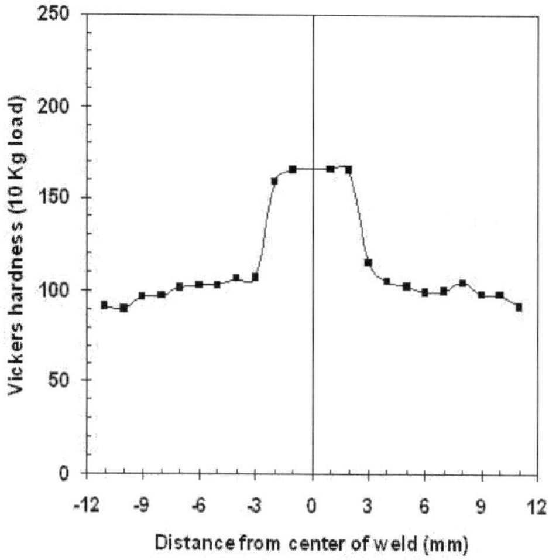

Fig. 8. Hardness profile through the powder metal Fe-Ni alloy using filler metal of mild steel.

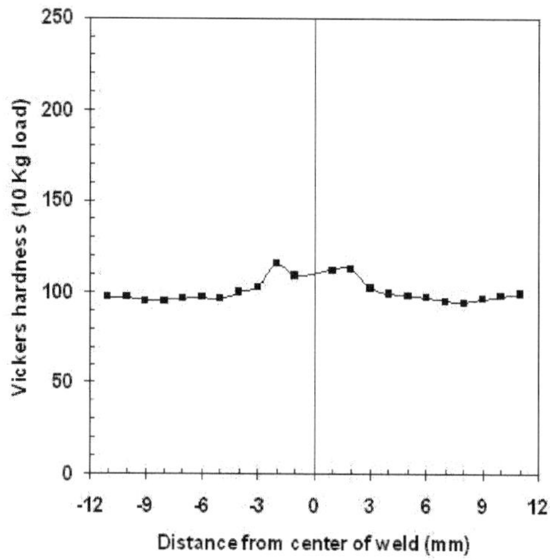

Fig. 9. Hardness profile through the powder metal Fe-Ni alloy using filler metal of Fe-Ni.

b. Powder metal Fe-Ni-P

According to author (Beiss, 1989), PM carbon steels with additions of 0,35% P may be fusion successfully welded without the occurrence of solidification cracks since the carbon content is lower than 0,2%. PM carbon steels with higher carbon contents tend to facilitate the segregation of the phosphorus and the formation of the M_3P eutectic.

Figure 6 shows the microstructure of the transverse section of weld joint of the alloy Fe-Ni-P using Fe-Ni (60% Fe-40% Ni) filler metal. Despite the good toughness of the Fe-Ni filler metal to absorb the shrink stresses during the weld metal cooling, the weld metal presented solidification cracks and pores after the pulsed GTA welding.

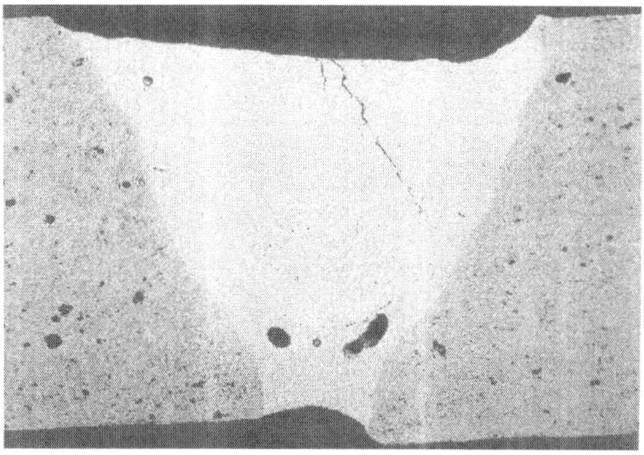

Fig. 10. Macrograph of weld of 7 mm powder metal Fe-Ni-P alloy using filler metal of Fe-Ni (60% Fe-40% Ni). Note the presence of solidification cracks and porosity in the fusion zone. Etching 2%. Magnification: 15x.

The weld solidification cracking can be mainly attributed to the presence of the low-melting eutectic Fe_3P and/or Ni_3P in the weld pool. According to the literature (Lancaster, 1987; Lippold & Kotecki, 2005), the excessive amount of phosphorus combines with nickel or iron forming the low-melting eutectic Ni_3P or Fe_3P. The continuous presence of the segregated Ni_3P or Fe_3P liquid film in the last stages of solidification o fthe weld pool combined with the higher shrinking stress due to faster cooling rates during fusion welding, may have contributed to the appearance of solidification cracking (Briskman, 1979). In addition, considering the potential of the nickel as an austenite stabilizer, the Fe-Ni filler metal solidifies in the austenitic mode, which increases the segregation of phosphorus in the weld pool and, consequently, the susceptibility to cracking (Lippold & Kotecki, 2005).

Figures 11 and 12 show the presence of phosphorus eutectic in the HAZ and fusion zone of the powder metal Fe-Ni-P alloy using Fe-Ni filler metal, which is characterized by the presence of small "islands" in the ferritic grains.

As can be seen in Figure 13, a complete elimination of the weld solidification cracking and porosity in the fusion zone of the Fe-Ni-P alloy was possible using the filler metal 309L stainless steel and adjusting the welding parameters to those values shown in table 2. According the literature (Lippold & Kotecki, 2005), the principal reason for the absence of solidification cracking in the weld metal of the 309L filler metal is the low carbon content of the 309L filler metal and, mainly, the presence of a two-phase austenite/ferrite mixture in the microstructure of the weld metal.(See Fig 14).

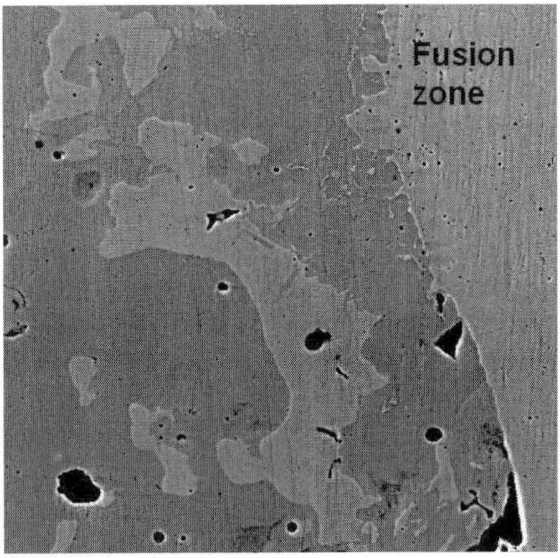

Fig. 11. Micrograph of the HAZ of the powder metal Fe-Ni-P alloy. Note the presence of the phosphorus eutectic islands in the ferritic grains and some pores. Etching: 2% nital. Magnification 800x.

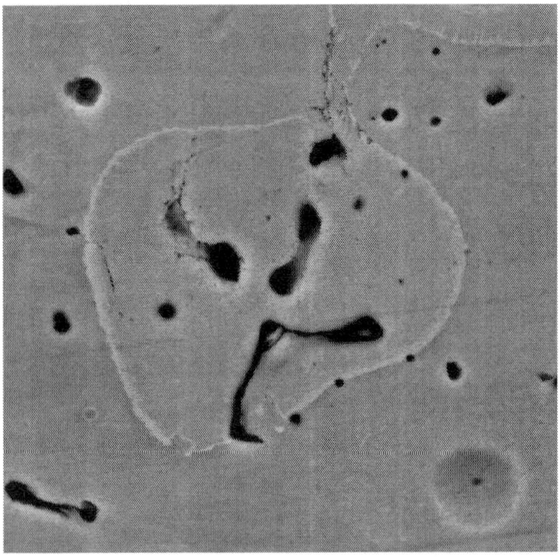

Fig. 12. Micrograph of the fusion zone of the powder metal Fe-Ni-P alloy. Note the presence of the phosphorus eutectic in the ferritic grains and some pores. Etching: 2% nital. Magnification: 2000x.

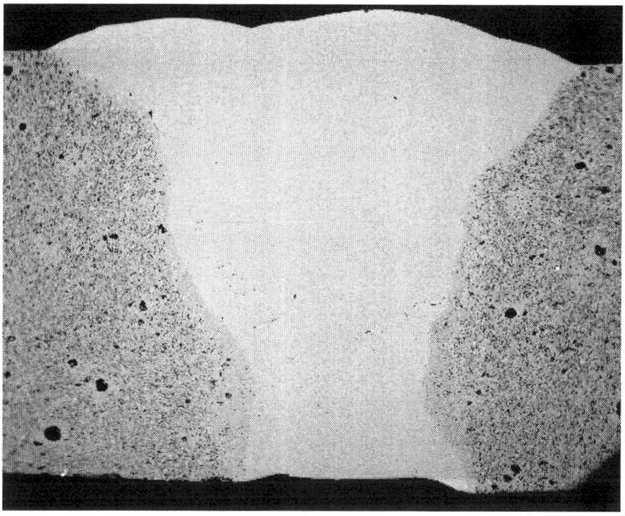

Fig. 13. Macrograph of weld of 7 mm powder metal Fe-Ni-P alloy using filler metal of 309L stainless steel. Note the absence of solidification cracks and pores in the fusion zone. Note also large pores in the base metal. Etching: nital 2%. Magnification: 15x.

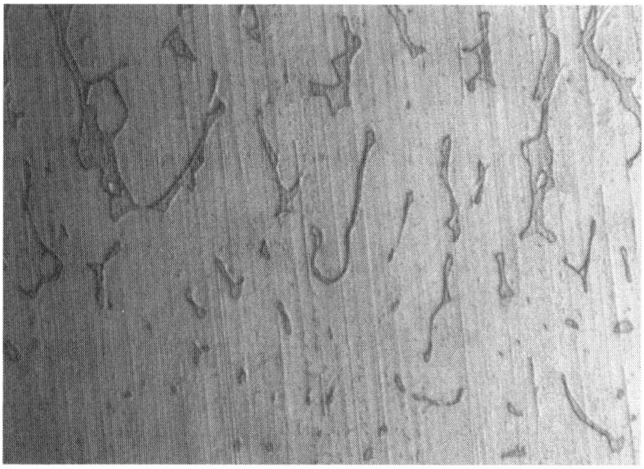

Fig. 14. Micrograph showing the two-phase austenite + vermicular delta ferrite mixture in the weld metal of 309L stainless steel. Etching: (HCl/HNO_3) reagent. Magnification: 1600x.

Image analysis results (Table 3) from optical micrographs, similar to that showed in Fig 14, together with observations in the WRC-1992 diagram, indicated that the ferrite number (FN) in the 309L stainless steel weld metal was approximately 7 FN. In general, above 3 FN, but less than 20 FN, solidification of austenitc stainless steels is most likely in the FA mode (Suutala, 1983).

309L stainless steel weld metal		
Statistical function	Phase A (Delta ferrite)	Phase B (Austenite)
Unity	%	%
Counts	15	15
Mean	6,77	93,23
Standard deviation	0.89	2.82

Table 3. Image analysis results of the delta ferrite volume fraction of the 309L weld metal.

In the FA mode, the duplex microstructure (delta ferrite + austenite) presents at the end of solidification, increase the amount of tortuous phase boundaries that resist wetting by liquid films and along which cracks might propagate. Thus, once the crack is nucleated, it becomes very difficult for it to propagate along to the nonplanar crack path generated to these tortuous bondaries (Briskman, 1979). Additionaly, as the solubility of the phosphorus in the ferrite is higher than that observed in the austenite, the delta ferrite in the weld metal is prone to absorb a significant amount of this element, which reduce the concentration of the phosphorus in liquid film, avoiding the permanence of the segragated low-melting eutectic until the last stages of solidification and, consequently, the solidification cracking.

It can be also seen in Fig. 14 the vermicular morphology of the delta ferrite. In general, this ferrite morphology is present when welding cooling is moderate and/or when the Creq/Nieq is low but still within the FA mode (Lippold & Kotecki, 2005).

It is worthwhile mentioning that other significant factors involving the pulsed GTA welding of teh powder metal Fe-Ni-P alloy may have also contributed to minimize the occurence of solidification cracking. These may be, for instance the utilization of pulsed current and multipass weld, which have the effect of refinement of the as-cast microstructure in the fusion zone (Balasubramanian et al., 2008).

Also, it can be noted that there was no presence of pores in the weld metal. However, th e pores in the base metal of this alloy (see Fig. 15) were large and rounded with higher densification of ferrite in their surroundings. The size of the Fe-P pre-alloyed particles added in this alloy probably is the cause of the pores characteristics (large and rounded) in the base metal.

Figure 16 shows that Fe-Ni-P alloy welded with 309L austenitic stainless steel filler metal did not evidence significant changes of hardness profile in the HAZ in comparison with the base metal, even though the phosphorus is prone to harden the ferrite by solid solution.

4.1.2 Tensile tests of the weldments

Tensile tests carried out in the welded samples of the pure Fe and Fe-Ni using AWS R 70S-6 and AWS R Fe-Ni filler metals (filler metal of mild steel and Fe-Ni alloy) and Fe-Ni-P powder metal alloy using AWS R 309L filler metal showed that the failures of samples occurred always in the base metal. Furthermore, the welded samples of these alloys presented ultimate tensile strength slightly higher than unwelded samples (see Table 4).

The higher tensile strength of the welded samples may be attributed to the residual stress in the samples due to their small dimensions (width and length) combined with the high heat input and the relatively rapid cooling of the weld metal during welding. It is worthwhile

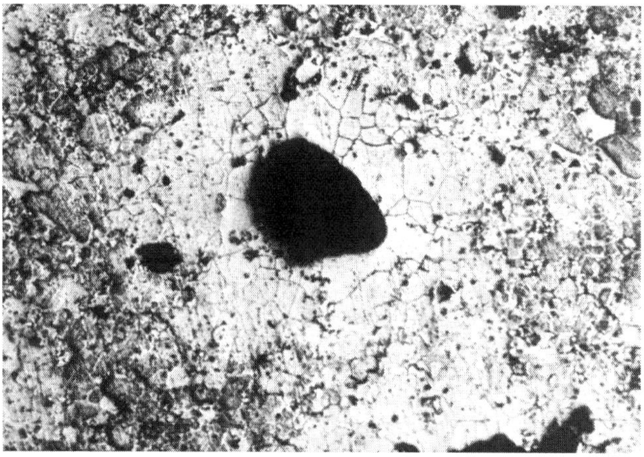

Fig. 15. Micrograph showing the large and rounded pore in the base metal. Note the high densification of ferrite in its surroundings . Etching: nital 2%. Magnification: 1600x.

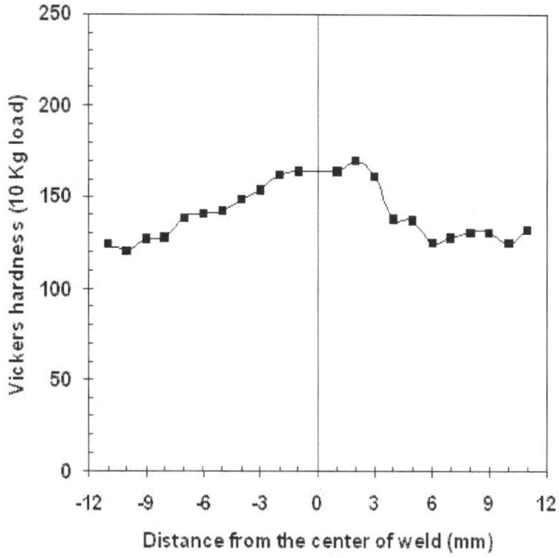

Fig. 16. Hardness profile through the powder metal Fe-Ni-P alloy using 309L stainless steel filler metal.

mentioning that the welded samples were not annealed after welding. However, the tensile tests results together with the hardness results indicated that weldments of these materials presented a good continuity of the mechanical properties with relation to the base metal, even when these materials are welded by a fusion welding process.

Tensile Properties	Unwelded samples			Welded samples		
	Fe	Fe-Ni	Fe-Ni-P	Fe	Fe-Ni	Fe-Ni-P
Ultimate tensile strength (MPa)	170	205	200	197	215	218
Elongation (in 25.4 mm) (%)	4.3	4.1	1.6	3.8	3.9	1.5

Table 4. Tensile properties of GTA welded alloys.

5. Conclusion

In this chapter was shown the benefits of the PM components and the advantages of their utilization in replacing casting, machined and forged materials. However, experimental information about joining these PM materials using fusion welding processes is still scarce.

Through of the study presented in this chapter, it was possible to verify that PM iron based alloys pure Fe and Fe-Ni may be successfully welded by pulsed GTAW process without additional techniques such as preheating, post-heating or special joint configuration.

Concerning to the Fe-Ni-P alloy, this alloy also may be successfully welded by pulsed GTAW process but a discerning selection of the filler metal and careful control over the welding parameters (heat input, peak current, base current, peak time, base time and travel speed) must be done. Due to the presence of phosphorus, the correct selection of the filler metal may avoid the presence of low-melting eutectic films at the end of the solidification and, consequently, the solidification cracking. A rigid control of the heat input, in turn, prevents the higher dilution of the base metal, which decreases the shrink stress in the fusion zone during the solidification of the weld pool leading to a lower susceptibility to cracking of the weldment.

Hardness results showed that no excessive hardening was observed in the weld metal and HAZ of the iron powder metal alloys studied. This is in agreement with the base metal fracture location in the tensile tests and with the slight increase of the tensile properties of welded samples in comparison with unwelded samples. In general, excessive hardening (higher Vickers hardness values) is prone to increase significantly the tensile strength and decrease the toughness.

6. Acknowledgment

The author acknowledges the Brazilian Government Agencies CNPq, CAPES and FAPEMIG for the financial support to carry out this study.

7. References

ASM Handbook, (1999). *Metallography and Microstructure,*, vol. 9, ASM, USA

Briskman, A.N. (1979). The effect of welding currente pulses on the susceptibility of weld metal to hot cracking during argon TIG welding, *Aut. Weld. (7)*, pp. 40-43.

Balasubramanian, M., Jayabalan, V. & Balasubramanian, V. (2008). Optimizing the Pulsed Current GTAW parameters to Attain Maximum Impact Toughness, *Materials and Manufacturing Processes*, Vol. 23, n. 1-2, pp. 69-73.

Beiss, P. (1989). Finishing Process in Powder Metllurgy, *Powder Metallurgy*, 32 (4), pp. 277-284.

Chawla, N. & Deng, X. (2005). Microstructure and Mechanical behaviour of porous sintered steel, *Materials Science and Enginnering A, 390,* pp. 98-112.

Correa, E.O., Costa, S.C. & Santos, J.N. (2008). Weldability of iron based powder metal using pulsed plasma arc welding process, *Journal of Materials Processing Technology*, 198, pp. 323-329.

D' Oliveira, A.S.C.M., Paredes, R.S.C., & Santos, R.L.C. (2006). Pulsed Current Plasma Transferred Arc Hardfacing, *Journal of Materials Processing Technology,* 171, pp. 167-174.

German, R.M. (2005). Powder Metallurgy and Particulate Materials Processing, *Metal Powder Industries Federation*, New Jersey, USA.

Hamill, J.A. (1993). Pwhat are the joining process, Materials and Techniques for Powder Metal Parts, *Welding Journal*, 2, pp. 37-45, USA.

Jayabharat, K., Ashafaq, M, Venugopal, P., Achar, D.R.G. (2007). Investigation on the continuous drive friction welding of sintered powder metallurgical (P/M) steel and wrought copper parts, *Materials Science and Engineering A,* 454-455, pp. 114-123.

Jenkins, I. & Wood, J.V. (1991). Powder Metallurgy: An Overview. *The Institute of Metals*, London, UK.

Juang, S.C. & Tarng, Y.S. (2002). Process Parameter Selection for Optmizing the Weld Pool Geometry in the Tig Welding of Stainless Steel, *Journal of Materials Processing Technology*, Vol. 22, No 1, pp. 33-37.

Kumar, T.S., Balusubramanian, V. & Sanavullah, M.Y. (2007). Influences of Pulsed Current Tungsten Inert Gas welding parameters on the Tensile properties of AA6061 Aluminum Alloy, *Materials Design*, 28, pp. 2080-2092.

Kurt, A.H., Ates, A., Durgutlu, A. & Karacif, K. (2004). Pexploring the Weldability of Powder Metal Parts, *Welding Journal*, 83 (12), pp. 34-37.

Lancaster, J.F. (1987). *Metallurgy of Welding*, London, UK.

Lenel, F.V. (1980). Powder Metallurgy: Principles and Applications, *Metal Powder Industries Federation*, New Jersey, USA

Lippold, J.C & Kotecki, D.J. (2005). *Welding Metallurgy and Weldability of Stainless Steels*, 5 ed., USA.

Metal Powder Industries Federation (2004). In : *Design Solutions Manual*, MPIF, New Jersey, USA

Pawan, K., Kolhe, K.P., Morey, S.J. & Datta, C.K. (2011). Process Parameters Optimization of an Aluminum Alloy with Pulsed Gas Tungsten Arc Welding (GTAW) Using Gas Mixtures, *Materials Science and Applications*, N.2, pp. 251-257.

Schatt, W. & Wieters, K-P. (1997). Powder Metallurgy: Processing and Materials, *European Powder Metallurgy Association*, Shrewbury.

Sudhakar, K.V., Sampathkumaran, P & Dwarakadasa, E.S. (2000). Dry Slinding wear in high density Fe-2% Ni based P/M alloys, *Wear,* 242, pp. 207-212

Suutala, N. (1983). Effect of solidification conditions on the solidification mode in austenitic stainless steels, *Metallurgical Transactions*, vol. 14, n. 2, pp. 191-197.

Thummler, F. & Oberacker, R. (1993). *Institute of Materials*, London, UK.

Wang, S.H., Chiu, P.K., Yang, J.R. & Fand, J. (2006). Gama (γ) phase transformation in Pulsed GTAW weld metal of Duplex Stainless Steel, *EurMaterials Science and Engineering A,* N. 420, pp. 26-33.

7

Assessment of Stress Corrosion Cracking on Pipeline Steels Weldments Used in the Petroleum Industry by Slow Strain Rate Tests

A. Contreras[1], M. Salazar[1], A. Albiter[1], R. Galván[2] and O. Vega[3]
[1]*Instituto Mexicano del Petróleo,*
[2]*Universidad Veracruzana,*
[3]*Centro de Investigación en Materiales Avanzados-CIMAV*
México

1. Introduction

The Stress Corrosion Cracking (SCC) is a local corrosion process which is characterized by the initiation and propagation of cracks. It takes place under the simultaneous action of sustained tensile stresses and specific corrosive environment on a susceptible material.
The formation of SCC occurs below the yield strength of the material and typically below the design stress and fatigue limit of an engineering structure. Since the first discovery of SCC on the exterior surface of a buried high pressure natural gas transmission pipeline in 1965 (Leis & Eiber, 1997), SCC has continued to make a significant contribution to the number of leaks and ruptures in pipelines.
Two forms of SCC can exist on buried steel pipelines (Beavers & Harle, 2001). The first discovered form of SCC propagates intergranularly and is associated with a concentrated alkaline electrolyte in contact with the steel surface, commonly called as high pH-SCC or classical SCC. A second form of SCC was discovered in Canada in the early 1980. This form of SCC propagates transgranularly and is associated with a dilute neutral pH electrolyte in contact with the steel surface, commonly called as low pH-SCC, non-classical, or near neutral pH-SCC. Currently, there are some mechanisms proposed to explain the SCC occurrence including the following: (1) a role for hydrogen in enhancing crack tip dissolution; (2) a possible synergistic growth by fatigue and corrosion.
For high pH-SCC it is observed that the mechanism involves anodic dissolution for crack initiation and propagation. In contrast, for low pH-SCC is associated with the dissolution of the crack tip and sides, accompanied by the ingress of hydrogen in the steel. Steels with high tensile strength are more susceptible to SCC. Cracks propagate as a result of anodic dissolution in front of their tip in SCC process, due to the embrittlement of their tip by hydrogen based mechanism. It was revealed that cracking behavior of pipeline steel in the soil environment depends of the cathodic protection applied. Applying different potentials levels the dominance of SCC process changes. At relatively low potential, the steel cracking is based primarily on the anodic dissolution mechanism. When the applied potential increases negatively, hydrogen is involved in the cracking process, resulting in a transgranular cracking mode (Liu *et* al, 2008).

SCC can occur in both gas and liquid pipelines but is more common and catastrophic in gas pipelines (Manfredi & Otegui, 2002). SCC is the most unexpected form of pipeline failure that can involve no metal loss and must not be confused with wall thinning rupture. SCC on pipelines begins with small cracks develop on the outside surface of the buried pipe. These cracks are initially not visible to the eye and are most commonly found in colonies, with all the cracks in the same direction, perpendicular to the stress applied.

This chapter describes the mechanical and environmental effects as well fracture characteristics on SCC susceptibility of steels used in the oil industry using slow strain rate tests (SSRT), which were carried out according to requirements of NACE TM-0198, ASTM G-129 and NACE TM-0177 standards (NACE TM-0198, 2004; ASTM G-129, 2006; NACE TM 0177, 2005). Some tests were supplemented by potentiodynamic polarization and hydrogen diffusion tests. SSRT were performed in samples which include the longitudinal and circumferential weld bead of pipeline steels. The weld beads were produced using the submerged arc welding (SAW) and shielded metal arc welding (SMAW) process.

The SCC susceptibility has been evaluated using the results of SSRT in air (as an inert environment), sour solution according to NACE TM 0177 (solution A from method A) and in some cases a simulated soil solution called NS4. The studies include the effect of pH, temperature, microstructure, effect of multiple welding repairs and mechanical properties. The steels studied are low carbon steels API X52, X60, X65 and X70.

2. Stress corrosion cracking phenomena

2.1 What is SCC and how is presented in pipelines?

SCC is the cracking of the steel as result of the combined effect of corrosive environment and tensile stresses on a susceptible material. SCC is a term used to describe service failures in engineering materials produced by environmentally induced crack propagation (Jones, 1992). The stress required to produce SCC can be residual, externally applied or operational. SCC on pipelines begins when small cracks develop on the external surface of buried pipelines. These cracks initially are no visible, but when the time pass, this individual cracks may growth and forms colonies, and many of them join together to form longer cracks.

The SCC phenomenon has four stages:

1. Cracks nucleation.
2. Slow growth of cracks.
3. Coalescence of cracks.
4. Crack propagation and failure.

This process can take many years depending on the conditions of steel, environment and stresses.

2.2 The conditions for stress corrosion cracking

The studies performed indicate that SCC initiate as a result of the interaction of three conditions as is shown in Figure 1. All the three conditions are necessary to SCC occurs. If any of these three conditions can be mitigate or eliminate to a point where cracking will not occur, the SCC can be prevented.

2.3 Types of stress corrosion cracking

Generally, there are two types of SCC related with the pH. The pH is measured from the environment in contact with the pipe surface. The type involving high pH (greater than 9) is

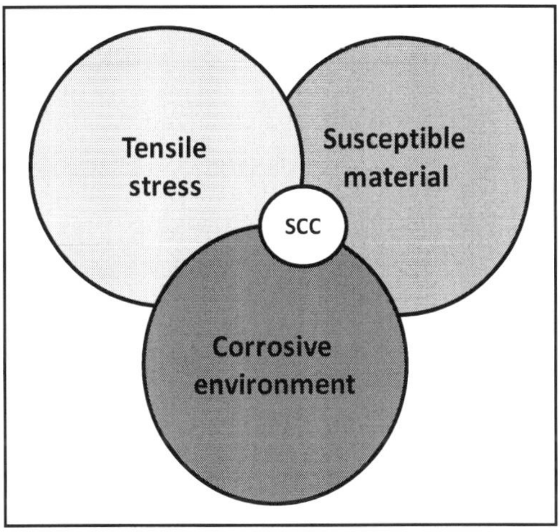

Fig. 1. Conditions necessary for SCC occur.

associated with intergranular cracking, while the type involving lower pH (<6) is associated mainly with transgranular cracking. High pH cracking generally involves rupture of passivating oxide films followed by dissolution in the crack tip (Krist & Leewis, 1998). Delanty *et* al. (Delanty & O'Beirne, 1992) found that the severity of the SCC increases with increasing bacterial concentration and the absence of oxygen.

High pH-SCC occurs only in a relatively narrow cathodic potential range (650-750 mV, $Cu/CuSO_4$) in presence of carbonate-bicarbonate environment and a pH greater than 9 (Stress Corrosion Cracking on Canadian Oil and Gas Pipelines, 1996). Growth cracks increases with temperature, and generally are cracks with no evidence of corrosion.

In contrast, it has been suggested that the low pH-SCC is associated with the dissolution of the crack tip and sides, accompanied by the ingress of hydrogen in the steel (Fang *et* al, 2003, 2010; Zhang *et* al, 1999). No apparent correlation with temperature was found; generally the pH range is between 5.5 to 7.5. A corrosion potential between 760 to 790 mV, $Cu/CuSO_4$ was observed, and wide cracks with evidence of corrosion. A comparative table of both SCC types was given elsewhere (Beavers & Harle, 2001).

2.4 Stress corrosion cracking evaluation techniques

2.4.1 Test methods

As was mentioned above to produce SCC in steel is necessary to apply tensile stresses on material susceptible exposed in a corrosive environment. To assess the SCC susceptibility there are many kinds of specimens including U-bends, bean beams, C-Ring and smooth tensile bars. More detail of these methods is given in NACE TM 0177.

The most common methods for testing these specimens are: 1) Constant Load Tests (CLT) using proof rings and 2) Constant Extension Rate Tests (CERT). The CLT gives few information due to only put the specimen in a specific solution for 30 days at tensile stress of 72% of yielding strength, if the specimen fail the steel no pass, if no fail the steel pass. The termination of the test shall be at tensile test specimen failure or after 720 hours, whichever occurs first.

The CERT method commonly is used with smooth tensile specimens through the SSRT. In these tests specimens are slowly strained in tension and simultaneously exposed to a corrosive environment. SCC susceptibility is evaluated by comparison of failure times, maximum stress, plastic elongation, strain or reduction in area to values obtained in tests conducted in an inert environment (Air). From this comparison of the mechanical properties mentioned above, a ratio is obtained. Additionally, scanning electron microscopy (SEM) observations of samples with low ratios (<0.8) should be carried out. More details of the evaluation of SCC susceptibility using SSRT is given on NACE TM-0198 and ASTM G-129.

2.4.2 Slow strain rate tests (SSRT)

In recent years the SSRT has become widely used and accepted for SCC evaluations to screen materials and to identify alloys that should not experience SCC in service. SSRT involves the slow straining of a specimen of the steel of interest in a solution in which will be in service. Typically a strain rate of the order of $1x10^{-6}$ in/sec is used, which is about four orders of magnitude slower than strain rate used in a standard tensile test. A major advantage of SSRT over CLT is that the test period is generally shorter. Thus, using SSRT we can screening or evaluate materials in a fast way within a few days to determine SCC susceptibility (Kane *et* al, 1997).

It is well known that the SSRT provides not only a useful information on SCC susceptibility of the materials in any corrosive environments, but also a relatively short experimental time to evaluate SCC susceptibility, where a maximum fracture time is that obtained at the lowest strain rate in an inert environment. For that reason, SSRT has been widely used for SCC assessment. However, to use SSRT for the SCC experiments, we need to compare the parameters (time to failure, maximum stress, strain, reduction area, plastic elongation, etc.) obtained in the corrosive environments with those in an inert environment. In addition, it must be kept in mind that the specimens are always fractured in both environments, by which in some cases it would be difficult to judge whether the fracture of the specimens takes place by SCC or not (Nishimura & Maeda, 2004).

2.5 Stress corrosion cracking assessment

The susceptibility to SCC is evaluated according to NACE TM-0177 for CLT, and according to NACE TM 0198 and ASTM G129 for SSR tests. To evaluate the SCC susceptibility through SSRT is expressed in terms of the percentage reduction in area (%RA) calculated by the following expression:

$$RA(\%) = \frac{(D_i^2 - D_f^2)x100}{D_i^2} \quad (1)$$

where D_f and D_i are the final and the initial diameters of the tensile specimen respectively. The reduction area ratio (RAR) after fracture for the specimen in the test environment (RA_e) to the corresponding value determined in the controlled environment (RA_c) was calculated according to the following expression:

$$RAR = \frac{RA_e}{RA_c} \quad (2)$$

Additionally, the SCC susceptibility using the time to failure ratio (TFR) can be evaluated according to the following equation:

$$TFR = \frac{TF_e}{TF_c} \quad (3)$$

where TF_e is the time to failure determined for the material in the test environment and TF_c is the time to failure to the corresponding value determined in the controlled environment. The similar way it can be assessed the SCC susceptibility using the plastic elongation according to the following expressions:

$$\%EP = \left[\frac{E_F}{L_I} - \left(\frac{\sigma_F}{\sigma_{PL}} \right) \left(\frac{E_{PL}}{L_I} \right) \right] x100 \quad (4)$$

$$EPR = \frac{EP_e}{EP_c} \quad (5)$$

where:
EP = Plastic strain to failure (%)
E_F = Elongation at failure (mm/in)
E_{PL} = Elongation at proportional limit (mm/in)
L_I = Initial gauge length (mm/in) (usually 25.4 mm/1 in)
σ_F = Stress at failure (MPa)
σ_{PL} = Stress at proportional limit (MPa)

Ratios in the range of 0.8-1.0 normally denote high resistance to environment assisted cracking (EAC), whereas low values (i.e.<0.5) show high susceptibility. Therefore, to maximize the SCC resistance, it is desirable to obtain values of ratios as close to unity as possible. Lower values of ratios generally indicate increasing susceptibility to SCC.
Complementary metallographic examination must be performed to establish whether or not there is SCC on the samples. Overalls when there is some uncertainty in the assessment of SCC susceptibility evaluating the mechanical properties (RAR, TFR, EPR). The presence of cracks must be evaluated on the longitudinal section of the gage. The SEM observation it is recommended when there is ratios lower than 0.8.

3. SCC susceptibility of low carbon steels

Pipelines of low carbon steel welded by electric arc have been used for many years and are widely used in the petroleum industry. However, frequent failures during operation over the years (Craig, 1998) have prompted several studies of the design, construction, operation and maintenance of equipment and metallic structures used in this industry. Oil and gas from Mexico contain entrained H_2O, CO_2 and H_2S mainly (Presage of production of the marine and south regions from México for a horizon of the 2000-2014). These constituents, when moving through the pipelines, induce failures, mainly in the weld bead. Studies of weld bead failures have demonstrated that these occur mainly in the heat affected zone (Greer, 1975). This is due to the heterogeneity in microstructure (grain growth) and residual stresses.

3.1 Materials

In this work the susceptibility to SCC and corrosion of the main pipeline steels used in the oil industry were investigated. API X52, X60, X65 and X70 pipeline steels were studied. The chemical compositions and equivalent carbon (C_{eq}) are showed in Table 1. Dimensions of the pipeline used in this study are showed in Table 2.

Steel	C	Mn	Si	P	S	Al	Nb	Ni	V	Mo	Ti	Cr	Cu	C_{eq}
X52-A	0.080	1.05	0.26	0.019	0.003	0.038	0.041	0.019	0.054	---	0.002	0.02	0.019	0.27
X52-B	0.090	0.89	0.30	0.006	0.0015	0.025	---	0.05	0.036	0.05	0.016	0.07	0.12	0.28
X60	0.020	1.57	0.14	0.013	0.002	0.046	0.095	0.17	0.004	0.05	0.014	0.26	0.30	0.32
X65	0.070	1.46	0.25	0.012	0.002	0.041	0.047	0.050	0.069	---	0.017	0.02	0.09	0.34
X70	0.027	1.51	0.13	0.014	0.002	0.035	0.093	0.16	0.004	0.004	0.011	0.27	0.28	0.36

Table 1. Chemical composition of the steels studied (wt.%).

Steel	Diameter (in)	Thickness (in)
X52-A	36	0.375
X52-B	8	0.437
X60	42	0.500
X65	24	0.562
X70	36	0.902

Table 2. Dimensions of the pipeline studied.

The pipeline steels with longitudinal welding acquired in PMT (Productora Mexicana de Tubería) were used in this study. The longitudinal and circumferential welding were carried out by the technique of submerged arc welding (SAW) and shielded metal arc welding (SMAW).

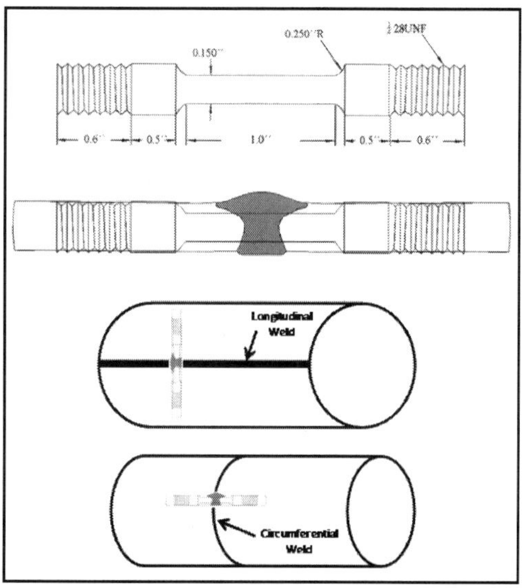

Fig. 2. Schematic representation where the specimens were obtained from the pipeline.

Cylindrical tensile specimens with a gauge length of 25.4 mm (1 inch) and 3.81 mm (0.150 inches) gauge diameter were machined from the pipeline perpendicular to weld bead as is shown in Figure 2.

3.2 Experimental set-up

To perform the SSR tests in the NACE solution saturated with H_2S and NS4 solution, a 500 mL glass autoclave as is shown in Figure 3(a) was used. The autoclave containing the specimen was externally heated by means of a heating element to the temperature required. A Solartron potentiostat controlled by a desktop computer was used for potentiodynamic polarization. The SSRT were performed in an Inter-Corr machine type M-CERT with load capacity of 44 kN and total extension of 50 mm as is shown in Figure 3(b).

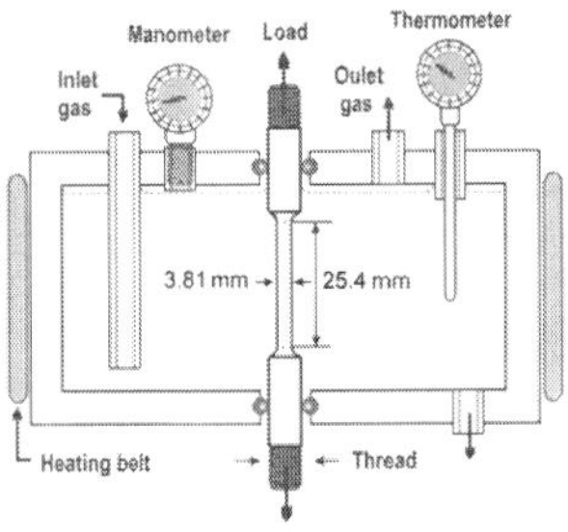

(a)

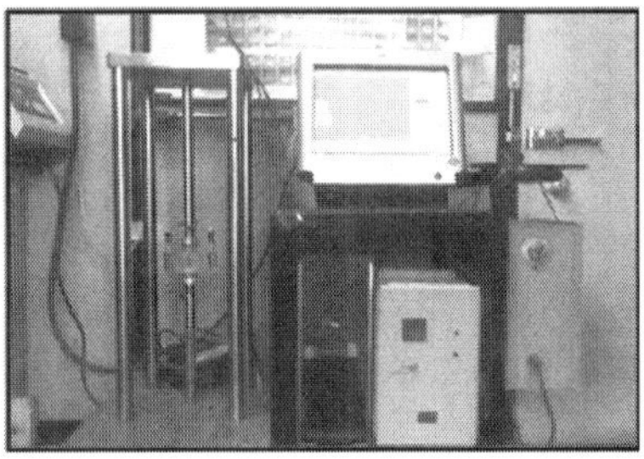
(b)

Fig. 3. (a) Autoclave and (b) M-CERT machine used to perform the SSRT.

3.3 SCC susceptibility of API X52 and X70 pipeline steels

The susceptibility to SCC in a sour solution saturated with H_2S (commonly called as sulphide stress corrosion cracking, SSCC), of API X52 and X70 steels was studied using SSRT. SCC tests were performed in samples which include the longitudinal weld bead of the pipeline steels. The SSRT tests were performed at room temperature in air and with the NACE solution saturated with H_2S at 50°C and at room temperature, using a strain rate of $1x10^{-6}$ in/sec. Cylindrical tensile specimens were machined from the tube according to the NACE TM-0198 standard as was shown in Figure 2.

The test solution according to the standard NACE TM-0177 consists of 50 g of NaCl and 5 g of glacial acetic acid dissolved in 945 g of distilled water. The solution was subsequently saturated with H_2S at a flux rate of 100 to 200 mL/min for 20 minutes.

Table 3 shows the ultimate tensile strength (UTS), elongation (EL), percentage of reduction in area (RA), reduction in area ratio (RAR) obtained from the SSRT curves. From Table 3 can be observed that specimens tested in air showed the maximum %RA, which indicate a high ductility compared with those tested in aggressive environment. The X52 specimens tested in the NACE solution saturated with H_2S at room temperature presented the maximum susceptibility to SCC. This is in agreement with results reported in the literature (Lopez *et* al, 1996; NACE TM 0177, 2005; Perdomo *et* al, 2002). Meanwhile, the maximal susceptibility to SCC in X70 specimens tested in NACE solution saturated with H_2S was at 50 °C.

SSRT tests are widely used to evaluate the susceptibility to stress corrosion cracking of various materials (Wang *et* al, 2001; Casales *et* al, 2000, 2004; Parking & Beavers, 2003; Chen *et* al, 2002; Wang *et* al, 2002; Zhang *et* al, 1999; Brongers *et* al, 2000; Park *et* al, 2002; Beavers & Koch, 1992; Liou *et* al, 2002). The failure time for samples tested in air, NACE solution at 20°C and NACE solution at 50°C was around 45, 20, 27 and 41, 23 and 39 hours for the X52 and X70 steels respectively. It is clear that specimens tested in the sour solution saturated with H_2S presented high susceptibility to SCC. Corrosion was found to be an important factor in the initiation of some of the cracks. The susceptibility to SCC was manifested as a decrease in the mechanical properties, e.g., strain values before failure, ultimate tensile strength, reduction in area, time to failure and in some cases the presence of secondary cracking along the gauge length of the specimen.

Steel	Environment	UTS (MPa)	EL (mm)	%RA	RAR	Failure zone
X52	Air	391	2.03	55.6	N/A	Base Metal
	NACE+H_2S at 25°C	249	1.42	13.8	0.248	Weld bead/HAZ
	NACE+H_2S at 50°C	233	1.88	7.25	0.130	Weld bead/HAZ
X70	Air	462	2.64	50.98	N/A	Weld bead
	NACE+H_2S at 25°C	213	1.21	6.91	0.135	Weld bead
	NACE+H_2S at 50°C	355	2.03	4.38	0.085	Weld bead/HAZ

Table 3. Summary of the SSRT results to evaluate the SCC susceptibility.

The susceptibility to SCC depends on many factors such as: alloying elements (Kim *et* al, 1998) microstructure (Wilhelm & Kane, 1984), applied stresses (Miyasaka *et* al, 1996), environments (Vangelder et al, 1987), temperature (Casales *et* al, 2002), strength (López *et* al, 1996), strain (Parkins, 1990), among others.

In the SSR tests, the specimens tested in air exhibited a ductile type of failure. Whereas, in the corrosive solution, the specimens shown a brittle fracture. Among all testing environments, the NACE solution saturated with H_2S had strong influence on SSRT results, reflected in the degradation of mechanical properties. Both steels presented a corrosive attack in the form of anodic dissolution of the material. The X52 steel showed the best resistance to the corrosive attack. All the cracks, primary and secondary were perpendicular to the applied tension axis, being indicative of SSCC.

These cracks were related to the diffusion of atomic hydrogen promoting the embrittlement damage. The failure in air occurred in some cases in the weld joint, but in presence of the NACE solution saturated with H_2S the failure always occurred in the HAZ. The path of the crack was very irregular with brittle appearance.

The failure in air occurred in some cases in the base metal (BM) for the X52 steel, meanwhile, for the X70 occurs in the weld bead (WB) as is observed in Figure 4(a). In presence of the NACE solution the failure always occurred in the heat affected zone (HAZ) (Figure 4b and 4c). The major occurrence of those fractures, is related to the higher stress concentration by the lost metal regions by anodic dissolution (pits and general corrosion) and the diffusion of atomic hydrogen into the entrapping sites during the welding process and also to microstructural trappings.

Additionally, the diffusion of hydrogen and the stresses focused in the surrounding of the crack initiations sites develop the initiation and the growing of cracks (Contreras *et* al, 2005). The mode of crack growth was discontinuous, observing little deformation and the formation of a neck in the failure zone was absent for the tests performed in the NACE solution, which it is related to the brittle fracture. The SCC of low strength steels is characterized by transgranular fracture, in contrast to intergranular fracture of high strength steels (Shim & Byrne, 1990; Asahi *et* al, 1988; Tsay *et* al, 2000).

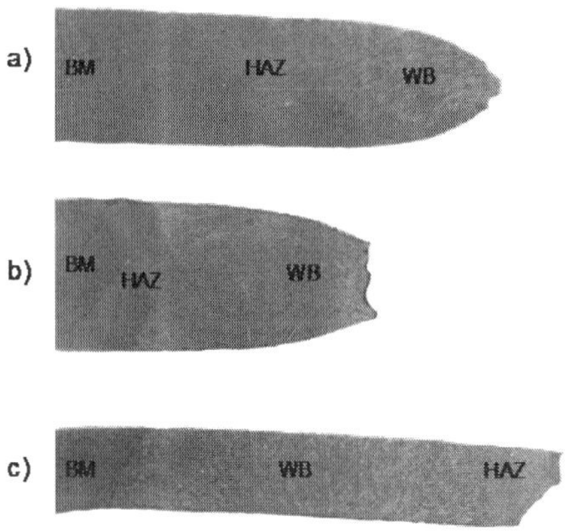

Fig. 4. Optical macrographs of the longitudinal sections (X70 steel) of the specimens after cracked, showing the zone where they failed: (a) tested in air, (b) tested in NACE solution at 25°C and (c) tested in NACE solution at 50°C

Mechanical fracture techniques have been used to quantify the stress effects and the sour environment effects in the cracking. Figure 5 shows SEM micrographs of the near surface cross-section micrographs in the failed SSRT specimens tested in the NACE solution saturated with H_2S. The path of the crack was very irregular with brittle appearance (Figure 5a). In addition, it is observed that X70 steel is more susceptible to the corrosion attack as shown in Figure 5(b), showing secondary cracks after being fractured in a NACE solution saturated with H_2S at 50°C. It was observed that at 25°C, the attack was in the pitting form and it was less severe than at 50°C. Meanwhile, at 50°C the attack was in the form of micro-cracks (Fig. 5b) and it was more homogeneous along the gauge section of the specimen.

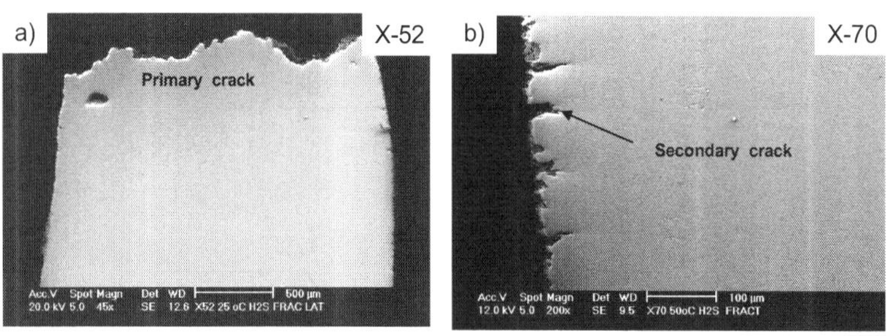

Fig. 5. SEM micrographs of tensile fractured samples (a) X52 steel showing the appearance of the fracture surface (primary crack) tested at 25°C, and (b) X70 steel showing the secondary cracks in the longitudinal gauge section, tested at 50°C.

3.4 Effect of multiple repairs in girth welds of API X52 pipeline on SCC
3.4.1 Slow strain rate tests carried out in NACE solution

Sulphide stress corrosion cracking (SSCC) susceptibility of four conditions of shielded metal arc welding repairs and one as welded specimen of the girth weld in seamless API X52 PSL2 steel pipe was evaluated using SSRT. The SSR tests were performed in air and in a sour solution saturated with H_2S both at room temperature to a constant elongation rate of $1x10^{-6}$ in/sec. The SSCC susceptibility was evaluated in function of the reduction in area ratio and elongation plastic ratio and also was manifested as a decrease in the mechanical properties. The dimensions of the line pipe were 8 inches (203.2mm) in diameter and 0.437 inches (11.1mm) in nominal wall thickness, with V-bevel at 30° in the welding. The chemical composition was showed in Table 1 (X52-B). The girth welds were obtained from the quality control department of the company Construcciones Maritimas Mexicanas (CMM-PROTEXA), carried out by qualified welders under a qualified welding procedure according to API 1104 standard (API 1104, 2005), using the SMAW process, with a 0.125 inch in diameter consumable filler rod E-6010 for the root and hot pass and with a 0.185 inches electrode E-7010G for the subsequent passes. The original weld (0R) was repaired by arc air and hand disc grinder to a depth between the root and hot pass and rewelding using the same welding procedure (1R). To simulate multiple welding repairs, the repaired weld was similarly removed and welded again, to obtain a second (2R), third (3R) and fourth (4R) welding repair. X-ray inspection was used to verify the quality of the welded unions after each welding repair according to the requirements of API-1104.

Several tests (SSRT) were carried out for each one of the welding repair conditions, for both, in air and NACE solution. Profiles obtained from the SSRT are shown in Figure 6.

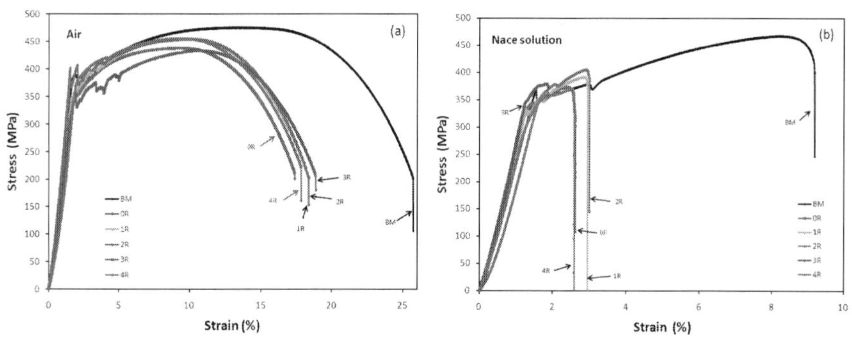

Fig. 6. Stress versus strain profiles obtained from SSRT for all repair conditions (a) Tested in air; (b) tested in NACE solution saturate with H_2S at room temperature.

SCC susceptibility was evaluated of the results obtained from SSRT according to NACE TM 0198 and ASTM G129. The results are showed in Table 4. The strength, elongation and reduction in area decreases significantly when the samples are exposed to the NACE solution saturate with H_2S. This behavior could be attributed to combined action between tensile stress and the specific corrosive environment. In fact, the synergic effect between the stress, chloride and sulphide (obtained of hydrogen sulphide) increased significantly the susceptibility to SSCC.

According to the results of RAR and EPR is clear that the specimens tested in the NACE solution saturated with H_2S exhibited high susceptibility to SSCC. It is suggested that decrease in mechanical properties is due to a hydrogen embrittlement mechanism, because the fractured samples showed little deformation in the gauged section and there is not formation of neck in the failure zone. From the different welding repair conditions, the second repair showed the best performance and behavior in the tests, for both, in air and in the NACE solution.

Condition	Environment	RA, %	EP, %	RAR	EPR	Failure zone
Base metal	Air	87.62	23.26			
As welded	Air	83.16	15.00			BM
First repair	Air	86.86	16.18			BM
Second repair	Air	86.88	16.69			BM
Third repair	Air	82.30	14.37			BM
Fourth repair	Air	86.34	15.51			BM
Base metal	NACE Solution	10.40	4.92	0.118	0.211	
As welded	NACE Solution	6.17	1.34	0.074	0.089	ICHAZ
First repair	NACE Solution	7.84	1.22	0.090	0.075	ICHAZ
Second repair	NACE Solution	9.30	1.65	0.107	0.098	ICHAZ
Third repair	NACE Solution	4.87	0.689	0.059	0.047	ICHAZ
Fourth repair	NACE Solution	5.26	0.984	0.060	0.063	ICHAZ

Table 4. Summary of the SSRT results to evaluate the SCC susceptibility of multiple repairs in girth welds of API X52

According to Lant *et* al. (Lant *et* al, 2001) the temper bead technique generates overlap beads producing grain refinement in the coarse grained heat affected zone (CGHAZ) of the previous bead and decreases the residual stresses due to the input of additional thermal energy. Considering this assumption, this it is the reason for which the second welding repair presents the best mechanical behavior, for both, in air and in NACE solution.

The microstructure as well as the mechanical properties obtained for the different weld repair conditions were shown elsewhere (Vega *et* al, 2008). Micrographs of the fracture surface specimens were shown elsewhere (Vega *et* al, 2009). In air, all the specimens exhibited a ductile type of failure (the presence of quasi-cleavage fracture mixed with microvoid coalescence was observed). Meanwhile, in NACE solution saturated with H_2S showed a brittle fracture with a transgranular appearance. The brittle fracture is related to the higher stress concentration by the lost metal regions due to anodic dissolution (pits and general corrosion) and the diffusion of atomic hydrogen into the entrapping sites during the welding process. Additionally, the diffusion of hydrogen and the stresses close to the crack initiations sites develop the initiation and the growing of cracks.

The specimens tested in air for the different conditions of repair, the failure occurred in the BM very close to the interface with the HAZ. For the samples tested in the NACE solution for all the repair conditions, the failure occurred mainly along the intercritical heat affected zone (ICHAZ) as is shown in Figure 7.

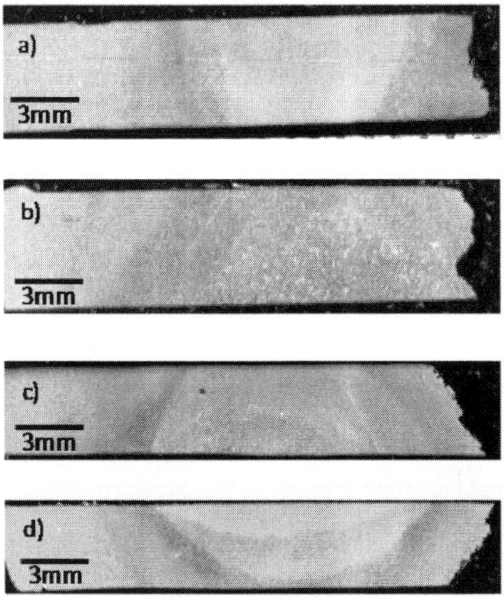

Fig. 7. Optical micrographs of longitudinal section of fracture samples from SSRT showing failure zone of specimens tested in NACE solution at room temperature, a) first repair; b) second repair; c) third repair; d) fourth repair

SSCC in steels and in welded line pipes of low and medium strength has also been denominated as stress oriented hydrogen induced cracking (SOHIC) (Takahashi, 1995, 1996a, 1996b; Carneiro, 2003). The SOHIC is divided in two stages: in the first stage, cracks parallel to applied stress are formed by the hydrogen induced cracking (HIC) mechanism; in the latter stage, the cracks link perpendicularly to applied stress like SSCC. All the found cracks in the different welding repair conditions agree with the theory proposed by the SOHIC mechanism; cracks formed parallel to the applied stress after link with cracks formed perpendicular to the direction of the applied stress, preferentially induced along the ICHAZ as is shown in Figure 8.

Most of the failures reported by SSCC or SOHIC in welded line pipes of low and medium strength have occurred in the ICHAZ (Kimura, 1989; Takahashi, 1996; Endo, 1994). According to Takahashi *et* al. (Takahashi, 1996) the applied stress acts to enhance shear stress around the first parallel formed cracks to the stress and the shear stress can cause local yielding which facilitates the vertical linking of the parallel cracks to break the specimen eventually.

McGaughy *et* al. (McGaughy, 1992; McGaughy & Boyles, 1990) evaluated the effects of SMAW repairs on the residual stress distribution of girth welds in API 5L X65 line pipe. The girth welds contained repairs which included a single, double and a full wall repair. The results showed that increasing number of welding repairs increased the level of axial and hoop residual stresses, inside and outside surface of the line pipe. This suggested that the contribution of the residual stresses together with the presence of discontinuities near to the fracture interface makes the WB/HAZ interface more susceptible zone to SCC.

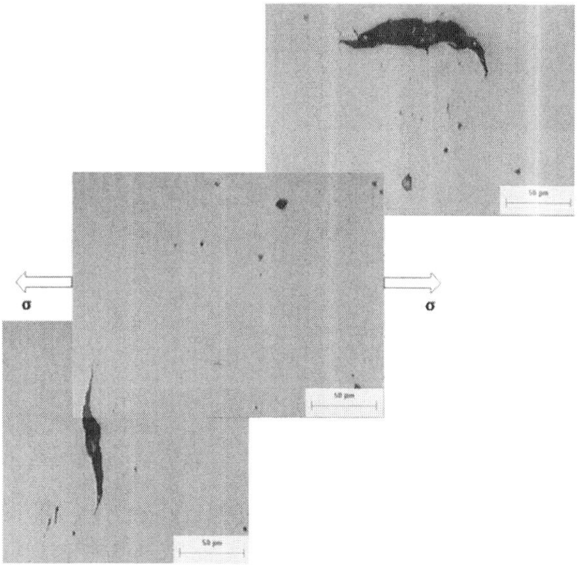

Fig. 8. Scanning electron microscopy images showing parallel and perpendicular cracks found along ICHAZ (sample with third repair).

3.4.2 Slow strain rate tests carried out in NS4 solution

The susceptibility to SCC in girth welds of seamless API X52 steel pipe containing multiple shielded metal arc welding (SMAW) repairs and one as-welded condition were evaluated using SSRT according to NACE TM-0198 standard. The SSRT were performed in air and NS4 solution at pH 10 (basic) and pH 3 (acid) at room temperature and at constant elongation rate of $1x10^{-6}$ in/sec. Cylindrical tensile specimens were transversal machined to the direction of the application to the girth weld. The specimens according to the number of repairs were identified as 0R (as-welded), 1R, 2R, 3R and 4R respectively.

The experimental research about SCC in pipelines considering the external environment have been studied using NS4 solutions (Elboujdaini *et* al, 2000; Pan *et* al, 2006; Bulger & Luo, 2000; Fang *et* al, 2007; Lu & Luo, 2006), this as results of investigations about the chemical composition of the solution on the surface of the pipeline failed by SCC. The main goal using NS4 solution is to simulate the chemical composition of the soil. The great majority of these studies were made in the base metal of pipeline steels; there are few studies on the longitudinal and circumferential welding but none reported in function of the number of repairs. The chemical composition of the soil solution use to carry out the SSRT is shown in Table 5.

Stress vs. Strain profiles obtained from the SSRT performed performed in air and in the NS4 solution both at room temperature for the different welding repair conditions are shown in Figure 9. Base metal (BM) presented the maximum strain for SSRT carried out in air (26%) and in NS4 solution (20% with pH 3 and 23% with pH 10) in comparison with the four weldments. Specimens tested in air showed a strain about 16-19% meanwhile the specimens tested in NS4 showed a strain between 14-19% with pH 3 and pH 10 respectively.

Compound	Composition (gr/L)
Sodium bicarbonate ($NaHCO_3$)	0.483
Calcium chloride ($CaCl_2$)	0.137
Magnesium sulfate ($MgSO_4$)	0.131
Potassium chloride (KCl)	0.122

Table 5. Chemical composition of the NS4 solution

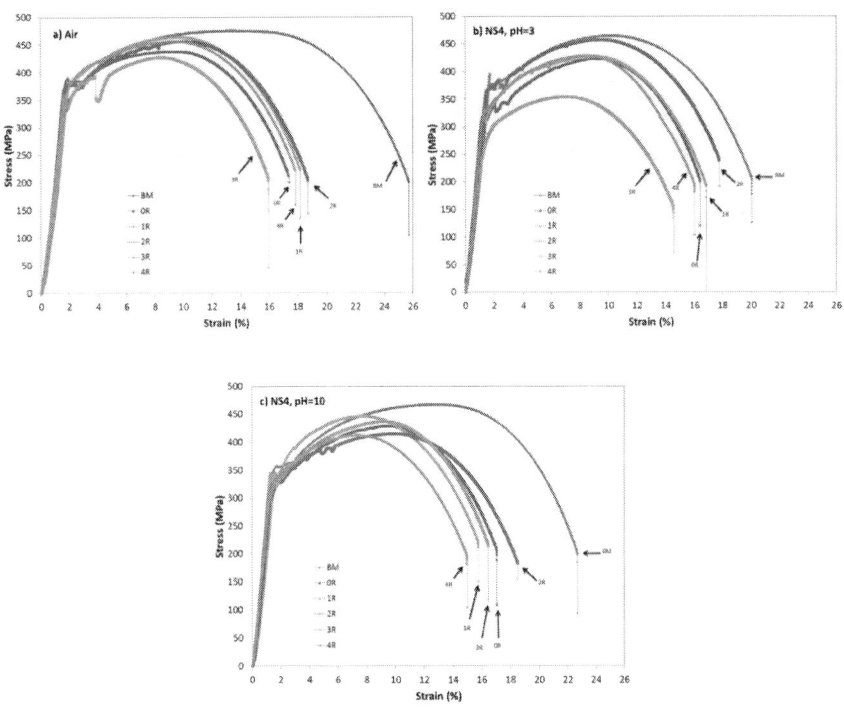

Fig. 9. Stress versus strain profiles obtained from the SSRT in function of pH and number of repairs, a) in air, b) in NS4 solution with pH 3 and c) in NS4 solution with pH 10.

From the analysis of these curves we obtain results of reduction in area (RA), plastic elongation (EP), reduction in area ratio (RAR) and elongation plastic ratio (EPR) which are shown in Table 6. RAR and EPR were calculated according to NACE TM 0198. Ratios in the range of 0.8-1.0 indicate that steel has high resistance to SCC, whereas low values (i.e. <0.5) show high susceptibility.

Condition	Environment	YS (MPa)	UTS (MPa)	RA (%)	E_p (%)	RAR	EPR
BM	Air	386.1	475.2	89.10	23.26		
0 rep		356.0	437.5	85.74	15.00		
1° rep		379.9	464.2	88.10	16.18		
2° rep		384.1	456.2	88.53	16.69		
3° rep		359.7	427.4	84.60	14.37		
4° rep		379.2	455.9	86.34	15.51		
BM	NS4, pH=3	396.8	464.8	88.13	19.72	0.98	0.84
0 rep		325.9	423.0	84.10	14.17	0.98	0.94
1° rep		322.5	427.5	86.84	15.47	0.98	0.95
2° rep		351.8	438.2	86.90	16.69	0.98	1.00
3° rep		318.1	354.9	83.91	13.78	0.99	0.95
4° rep		329.7	428.8	85.01	15.00	0.98	0.96
BM	NS4, pH=10	357.0	467.7	87.62	18.34	0.98	0.78
0 rep		316.0	428.4	83.16	14.98	0.96	0.99
1° rep		324.9	446.3	86.86	15.51	0.98	0.95
2° rep		340.6	415.2	86.88	16.22	0.98	0.97
3° rep		344.6	436.8	82.30	12.99	0.97	0.90
4° rep		318.0	412.1	86.10	14.60	0.99	0.94

Table 6. Assessment of the susceptibility to SCC obtained from the SSR tests

According to these results, it is clear that the specimens tested in the NS4 solution not exhibited susceptibility to SCC. In addition, secondary cracks or corrosion were not observed in the specimens after made the SSR tests. The strength, elongation and reduction in area decreases slightly when the samples are exposed to the NS4 solution.

Passive film rupture, anodic dissolution and repassivation are the generally accepted mechanism on SCC, but this is dependent on the strain rate. If the strain rate is too high, the material fails predominantly under mechanical loading, and the environment did not have time to damage the material. By other hand, at too slow strain rate the passive film formed over a longer period of time will be too dense to be ruptured by the slow strain rate.

SEM observations of the fracture surfaces of specimens tested in air and NS4 solution with pH of 3 and 10 showed microvoids coalescence for all the conditions studied which is characteristic of a ductile type of fracture. Figure 10 shows SEM images of the fracture surface after SSRT were performed.

As the strain increases the neck formation in the gauge section before the samples failed was observed. This was reflected in the assessment of reduction area on the fracture surface. On the edge of the fracture surface an intergranular type of fracture was observed, towards the centre of the fracture change to a ductile type.

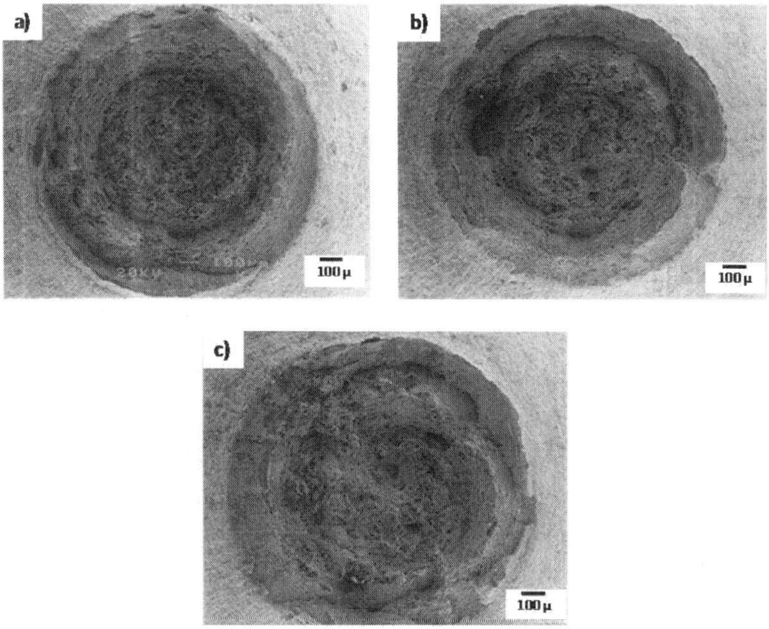

Fig. 10. Fracture surfaces after made the SSRT, a) in air, b) NS4 solution with pH 3 and c) NS4 solution with pH-10.

Optical micrographs of the longitudinal sections for specimens tested in NS4 solution with pH-10 showing the failure zone are shown in Figure 11. The specimens tested in NS4 solution with pH of 3 and 10 for the different conditions of repair, the failure occurred in the BM/HAZ interface without presence of secondary cracks in the gauge section.

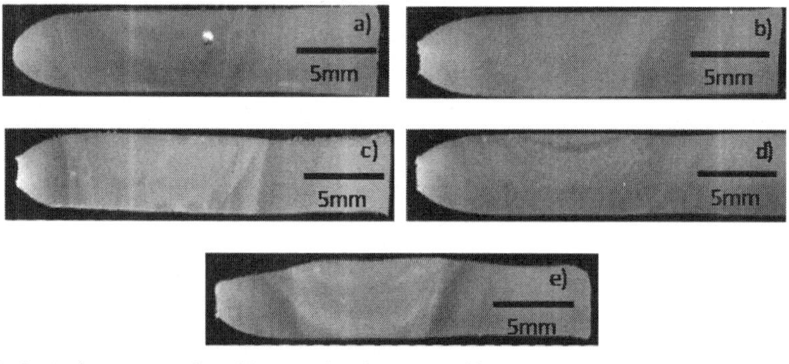

Fig. 11. Optical micrographs of longitudinal section of fractured samples from SSRT performed in NS4 solution with pH-10 showing failure zone, a) as welded; b) first repair; c) second repair; d) third repair; e) fourth repair.

The results showed that increasing number of welding repairs increased the level of axial and hoop residual stresses, inside and outside surface of the line pipe. The SCC susceptibility was expressed in function of the reduction in area ratio and elongation ratio. The yield strength, tensile strength and ductility of the welded joints shown a decrease when they are exposed to the NS4 solution. The metallographic observations of the fractured specimens show that the most susceptibility area to SCC was the BM/HAZ interface.

3.5 SCC susceptibility of API X60 and X65 pipeline steels

The SCC susceptibility of API X60 and X65 longitudinal weld beads was evaluated using SSRT in a brine solution saturated with H_2S at room temperature (25°C), 37°C and 50°C. The tests were supplemented by potentiodynamic polarization curves and hydrogen permeation measurements. Longitudinal weld beads produced by SMAW process were analyzed. The chemical composition of these steels was shown in Table 1. These types of pipes are typically used in the Mexican pipeline systems for transporting hydrocarbons.
Figure 12 shows micrographs of the X60 and X65 steels weldments. These figures clearly show the different microstructures found in a weldment, which consists mainly of polygonal and coarse acicular ferrite. This microstructure optimizes the strength and the toughness of the weld beads (Bhatti, 1984; Dolby, 1976; Kirkwood, 1978; Asahi, 1994).

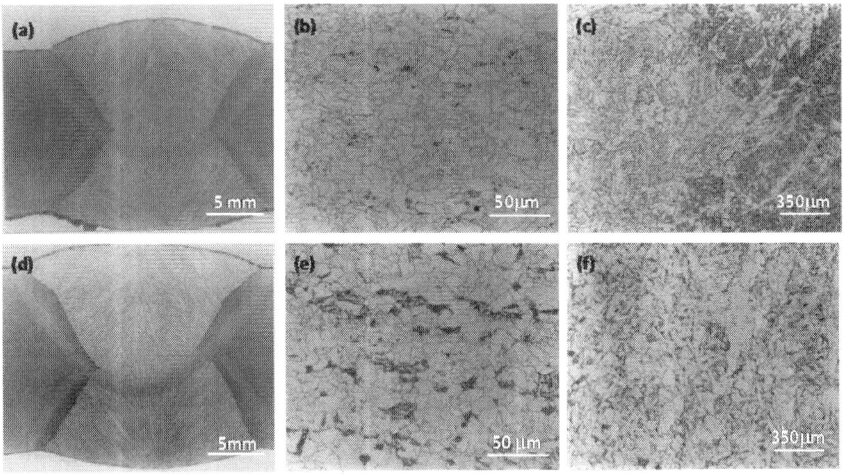

Fig. 12 Microstructures obtained by optical microscopy of the weld bead: (a-c) API X60 steel, (d-f) API X65 steel.

The welding industry has recognized that weld induced stresses play an important role in SCC phenomena. Each year, tens of millions of dollars are expended to replace or repair pipes and vessels that suffer SCC or hydrogen embrittlement (HE). When H_2S is present in the pipelines transporting hydrocarbons, this type of brittle failure is known as sulfide stress corrosion cracking, and it has been established as a particular case of hydrogen embrittlement (Tsay et al, 2000). The transport of these types of products always induces failures in the pipeline systems, and is very frequently in the weld beads.

The development of multi-phase microstructures is important for the attainment of certain mechanical properties, but it can be detrimental for resistance to SSCC. Carbon-rich phases such as pearlite, bainite, or martensite can be particularly susceptible to this mode of HE. The susceptibility toward SSCC was measured with the I_{SSCC} index according to:

$$I_{SSCC} = \frac{\%RA_{AIR} - \%RA_{NACE}}{\%RA_{AIR}} \tag{6}$$

where $\%RA_{AIR}$ and $\%RA_{NACE}$ are the percentage reduction in area values in air and in the NACE solution saturated with H_2S. The results are plotted in Figure 13(a). Values close to the unit mean that the steel is highly susceptible to SCC, whereas values close to zero mean that the steel is immune to SSCC. Thus, Figure 13(a) clearly shows that, in all cases, regardless of the temperature, both steels are highly susceptible to SCC, and the effect of the temperature is negligible, although the tendency is that this susceptibility increases with increasing temperature. X60 pipeline steel was more susceptible to SCC than the X65 steel, although this difference seems to be negligible.
Hydrogen permeation tests were carried out using the two-component Devanathan-Stachurski cell (Devanathan & Stachurski, 1962). Figure 13(b) shows the effect of the temperature on the hydrogen uptake (C_0) for both steels. It is clear that the amount of hydrogen uptake increases with temperature for both steels, being always higher in the X60 than in the X65 steel. All these results are consistent with those found in the literature (Asahi *et* al., 1994). The corrosion rate, taken as the corrosion current density, I_{corr}, the amount of hydrogen uptake for the weldments, C_0, and the SSCC susceptibility increased with an increase in the temperature from 25°C to 50°C. Although anodic dissolution seems to play an important role in the cracking mechanism, the most likely mechanism for the cracking susceptibility of X60 and X65 weldments in H_2S solutions seems to be hydrogen embrittlement (Natividad *et* al., 2006).

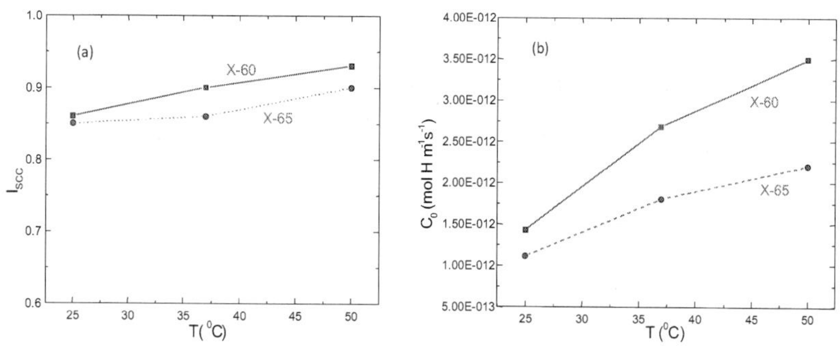

Fig. 13. (a) Effect of temperature on the I_{SSCC} values for both steels and (b) on the hydrogen uptake for both steels.

Potentiodynamic polarization curves were performed at a sweep rate of 1.0 mV/s using a fully automated potentiostat controlled with a desktop computer. The scanning started at -500 mV, with respect to E_{corr}, and finished at 300 mV more positive than E_{corr}, at a scanning rate of 1 mV/s. Corrosion current values (I_{corr}) were calculated using Tafel extrapolation. Figure 14 show the effect of temperature on the polarization curves for X60 and X65 pipeline steels. As expected in these solutions, there is no passive region in any of the cases, only active dissolution. For both steels, the E_{corr} decreases as the temperature is increased, with the most noble value at 25°C and -600 mV, and the most active at 50°C and around -800 mV. The E_{corr} value for the X60 steel at 37°C was -650 mV whereas for the X65 steel, it was -700 mV.

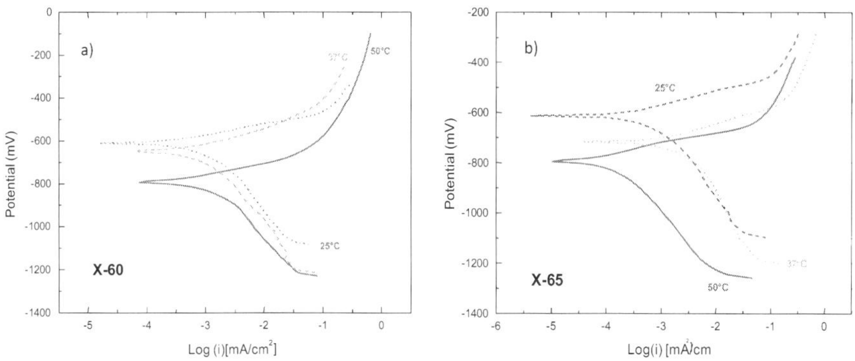

Fig. 14. Polarization curves obtained from weld bead at different temperatures in the NACE solution saturated with H_2S, (a) X60; (b) X65 steel.

Metallographic cross sections of X60 and X65 steels are shown in Figure 15. The cracks were transgranular in nature predominantly, and, just as indicated by the polarization curve that the corrosion rate increased as the temperature increased, the amount of corrosion products inside the cracks is more pronounced at 50 than at 25°C. Cross sections of the gage section of X60 steel showed secondary cracks as is shown in Figure 15(a). For X65 steel, no cracks were observed, only pits, as is shown in Figure 15(b).

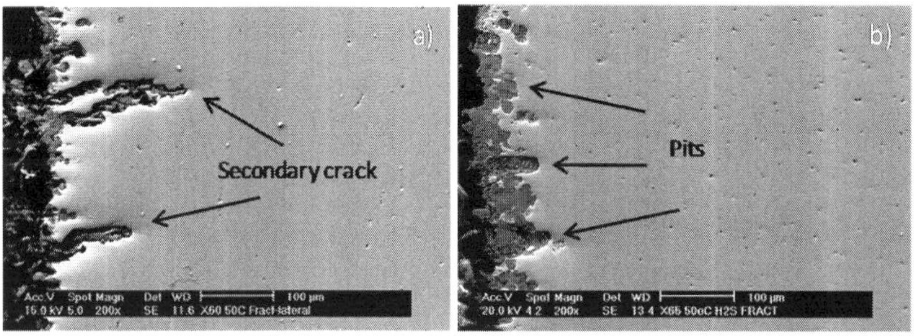

Fig. 15. Cross-sections of weldments tested in the NACE solution saturated with H_2S. (a)X60 steel at 50°C, (b)X65 steel at 50°C.

4. Conclusions

This chapter presents the mechanical and environmental effects as well as fracture characteristics on SCC susceptibility of steels used in the oil industry using SSRT. The tests were performed in samples which include the longitudinal and circumferential weld bead of pipeline steels. The steels studied are low carbon steels API X52, X60, X65 and X70. These steels were evaluated in a brine solution saturated with H_2S according to NACE TM 0177 (to evaluate sulphide stress corrosion cracking susceptibility). Additionally, some circumferential weldments of API X52 were evaluated in NS4 solution.

The results of the SSRT carried out in API X52 and X70 steels to evaluate the SCC susceptibility revealed that specimens tested in the NACE solution saturated with H_2S at room temperature and 50°C, presented high susceptibility to SCC, reflected in the degradation of mechanical properties. The specimens tested in air exhibited a ductile type of fracture. Whereas, in the corrosive solution, the specimens shown a brittle fracture. Both steels presented a corrosive attack in the form of anodic dissolution of the material. The X52 steel showed best resistance to SCC. All the cracks, primary and secondary were perpendicular to the applied tension axis, being indicative of SCC. These cracks were related to the diffusion of atomic hydrogen promoting the embrittlement damage. The failure of samples tested in air occurred in some cases in the weld joint, but in presence of the NACE solution the failure always occurred in the HAZ.

SCC susceptibility of API X52 with multiple welding repairs in girth welds of pipelines was evaluated by means of SSRT in NACE solution saturated with H_2S. According to the results of RAR and EPR, it is clear that welding joints are susceptible to SCC. The specimens tested in air for the different conditions of repair, the failure occurred in the BM very close to the interface with the HAZ. In presence of NACE solution saturated with H_2S at room temperature, the most susceptible zone to SCC was the ICHAZ following the SOHIC. Although in presence of imperfections like pores and non-metallic inclusions near to the interface with the HAZ, fusion line becomes the most susceptible zone to SCC. Using temper bead technique contributes to the reinforcement of the mechanical behavior, improving resistance to SCC as was observed in the second welding repair. The temper bead technique generates overlap beads producing grain refinement in the coarse grained heat affected zone of the previous bead and decreases the residual stresses due to the input of additional thermal energy.

The susceptibility to SCC in girth welds of seamless API X52 steel pipe containing multiple welding repairs was evaluated using SSRT in soil solution. The SSRT were performed in air and NS4 solution at pH 10 (basic) and pH 3 (acid) at room temperature. The main goal using NS4 solution is to simulate the chemical composition of the soil. According to the RAR and EPR results, it is clear that the specimens tested in the NS4 solution not exhibited susceptibility to SCC. In addition, secondary cracks or corrosion were not observed in the specimens after perform the SSRT. The strength, elongation and reduction in area decreases slightly when the samples are exposed to the NS4 solution. SEM observations of the fracture surfaces of specimens tested in air and NS4 solution with pH of 3 and 10 showed microvoids coalescence for all the conditions studied which is characteristic of a ductile type of fracture. The metallographic observations of the fractured specimens show that the most susceptible area to SCC was the BM/HAZ interface.

The SCC susceptibility considering the effects of the temperature on the corrosion rate, and hydrogen uptake of API X60 and X65 weldments through SSRT was carried out. The SCC susceptibility and corrosion rate, taken as I_{corr}, for both weldments, increased with an increase in the temperature from 25 to 50°C, as well as the amount of hydrogen uptake for the weldments. The most likely mechanism for the cracking susceptibility of X60 and X65 weldments in sour solutions seems to be hydrogen embrittlement, but anodic dissolution seems to play an important role in the cracking mechanism. Specimens tested in NACE solution showed brittle type of fracture, with a very small percentage reduction in area values, and the gage section shown a large number of secondary cracks.

5. References

API STD 1104-2005 (R 2010) Welding of Pipelines and Related Facilities - Twentieth Edition; July 2007, 2: December 2008

Asahi, H.; Sogo, Y.; Ueno, M. & Higashiyama, H. (1988). Effects of Mn, P, and Mo on sulfide stress cracking resistance of high-strength low-alloy steels, *Metallurgical Transactions A-Physical metallurgy and materials science*, Volume 19, Issue 9 (Sep 1988), pp. 2171-2177, ISSN 0360-2133

Asahi, H.; Ueno, M. & Yonezawa, T. (1994). Prediction of Sulfide Stress Cracking in High-Strength Tubular. *Corrosion*, Volume 50, Issue 7, (July, 1994), pp. 537-545. ISSN 0010-9312

ASTM G129-2000 (R 2006) Standard practice for slow strain rate testing to evaluate the susceptibility of metallic materials to environmentally assisted cracking

Beavers, J. & Koch, G. (1992). Limitations of the slow strain rate test for stress-corrosion cracking testing, *Corrosion*, Volume 48, Issue 3 (Mar 1992), pp. 256-264, ISSN 0010-9312

Beavers, J.; & Harle, B. (2001). Mechanisms of high-pH and near-neutral-pH SCC of underground pipelines, *Journal of Offshore Mechanics and Arctic Engineering-Transactions of the ASME*, Volume 123, Issue 3 (Aug 2001), pp. 147-151, ISSN: 0892-7219

Bhatti, A.; Saggese M.; Hawkins D.; Whiteman, M. & Golding M. (1984). Analysis of Inclusions in Submerged-Arc Welds in Microalloyed Steels. *Welding Journal*, Volume 63, Issue 7, (1984), pp. S224-S230), ISSN 0043-2296

Brongers, M.; Beavers, J.; Jaske, C. & Delanty, B. (2000). Effect of hydrostatic testing on ductile tearing of X65 line pipe steel with stress corrosion cracks, *Corrosion*, Volume 56, Issue 10 (Oct 2000), pp. 1050-1058, ISSN 0010-9312

Bulger, J. & Luo, L. (2000). Effect of microstructure on near-neutral pH SCC. *Proceedings in the International Pipeline Conference ASME 2000*, Calgary, Canada, October 1-5, 2000, ISBN 10 0791816664

Carneiro, R.; Ratnapuli, R. & Lins, V. (2003). The influence of chemical composition and microstructure of API line pipe steels on hydrogen induced cracking and sulfide stress corrosion cracking, *Materials science and engineering A-Structural materials properties microstructure and processing*, Volume 357, Issue 1-2 (Sep 2003), pp. 104-110, ISSN 0921-5093

Casales, M.; Espinosa-Medina, M.; Martinez-Villafañe, A.; Salinas-Bravo, V. & Gonzalez-Rodríguez, J. (2000). Predicting susceptibility to intergranular stress corrosion cracking of Alloy 690, *Corrosion*, Volume 56, Issue 11 (Nov 2000), pp. 1133-1139, ISSN 0010-9312

Casales, M.; Salinas-Bravo, V.; Espinosa-Medina, M.; Martinez-Villafañe, A. & Gonzalez-Rodríguez, J. (2004). Electrochemical noise generated during the stress corrosion cracking of sensitized alloy 690, *Journal of Solid State Electrochemistry*, Volume 8, Issue 5 (Apr 2004), pp. 290-295, ISSN 1432-8488

Casales, M.; Salinas-Bravo, V.; Martinez-Villafañe, A. & Gonzalez-Rodríguez, J. (2002). Effect of heat treatment on the stress corrosion cracking of alloy 690, *Materials Science and Engineering A-Structural materials properties microstructure and processing*, Volume 332, Issue 1-2 (Jul 2002), pp. 223-230, ISSN 0921-5093

Chen, W.; King, F. & Vokes, E. (2002). Characteristics of near-neutral-pH stress corrosion cracks in an X65 pipeline, *Corrosion*, Volume 58, Issue 3 (Mar 2002), pp. 267-275, ISSN 0010-9312

Contreras, A.; Albiter A., Salazar M., Pérez R. (2005). Slow strain rate corrosion and fracture characteristics of X52 and X70 pipeline steels, *Materials Science Engineering A,* Volume 407 (2005) pp. 45-52, ISSN 0921-5093

Craig, B. (1998). Calculating the lowest failure pressure for electric resistance welded pipe, *Welding Journal,* Volume 77, Issue 1 (Jan 1998), pp. 61-63, ISSN: 0043-2296

Delanty, B. & O'Beirne, J. (1992). Major field-study compares pipeline SCC with coatings, *Oil & Gas Journal,* Volume 90, Issue 24, (Jun 1992), pp. 39-44, ISSN: 0030-1388

Devanathan, M.A.V., Stachurski Z., Proc. R. Soc. London A 270 (1962): p. 90.

Dolby, R. (1976). Factors Controlling Weld Toughness-The Present, Position, Pt. II-Weld Metals. The Welding Institute Members Report 14/1976/M (May, 1976). London

Elboujdaini, M.; Wang, Y. & Revie, R. (2000). Initiation of stress corrosion cracking on X65 line pipe steels in near-neutral pH environment, *Proceedings of the International Pipeline Conference ASME 2000,* Calgary, Canada, October 1-5, 2000, ISBN 10 0791816664

Endo, S.; Nagae, M.; Kobayashi, Y. & Ume, K. (1994). Sulfide stress-corrosion cracking in welded-joints of welded line pipes, *ISIJ International,* Volume 34, Issue 2 (1994), pp. 217-223, ISSN 0915-1559

Fang, B.; Atrens, A.; Wang, J.; Han, E.; Zhu, Z. & Ke, W. (2003). Review of stress corrosion cracking of pipeline steels in "low" and "high" pH solutions, *Journal of Materials Science,* Volume 38, Issue 1 (Jan 2003), pp. 127-132, ISSN: 0022-2461

Fang, B.; Eadie, R.; Chen, W. & Elboujdaini, M. (2010). Pit to crack transition in X52 pipeline steel in near neutral pH environment Part 1-formation of blunt cracks from pits under cyclic loading, *Corrosion Engineering, Science and Technology,* Volume 45, Issue 4 (Aug 2010), pp. 302-312, ISSN: 1478-422X

Fang, B.; Han, E.; Wang, J. & Ke, W. (2007). Mechanical and environmental influences on stress corrosion cracking of an X70 pipeline steel in dilute near-neutral pH solutions, *Corrosion,* Volume 63, Issue 5 (May 2007), pp. 419-432, ISSN 0010-9312

Greer, J. (1975). Factors affecting sulfide stress cracking performance of high-strength steels, *Materials Performance,* Volume 14, Issue 3 (1975), pp. 11-22, ISSN 0094-1492

Jones, R. (1992). *Stress Corrosion Cracking,* Edited by Russell H. Jones, ASM International, 1-40, ISBN 0-87170-441-2, Ohio, U.S.A.

Kane, R.D.; Joia C.J.B.M.; Small A.L.L.T. & Ponciano J.A.C (1997). Rapid Screening of Stainless Steels for Environmental Cracking, *Materials Performance,* Volume 36, Issue 9 (Sep. 1997), pp. 71-74, ISSN 0094-1492

Kim, K.; Zhang, P.; Ha, T. & Lee, Y. (1998). Electrochemical and stress corrosion properties of duplex stainless steels modified with tungsten addition, *Corrosion,* Volume 54, Issue 11 (Nov 1998), pp. 910-921, ISSN: 0010-9312

Kimura, M.; Totsuka, N.; Kurisu, T.; Amano, K.; Matsuyama, J. & Nakai, Y. (1989). Sulfide stress-corrosion cracking of line pipe, *Corrosion,* Volume 45, Issue 4 (Apr 1989), pp. 340-346, ISSN 0010-9312

Kirkwood, P. (1978). Microstructural and Toughness Control in Low-Carbon Weld Metals. *Metal Construction,* Volume 10, Issue 5, (1978), pp. 260-264, ISSN 0307-7896

Krist, K & Leewis, L. (1998). GRI research - Stress corrosion cracking mechanisms in pipelines. *Pipeline & Gas Journal,* Volume 225, Issue 3, (Mar, 1998), pp. 49-52, ISSN 0032-0188

Lant, T.; Robinson, D.; Spafford, B. & Storesund (2001). Review of weld repair procedures for low alloy steels designed to minimize the risk of future cracking, *International journal of pressure vessels and piping,* Volume 78, Issue 11-12 (Nov-Dec 2001), pp. 813-818, ISSN 0308-0161

Leis, B.N. & Eiber R.J. (1997). Stress-Corrosion Cracking on Gas-Transmission Pipelines: History, Causes and Mitigation, Proceedings of First International Business Conference on Onshore Pipelines, (Berlin, December 1997).

Liou, H.; Hsieh, R. & Tsai, W. (2002). Microstructure and stress corrosion cracking in simulated heat-affected zones of duplex stainless steels, *Corrosion Science*, Volume 44, Issue 12 (Dec 2002), pp. 2841-2856, ISSN 0010-938X

Liu, Z.Y., Li X.G., Du C.W., Zhai G.L. & Cheng Y.F. (2008). Stress corrosion cracking behavior of X70 pipe steels in an acid soil environment, *Corrosion Science,* Volume 50, (2008), pp. 2251-2257, ISSN 0010-938X

López, H.; Raghunath, R.; Albarran, J. & Martinez, L. (1996). Microstructural aspects of sulfide stress cracking in an API X-80 pipeline steel, *Metallurgical and materials transactions A-Physical metallurgy and materials science,* Volume 27, Issue 11 (Nov 1996), pp. 3601-3611, ISSN 1073-5623

Lu, B. & Luo, J. (2006). Relationship between yield strength and near-neutral pH stress corrosion cracking resistance of pipeline steels - An effect of microstructure *Corrosion*, Volume 62, Issue 2 (Feb 2006), pp. 129-140, ISSN 0010-9312

Manfredi C. & Otegui J.L. (2002). Failures by SCC in buried pipelines, *Engineering Failure Analysis,* Volume 9, (2002), pp. 495-509, ISSN 1350-6307

McGaughy T. and L. Boyles. Significance of changes in residual stresses and mechanical properties due to SMAW repair of girth welds in line pipe, Technical Report PR-185-905, Edison Welding Institute, Columbus, OH, USA, 1990, 1–18.

McGaughy T. The influence of weld repairs on changes in residual stress and fracture toughness in pipeline girth welds. Recent Advances in Structural Mechanics, ASME 1992; PVP, Volume 248, pp. 81–86.

Miyasaka, A.; Kanamaru, T. & Ogawa, H. (1996). Critical stress for stress corrosion cracking of duplex stainless steel in sour environments, *Corrosion,* Volume 52, Issue 8 (Aug 1996), pp. 592-599, ISSN 0010-9312

NACE TM 0177-2005 Laboratory testing of metals for resistance to sulfide stress cracking and stress corrosion cracking in H_2S environments - Item No: 21212

NACE TM 0198-2004 Slow strain rate test method for screening corrosion-resistant alloys (CRAs) for stress corrosion cracking in sour oilfield service - Item No: 21232

Natividad, C.; Salazar, M.; Contreras, A.; Albiter, A.; Pérez, R. & Gonzalez-Rodríguez, J.G. (2006). Sulfide Stress Cracking Susceptibility of Welded X60 and X65 Pipeline Steels. *Corrosion,* Vol. 62, Issue 5, (May 2006), pp. 375-382, ISSN 0010-9312

Nishimura, R. & Maeda, Y. (2004). SCC evaluation of type 304 and 316 austenitic stainless steels in acidic chloride solutions using the slow strain rate technique, *Corrosion Science,* Volume 46, Issue 3 (Mar 2004), pp. 769-785, ISSN 0010-938X

Pan, B.; Peng, X.; Chu, W.; Su, Y. & Qiao, L. (2006). Stress corrosion cracking of API X60 pipeline in a soil containing water, *Materials science and engineering A*, Volume 434, Issue 1-2 (Oct 2006), pp. 76-81, ISSN 0921-5093

Park, J.; Pyun, S.; Na, K.; Lee, S. & Kho, Y. (2002). Effect of passivity of the oxide film on low-pH stress corrosion cracking of API 5L X65 pipeline steel in bicarbonate solution, *Corrosion*, Volume 58, Issue 4 (Apr 2002), pp. 329-336, ISSN 0010-9312

Parkins, R. & Beavers, J. (2003). Some effects of strain rate on the transgranular stress corrosion cracking of ferritic steels in dilute near-neutral-pH solutions, *Corrosion*, Volume 59, Issue 3 (Mar 2003), pp. 258-273, ISSN 0010-9312

Parkins, R. (1990). Strain rate effects in stress-corrosion cracking - 1990 Plenary Lecture, *Corrosion*, Volume 46, Issue 3 (Mar 1990), pp. 178-189, ISSN 0010-9312

Perdomo, J.; Morales, j.; Viloria, A. & Lusinchi, A. (2002). Carbon dioxide and hydrogen sulfide corrosion of API 5L grades B and X52 steels, *Materials Performs,* Volume 41, Issue 3 (Mar 2002), pp. 54-58, ISSN 0094-1492

Presage of production of the marine and south regions from México for a horizon of the 2000-2014

Shim, I. & Byrne, J. (1990). A study of hydrogen embrittlement in 4330 steel. 1. Mechanical aspects, *Materials science and engineering A*- Structural materials properties microstructure and processing, Volume 123, Issue 2 (Feb 1990), pp. 169-180, ISSN 0921-5093

Stress Corrosion Cracking on Canadian Oil and Gas Pipelines, (1996), Edited by National Energy Board (NEB), (December 1996), p. 16. ISBN 0-662-81679-X.

Takahashi, A. & Ogawa, H. (1995). Influence of softened heat-affected zone on stress oriented hydrogen-induced cracking of a high-strength line pipe steel, *ISIJ International,* Volume 35, Issue 10 ((1995), pp. 1190-1195, ISSN 0915-1559

Takahashi, A. & Ogawa, H. (1996). Influence of microhardness and inclusion on stress oriented hydrogen induced cracking of line pipe steels, *ISIJ International,* Volume 36, Issue 3 (1996), pp. 334-340, ISSN 0915-1559

Takahashi, A; Hara, T. & Ogawa, H. (1996). Comparison between full scale tests and small scale tests in evaluating the cracking susceptibility of line pipe in sour environment, *ISIJ International,* Volume 36, Issue 2 (1996), pp. 229-234, ISSN 0915-1559

Tsay, L.; Lin, Z.; Shiue, R. & Chen, C. (2000). Hydrogen embrittlement susceptibility of laser-hardened 4140 steel, *Materials science and engineering A-Structural materials properties microstructure and processing,* Volume 290, Issue 1-2 (Oct 2000), pp. 46-54, ISSN 0921-5093

Vangelder, K.; Erlings, J.; Damen, J. & Visser, A. (1987). The stress-corrosion cracking of duplex stainless-steel in $H_2S/CO_2/Cl^-$ environments, *Corrosion Science,* Volume 27, Issue 10-11 (1987), pp. 1271-1279, ISSN 0010-938X

Vega, O.; Hallen, J.; Villagomez, A. & Contreras, A. (2008). Effect of multiple repairs in girth welds of pipelines on the mechanical properties, *Materials Characterization,* Volume 59, Issue 10 (Oct 2008), pp. 1498-1507, ISSN 1044-5803

Vega, O.; Villagomez, A.; Hallen, J. & Contreras, A. (2009). Sulphide stress corrosion cracking of multiple welding repairs of girth welds in line pipe, *Corrosion Engineering, Science and Technology,* Volume 44, Issue 4 (Aug 2009), pp. 289-296, ISSN 1478-422X

Wang, S.; Chen, W.; King, F.; Jack, T. & Fessler, R. (2002). Precyclic-loading-induced stress corrosion cracking of pipeline steels in a near-neutral-pH soil environment, *Corrosion,* Volume 58, Issue 6 (Jun 2002), pp. 526-534, ISSN 0010-9312

Wang, S.; Zhang, Y. & Chen, W. (2001). Room temperature creep and strain-rate-dependent stress-strain behavior of pipeline steels, *Journal of Materials Science,* Volume 36, Issue 8 8Apr 2001), pp. 1931-1938, ISSN 0022-2461

Wilhelm, S. & Kane, R. (1984). Effect of heat-treatment and microstructure on the corrosion and SCC of duplex stainless-steels in H_2S/Cl^- environments, *Corrosion,* Volume 40, Issue 8 (1984), pp. 431-439, ISSN 0010-9312

Zhang, X.; Lambert, S.; Sutherby, R. & Plumtree, A. (1999). Transgranular stress corrosion cracking of X60 pipeline steel in simulated ground water, *Corrosion,* Volume 55, Issue 3 (Mar 1999), pp. 297-305, ISSN 0010-9312

8

Evaluation of the Shielding Gas Influence on the Weldability of Ferritic Stainless Steel

Demostenes Ferreira Filho[1], Ruham Pablo Reis[1] and Valtair Antonio Ferraresi[2]
[1]*Federal University of Rio Grande/School of Engineering*
[2]*Federal University of Uberlândia/Faculty of Mechanical Engineering*
Brazil

1. Introduction

The use of stainless steels has been nowadays widespread in a number of industrial sectors. They usually offer exceptional performance regarding mechanical and corrosion properties, but according to Lee et al. (2008) stainless steels are considered as high cost materials as far as solutions for structural engineering are concerned. However, this material group can provide aesthetic characteristics as well as outstanding versatility, easy cleaning and maintenance conditions. Nevertheless, there are still plentiful possibilities for applying stainless steels in new situations or improving their use in current applications due to their appealing visual aspect and durability.

In the automotive industry, for instance, parts of the exhaustion system are in general composed of tubes and blanks (stamped metal sheets) that usually are welded and have ferritic stainless steels as the main base material. According to Alves et al. (2002), the main ferritic stainless steels used in the hot portion of automotive exhaustion systems are the AISI 409 and 441. On the other hand, in the cold portion the AISI 409, 439 and 436 are normally utilized.

Faria (2006) states that automotive exhaustion systems went through a number of changes along the last 20 years as a consequence of more restrict pollution policies, needs for longer durability and higher engine efficiencies as well as requirements for reduction in weight and costs. Stainless steels used in the hot parts of automotive exhaustion systems, according to Sekita et al. (2004), must be refractory, which can be accomplished by niobium additions, high levels of molybdenum and optimized silicon presence. The same authors also mention the importance of having a good formability in such hot parts.

The market for stainless steels has experienced constant growth because of their excellent properties and continuous improvement in manufacturing of these materials, especially when issues like increase in process productivity and reduction in costs are taken into consideration. However, recently there has been a sharp increase in the international prices of alloying elements largely used in stainless steels, mainly nickel and molybdenum. As a result, the most traditional stainless steels class (austenitic) went through severe price rise worldwide. Fortunately, the ferritic class, which contains no nickel, emerges as an alternative for some applications, but sometimes some drawbacks have to be figured out before replacing the austenitic class.

One of the main problems found in certain ferritic stainless steels applications is related to their weldability, but, according to Schwarz & Tessin (2003), advantages concerning fatigue strength and general corrosion behavior can be achieved with ferritic filler materials. According to Reddy & Mohandas (2001), ferritic class base materials welded with ferritic wires exhibit greater resistance to stress corrosion cracking when compared with weldments produced with austenitic wires. However, grain coarsening in the fusion zone often takes place. Renaudot et al. (2000) state that the welding of ferritic stainless steels with filler metals also made of ferritic stainless steels minimizes the metallurgical discontinuity around the weld bead and promotes better metallurgical compatibility between the base metal and molten zone due to small differences in microstructure and thermal dilatation. The same authors point out that a ferritic class, namely the ER409Nb, filler metal has been utilized since the 90's for welding low-chromium ferritic stainless steels. Tests carried out with this wire resulted in welds with good geometry quality, ductility and resistance to intergranular corrosion. This result is also cited by Inui et al. (2003). These authors also mention that the weld metal of ferritic stainless steels has a large columnar structure, often leading to a decrease in cracking resistance and high temperature strength. Furthermore, they also reported that large grains deteriorate oxidation resistance and corrosion resistance of the weld metal. Balmforth & Lippold (2000), mention that the mechanical properties of the weld zone of ferritic stainless steels are very sensitive to the microstructure constituents, and poor microstructure control, like martensite presence in the fusion zone, can limit their application.

Ferritic wires might contain different elements in their chemical composition such as Titanium, Niobium and Aluminum, as a way of improving mechanical properties and resistance to corrosion of the welded joints. Considering thus the variability of ferritic wires available, Inui et al. (2003) tested three types of non-commercial filler metals made of stabilized ferritic stainless steels to weld plates also made of ferritic stainless steels; one wire stabilized with titanium, one stabilized with niobium and aluminum and another one with niobium, titanium and aluminum. The authors verified that the presence of Aluminum, titanium and niobium in adequate fractions was able to produce fusion zones with fine grains and, therefore, better mechanical properties.

Madeira & Modenesi (2007), claim that the addition of niobium and/or titanium, and the consequent stabilization, can reduce the formation of martensite, maintaining a ferritic structure, and decrease the grain growth in the fusion zone. Wang & Wang (2008), cited, that Titanium carbonitrates have high thermal stability during welding, especially after high heat inputs. Another important fact is that with the stabilization of ferritic stainless steel wires there is also an inhibition in the formation of chromium carbides and nitrates, which are directly related to intergranular corrosion in welded structures. The main stabilizing elements are niobium and titanium. Madeira (2007) compared the results of the ER430Ti and ER430LNb wires using Ar+2%O_2 as shielding gas in GMA welding. A higher penetration in the weld beads was observed when the ER430Ti was used. This took place for the same welding setting (voltage and wire feed rate) in the power source, but higher current levels were needed for the same fusion rate compared to the other wire. It was concluded that the increase in penetration is related to the higher electrical resistivity of the ER430Ti wire in relation to that one found in the ER430LNb wire. Electrical resistivity values were measured by Resende (2007) in a comparative manner for the ER308LSi, ER430Ti and ER430LNb filler metals. He noticed that the first two wires had relatively similar resistivity values whilst the third one had a lower level. From the same reference, a comparison was carried out regarding the weld beads produced with the ER430Ti and ER430LNb wires. The weld bead appearance resulted from the ER430Ti utilization was significantly inferior (lower wettability and poor superficial quality). This fact was mainly linked to the ER430Ti superficial roughness, which resulted in an irregular feeding. In spite of the references presented so far, there is still lack of information concerning the weldability of ferritic stainless steels.

Concerning the welding of the ferritic stainless steels, Stenbacka & Persson (1992) mention that GMA welding of stainless steels is commonly carried out shielded by argon with low levels of an oxidant (O_2 or CO_2). According to them, the presence of oxidizing components blended with argon promotes arc stability and metal transfer improvement. The advantage of the CO_2 addition in the shielding gas mixture would be cost reduction related, yet metal transfer is strongly affected. The same authors believe that a small amount of CO_2 added to argon should be used for short-circuiting transfer. According to Chae et al. (2008), the addition of CO_2 to argon improves the wettability of the weld bead, improving the weld quality. However, Strassburg (1976) claims that an increase in the proportion of oxidizing elements in the shielding gas increases the loss of molybdenum, chromium and niobium. Lundvist (1980), states that the addition of CO_2 in the shielding gas result in carbon absorption and also in oxidation of the deposited metal. The negative aspect related to carbon incorporation into the weld pool is that the content of ferrite in the weld bead might get lower; as carbon is a strong austenite producer at high temperatures, during cooling, martensite might form along the ferritic grain boundaries, impairing the tenacity of the welded joint.

There have been some studies about the shielding gas used for welding austenitic stainless steels. Tusek & Suban (2000), for instance, studied the effect of hydrogen in argon as the shielding gas for arc welding of this stainless steel class. When GMA welding was used, the hydrogen addition to argon increases melting rate and melting efficiency of the arc, but the increase is much smaller than in GTA welding. Durgutlu (2004) reported the effect of hydrogen added to argon during GTA welding of 316L austenitic stainless steel plates. In this case, the mean grain size in the weld bead increased with rising hydrogen contents. In addition, the weld bead penetration depth and width increased as the hydrogen content was raised. Gülenç et al. (2005) studied GMA welding of 304L stainless steel samples and observed that the toughness of the weld beads increases with rising Hydrogen amount added to argon and with increase in the welding current level. Liao & Chen (1998) examined how the miscrostructure and mechanical properties of 304 stainless steel welds are influenced by mixtures of carbon dioxide (2 to 20%) in argon. They detected that spattering increases, notch toughness is affected by the delta-ferrite amount and oxidation potential, specially at room temperature with increase in the CO_2 content.

Despite the facts discussed so far, the volume of information available in the current literature about GMA welding with ferritic stainless steels wires is still very scarce, mainly in relation to the shielding gas effect on the welded joints. Thus, this manuscript aims to analyze the metallurgical characteristics of a ferritic stainless steel weldments by studying the influence of the shielding gas (argon by itself and blended with O_2 or CO_2) on the chemical composition, microstructure, hardness and ductility of the weld beads. It is expected that a broader insight of the subject can help users and developers in the pursuit of more productive and safe welded structures (optimized filler metal and gas selection).

2. Materials and experimental procedure

In this study two different types of wires for GMA welding, namely ER430Ti and ER430LNb (both with a diameter of 1.0 mm), were used to assess the influence of stabilizing elements on the weld bead microstructure produced. The chemical composition of each wire is shown in Table 1. Each wire was combined with different shielding gases to evaluate any influence of the arc atmosphere on the weld bead chemical composition. The welded samples consisted of beads deposited side by side and in layers on the surface of 6-mm-thick plates previously cut in 40x40 mm from a UNS 43932 bi-stabilized ferritic stainless steel. The chemical composition of the base metal is presented in Table 2. This ferritic stainless steel is usually applied in automotive plants.

Wire	C	Cr	Mn	Mo	Nb	Ni	Si	Ti
ER430Ti	0.108	17.45	0.65	0.036	-	0.4	1.04	0.35
ER430LNb	0.027	17.66	0.425	0.034	0.44	0.215	0.43	0.004

Table 1. Chemical composition of the wires (weight, %).

Element	C	Cr	Mn	N	Nb	Ni	Si	V	Ti
Weight %	0.010	17.128	0.143	0.008	0.201	0.178	0.403	0.051	0.198

Table 2. Chemical composition of the UNS43932 ferritic stainless steel (weight, %).

The effectiveness of metal transfer from the wire to the weld pool and so the influence of the wire on the weld bead microstructure are mainly governed by the wire and shielding gas chemical compositions and by the type and stability of the metal transfer. The microstructure, and consequent properties, of the resultant weld bead also dependents on the volume of the weld pool and on the heat input produced, which in turn will govern the thermal cycle (weld zone cooling rate from the fusion temperature). Thus, in order to have a fair comparison of welding conditions under different wire and shielding gas combinations, first of all it is necessary to find welding settings as similar as possible to each other for all the experiments. These settings are not usually the same, since the optimized situation is different for each combination of wire and shielding gas used and their pursuit is typically a complicate task due to the quantity of variables involved in GMA welding.

Thus, to have such similar welding conditions, some approaches were applied to the experiments. In order that the same metal transfer mode was achieved for all the situations, in this case the short-circuit one, the arc was kept short by using a constant voltage power supply and by setting the arc voltage always at 20 V.

The welding current is probably the most important parameter that controls heat in the arc and its delivery to the weld pool. Consequently, the current level should be virtually the same for all the samples. This intent was accomplished by appropriately setting the wire feed speed and varying the contact tip-to-work piece distance between 12 and 18 mm. The welding current accepted range was 170 ± 2 A. The WFS values set for both ER430Ti and ER430LNb wires was 7.6 m/min. As a way to reach approximately the same weld pool volume and the same heat input for all the samples, the deposition rates, for a given current level, were managed to be equivalent. This was possible by making the ratio between the wire feed speeds and the welding travel speeds constant throughout the tests. The welding travel speed was set at 20 cm/min. Filho et al. (2010) and Ferreira Filho and Ferraresi (2008, 2010) show in details that the welding conditions and the weld bead shapes were quite similar for all the situations evaluated. The equivalency in welding conditions and weld bead shapes is crucial to avoid influence on the metallurgical formations other than that one exerted by the wire and shielding gas combination. More straightforward approaches for the experimental procedure, like just setting the same welding parameters for each combination of gas and wire, would be unfair and also lead to misjudgments.

With the welding conditions already defined, welding samples, for each combination of wire and shielding gas, were finally produced with four layers of weld beads. The characterization of the deposited molten wire in each welded sample was carried out in the central part of the cross section taken from the last weld bead of the last deposited layer. Chemical analysis of the deposits was carried out with a Solaris CCD optical emission spectrometer (one measurement for each specimen). The resultant microstructures were characterized by using a Leica DMRXP optical microscope (cross sections etched by Vilella's reagent). Samples consisted of 2-mm-thick sheets previously cut in 50x100 mm from a UNS 43932 bi-stabilized ferritic stainless steel were square butt welded with a 3 mm gap in addition as an approach to assess a situation commonly found in automotive exhaustion systems. These complementary samples allowed taking into account the effect of weld bead dilution on the resultant microstructures and mechanical properties for the different shielding gas and wire combinations evaluated.

3. Results and discussions

3.1 Weld bead chemical composition

Table 3 shows the chemical composition of the weld beads produced with the different combinations of wires and shielding gases. As seen, using only argon, the weld bead composition was similar to that one of the wire in each case (Table 1).

Wire	Element	Ar	Ar+2%O_2	Ar+2%CO_2	Ar+4%CO_2	Ar+8%CO_2
ER430Ti	C	0.087	0.082	0.087	0.091	0.094
	Cr	17.391	17.489	17.444	17.515	17.500
	Mn	0.564	0.559	0.562	0.563	0.558
	Mo	0.053	0.053	0.053	0.053	0.053
	Ni	0.486	0.488	0.484	0.488	0.487
	Si	0.902	0.893	0.892	0.878	0.875
	Ti	0.323	0.305	0.298	0.273	0.247
ER430LNb	C	0.018	0.017	0.027	0.038	0.046
	Cr	18.09	18.058	18.144	18.035	18.042
	Mn	0.362	0.351	0.357	0.340	0.337
	Mo	0.050	0.049	0.050	0.049	0.049
	Nb	0.469	0.463	0.464	0.455	0.451
	Ni	0.253	0.252	0.250	0.251	0.252
	Si	0.367	0.360	0.365	0.356	0.348
	Ti	0.009	0.010	0.010	0.010	0.010

Table 3. Chemical compositions of the weld beads (weight, %).

This was expected to happen considering the fact that argon is an inert gas and the small difference found is probably related to sampling and measurement intrinsic errors. On the other hand, perceptible differences in the content of carbon, silicon, manganese, titanium and niobium took place when O_2 or CO_2 were present in the shielding gas.

Figures 1 and 2 graphically shows the changes in the percentage of carbon in the metal deposit according to the shielding gas used respectively to ER430Ti and ER430LNb. It can be noticed that with the increase in the CO_2 gas content there was a proportional increase in carbon in the weld bead, fact also observed by Lundvist (1980) and by Liao & Chen (1998). It is worth noting that with the ER430LNb wire the carbon presence was significantly larger (0.094% for 8% of CO_2 content, for instance). The ER430Ti wire led to a much lower value (0.046% for 8% of CO_2 content, for instance).

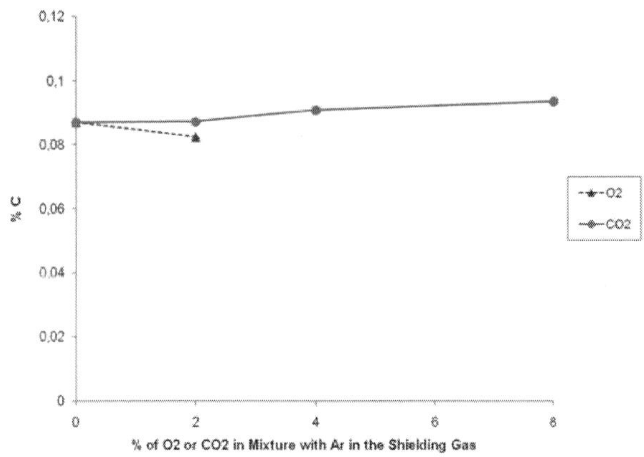

Fig. 1. Carbon presence in the weld beads versus shielding gas used for the ER430Ti wire.

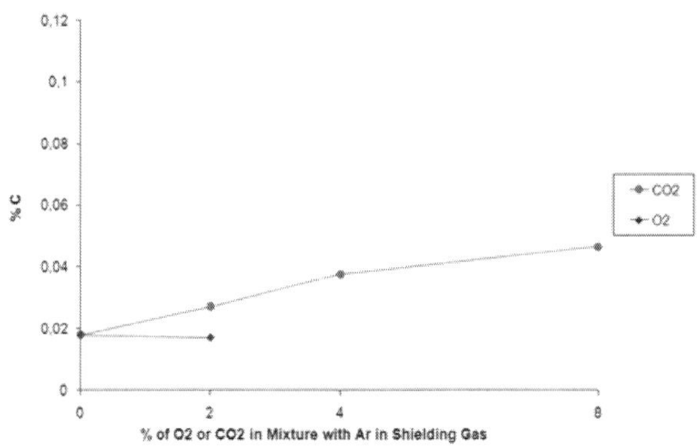

Fig. 2. Carbon presence in the weld beads versus shielding gas used for the ER430LNb wire.

Figures 3 and 4 respectively show the changes in manganese percentage in the weld bead depending on the shielding gas used to ER430Ti and ER430LNb. Figures 5 and 6 respectively show the changes in silicon percentage in the weld bead depending on the shielding gas used to ER430Ti and ER430LNb. Note that the amount of these two elements decreased in the weld beads produced with both wires as the O_2 or CO_2 content in the gas mixture was raised. This fact can be explained by the deoxidizing function of such elements. Once more the effect with the ER430Ti wire was less evident. The contents of manganese and silicon in the weld bead were much higher for the titanium alloyed wire than for the other one, which is a direct consequence of the larger presence of such elements in this wire.

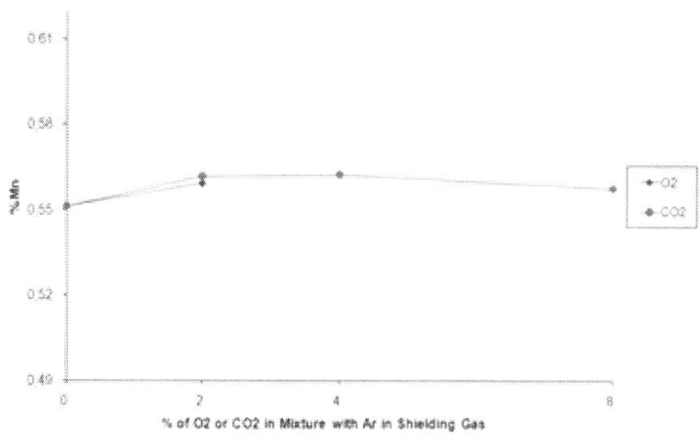

Fig. 3. Manganese presence in the weld beads versus shielding gas used for the ER430Ti wire.

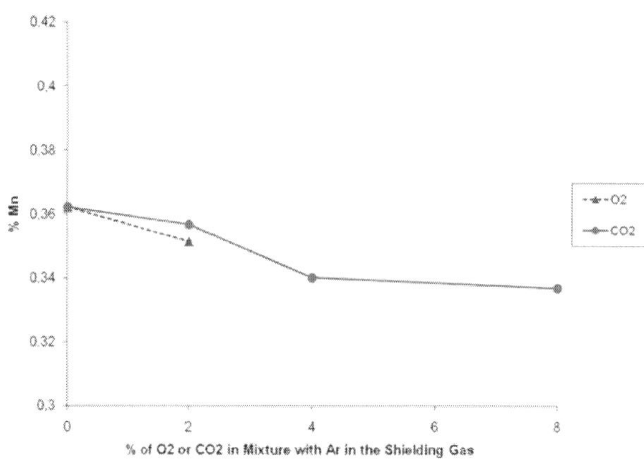

Fig. 4. Manganese presence in the weld beads versus shielding gas used for the ER430LNb wire.

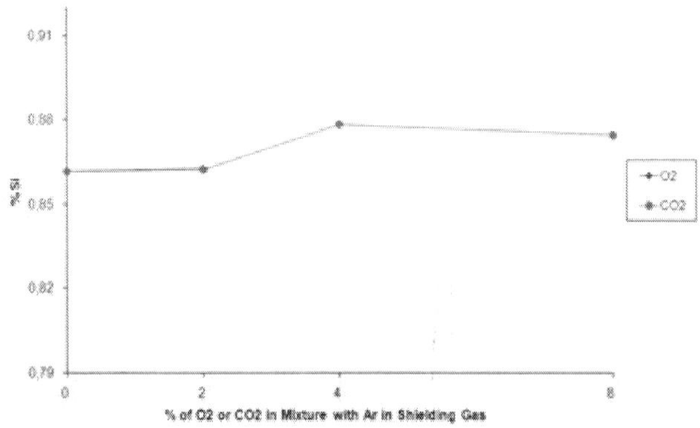

Fig. 5. Silicon presence in the weld beads versus shielding gas used for the ER430Ti wire.

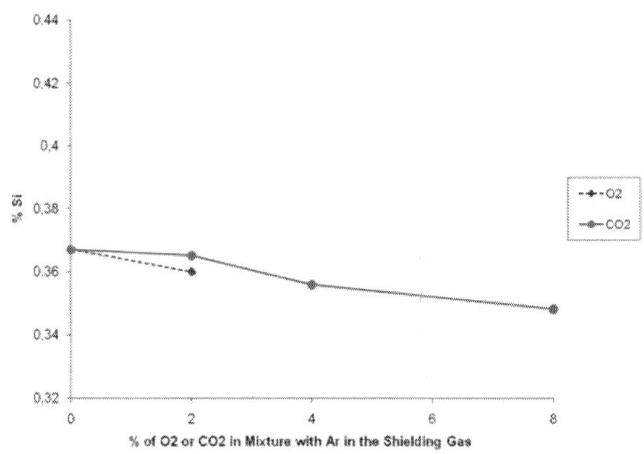

Fig. 6. Silicon presence in the weld beads versus shielding gas used for the ER430LNb wire.

Finally, Figure 7 shows the changes in the percentage of Ti in the weld bead when the ER430Ti wire was used and Figure 8 illustrates the changes in the presence of Nb in the weld bead with the use of the ER430LNb wire. The loss of Ti in the weld bead composition is more significant than the loss of Nb, especially when the weld pool is shielded by Ar+8%CO_2.

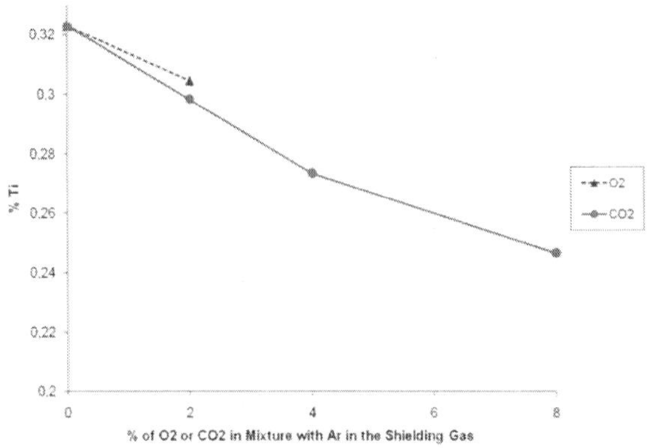

Fig. 7. Titanium presence in the weld beads versus shielding gas used for the ER430Ti wire.

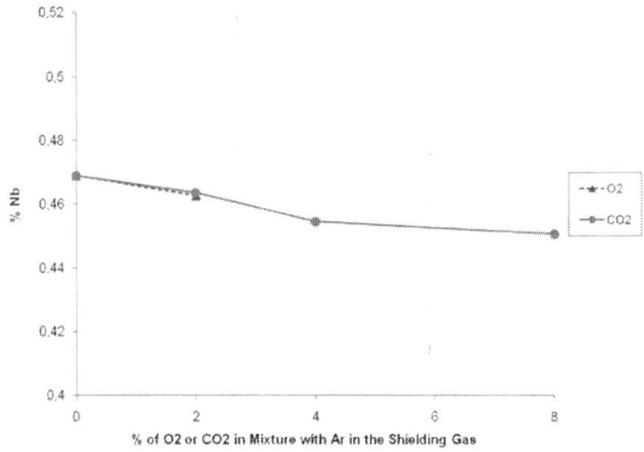

Fig. 8. Niobium presence in the weld beads versus shielding gas used for the ER430LNb wire.

3.2 Weld microstructure without dilution

Figures 9 to 13 show the microstructure of the fusion zones produced for the different combinations of wire and gas. The weld beads obtained with the ER430Ti wire (Figures 9(a) to 13(a)) contain a ferritic matrix, with columnar and coarse grains. This fact was expected, since the presence of stabilizer elements should retain grain growth as stated by Modenesi (2001). The presence of martensite and precipitates inside the grains increased as the shielding gas became more oxidizing. As seen before in Figure 1, there was an increase in carbon content in the weld pool with the presence of CO_2. This fact suggests that the severe drop in titanium amount with the increase of CO_2 contents in the shielding gas (Figure 7) is the responsible for martensite formation (it is known that titanium picks up carbon to form precipitates, retaining ferritic phases and avoiding by that martensite formation).

The microstructures of the weld beads with ER430LNb, Figures 9(b) to 13(b), were also composed of a ferritic matrix, with columnar and coarse grains. However, no presence of martensite is observed, regardless of the content of CO_2 in the shielding gas. This might be explained, despite the carbon absorption when CO_2 was used (Figure 2), by a lower content of carbon in the wire. Even so, the grains were as coarse as those found with ER430Ti wire. This fact can be explained by the Titanium precipitates high thermal stability during welding, especially after high heat inputs, as verified by Wang & Wang (2008). Niobium does not present the same grain growth effect as titanium does. It was not possible to observe changes in the precipitate quantities as the shielding gas was varied.

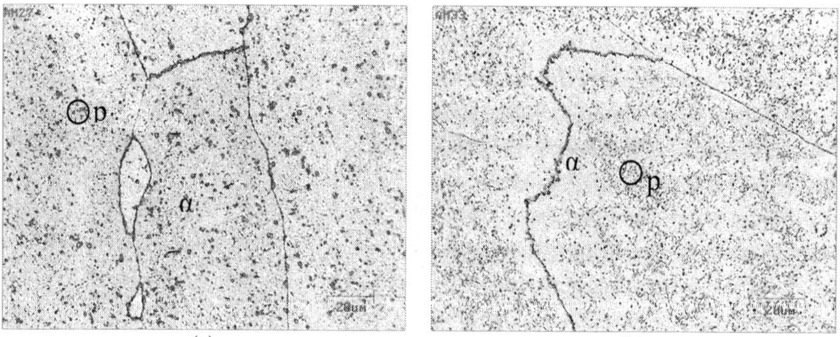

(a) (b)

Fig. 9. Fusion zone microstructure with argon as shielding gas for the ER430Ti wire (a) and for the ER430LNb wire (b) ("α" represents ferritic matrix and "p" precipitates).

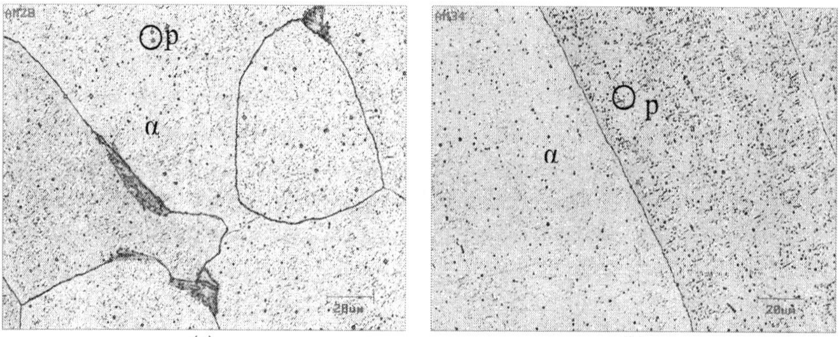

(a) (b)

Fig. 10. Fusion zone microstructure with Ar+2%O_2 as shielding gas for the ER430Ti wire (a) and for the ER430LNb wire (b) ("α" represents ferritic matrix and "p" precipitates).

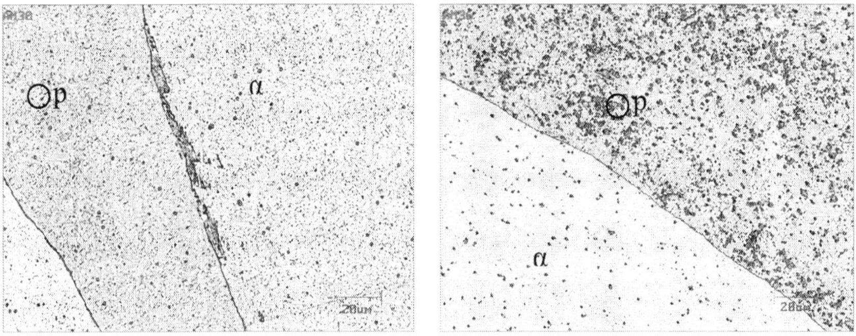

(a) (b)

Fig. 11. Fusion zone microstructure with Ar+2% CO_2 as shielding gas for the ER430Ti wire (a) and for the ER430LNb wire (b) ("α" represents ferritic matrix and "p" precipitates).

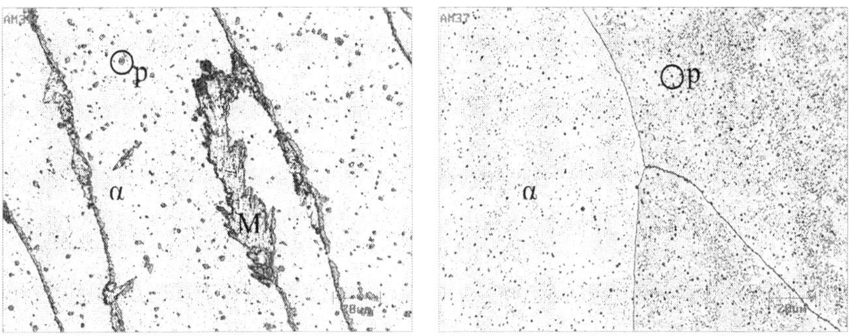

(a) (b)

Fig. 12. Fusion zone microstructure with Ar+4% CO_2 as shielding gas for the ER430Ti wire (a) and for the ER430LNb wire (b) ("α" represents ferritic matrix, "M" martensite and "p" precipitates).

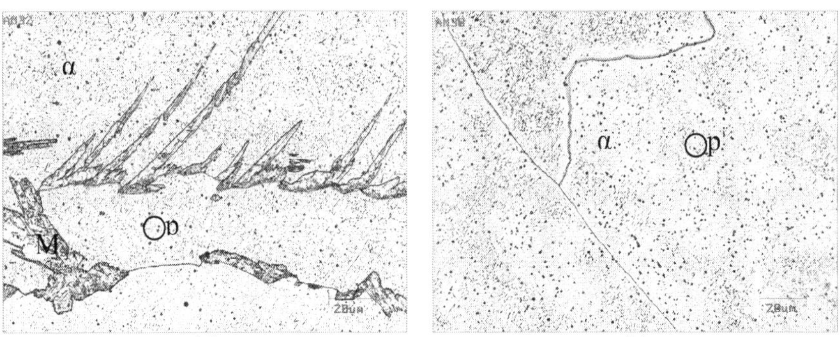

(a) (b)

Fig. 13. Fusion zone microstructure with Ar+8% CO_2 as shielding gas for the ER430Ti wire (a) and for the ER430LNb wire (b) ("α" ferritic matrix, "M" martensite and "p" precipitates).

3.3 Weld microstructure in the square butt joints

Based on the base and filler metals chemical compositions (Table 1 and 2, respectively) and in a dilution level of 39% (mean level for all shielding gases used), the microstructures and chemical compositions were estimated through Schaeffler diagrams for the case of weld beads produced using the ER430Ti wire (Figure 14 and 15).

As seen in Figure 14, the ER430Ti filler metal is located in the ferrite-martensite region, whilst the base metal falls in the ferritic region. The weld bead is located on the border between the ferritic and ferritic-martensitic regions, that is, carbon additions might lead to martensite formation in the fusion zone. Also from the diagram, the weld beads are estimated to have 0.070% of carbon, 0.79% of silicon, 0.46% manganese, 17.29% of chromium, 0.33% of nickel and 0.07% of niobium.

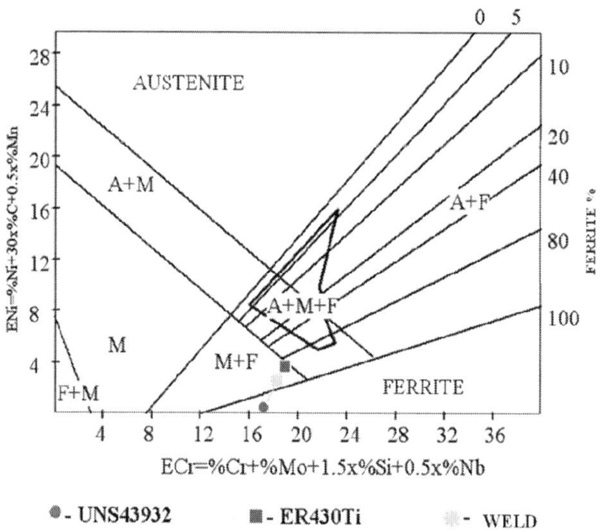

Fig. 14. Schaeffler diagram applied to the samples produced with the ER430Ti wire.

Applying the Schaeffler diagram methodology also to the samples welded with the ER430LNb wire (Figure 15), just ferritic microstructure was estimated to take place in the fusion zone. Concerning the chemical composition, the weld beads in this case are expected to have 0.023% of carbon, 0.12% of silicon, 0.35% of manganese, 17.49% of chromium, 0.38% of nickel and 0.37% of niobium.

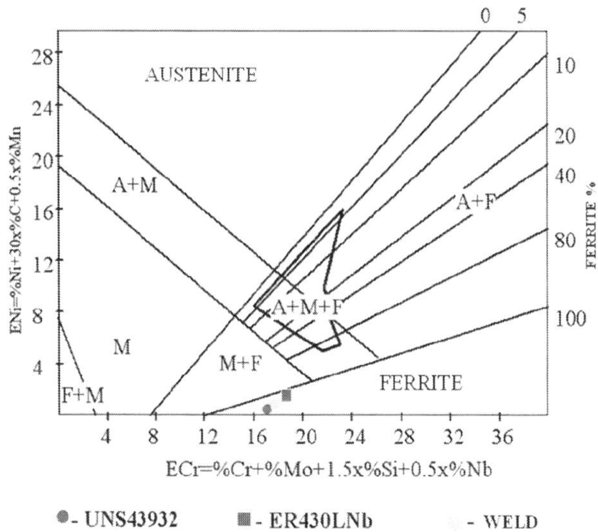

Fig. 15. Schaeffler diagram applied to the samples produced with the ER430LNb wire.

Figures 16 to 19 present the fusion zone microstructure produced by the ER430Ti and ER430LNb wires for, respectively, Ar, Ar+2%O2, Ar+4%CO2 and Ar+8%CO2 as shielding gases.

In the ER430Ti wire case (Figures 16(a) a 19 (a)), a ferritic matrix with titanium precipitates was observed in all grains, structure also observed by Madeira (2007) using SEM. Throughout the images it is possible to notice the increase in the amount of precipitates all around as the CO_2 content in the gas was raised. This fact is due to the increase in the presence of carbon in the fusion zone. It is worth saying that large carbon amounts can also form chromium precipitates, which decrease the corrosion resistance of the weld beads.

The estimations carried out by Schaeffler diagrams, in which titanium is not taking into account, showed the presence of ferrite and martensite along the grain boundaries in the microstructures. However, this was not the case of the weld beads produced using up to 8% of CO_2. This fact shows that titanium, up to this level of CO_2, was successful as a stabilizing element.

In the ER430LNb wire case (Figures 16 (b) a 19 (b)), a ferritic matrix with probably Nb precipitates formed in all grains. Despite it was not measured, an increase in the number of precipitates is visually verified as the CO_2 content in the gas was raised, but not as remarkable as for the ER430Ti wire case. The increase in the presence of C in the fusion zone is likely the reason for the increasing number of precipitates in the ER430LNb case as well.

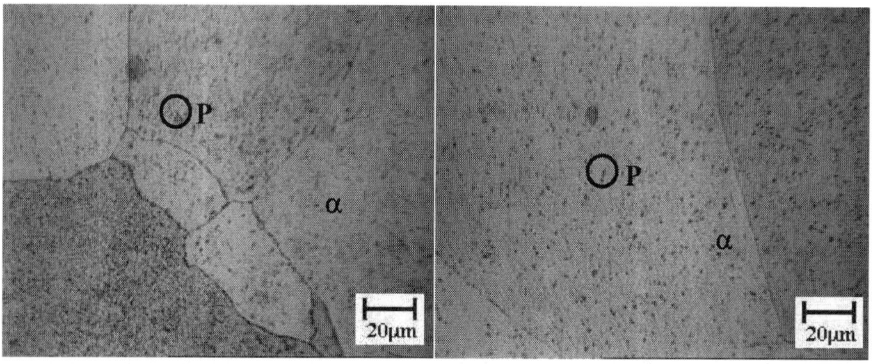

a b

Fig. 16. Fusion zone produced with Ar as shielding gas and with the ER430Ti wire (a) and with the ER430LNb wire (b) ((α) ferritic matrix, (P) titanium or niobium precipitates).

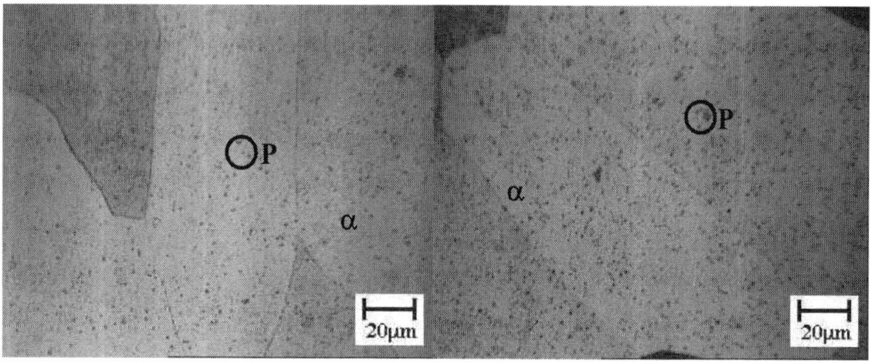

a b

Fig. 17. Fusion zone produced with Ar+2%O_2 as shielding gas and with the ER430Ti wire (a) and with the ER430LNb wire (b) ((α) ferritic matrix, (P) titanium or niobium precipitates).

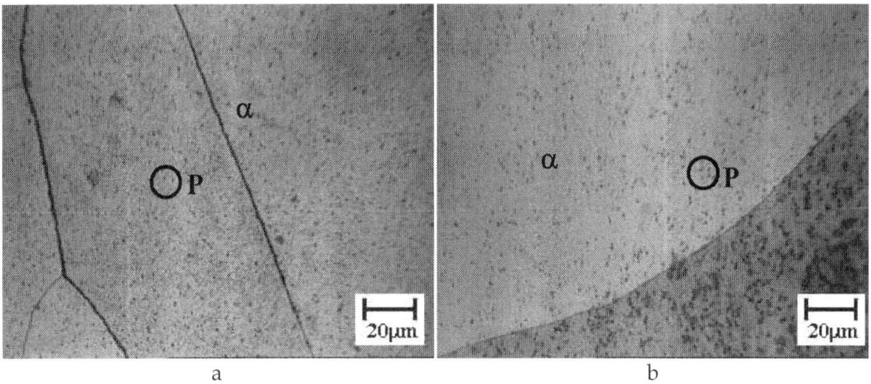

Fig. 18. Fusion zone produced with Ar+4%CO_2 as shielding gas and with the ER430Ti wire (a) and with the ER430LNb wire (b) ((α) ferritic matrix, (P) titanium or niobium precipitates).

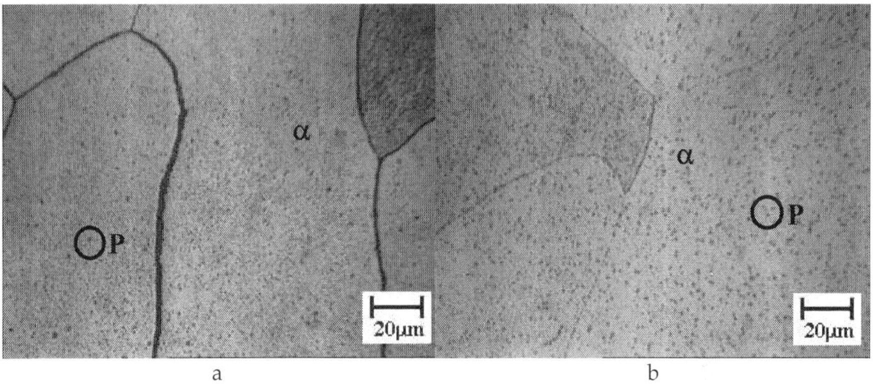

Fig. 19. Fusion zone produced with Ar+8%CO_2 as shielding gas and with the ER430Ti wire (a) and with the ER430LNb wire (b) ((α) ferritic matrix, (P) titanium or niobium precipitates).

As martensite was not observed along the grain boundaries with shielding gases with up to 8% of CO_2 in the weld beads produced in the square butt joints (with dilution), but it was found in the weld beads with dilution from 4% of CO_2 and using the ER430Ti wire, an extra shielding gas containing 25% of CO_2 was assessed for both wires.

As showed in Figure 20 (a), for the ER430Ti case, besides the formation of ferritic matrix and precipitates, there is also martensite formed along the grain boundaries when 25% of CO_2 is used in the shielding gas. In this case, the titanium present in the wire was unable to cause an adequate stabilization and the free carbon leads to transformation of austenite into martensite during the weld bead cooling. For the ER430LNb wire case (Figure 20 (b)), there was not martensite present in the weld beads even with the use of such level of CO_2. In this case, the niobium in this wire was able to stabilize the carbon for all the shielding gases.

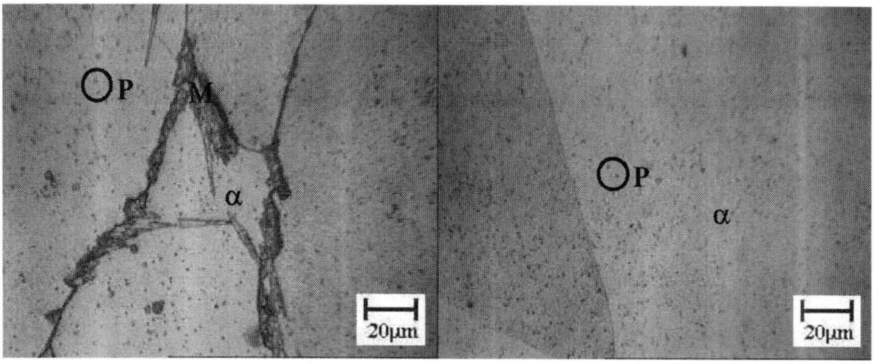

a b

Fig. 20. Fusion zone produced with Ar+25%CO_2 as shielding gas and with the ER430Ti wire (a) and with the ER430LNb wire (b) ((α) ferritic matrix, (P) titanium or niobium precipitates, (M) martensite).

3.4 Welded joints microhardness

Table 4 presents the microhardness values measured in the ferritic matrix for the ER430Ti and ER430LNb wires. Table 5 shows the microhardness values measured in the martensite formations in the weld beads produced with the ER430Ti wire and with Ar+25%CO_2 as the shielding gas.

Wire	Shielding gas	Microhardness ($HV_{2,5}$)	
		Mean value	Standard deviation
ER430Ti	Ar	180.3	1.5
	Ar+2%O_2	172.0	10.6
	Ar+4%CO_2	187.7	4.7
	Ar+8%CO_2	208.0	4.6
	Ar+25%CO_2	203.0	1.0
ER430LNb	Ar	163.0	2.6
	Ar+2%O_2	173.0	1.0
	Ar+4%CO_2	160.3	7.2
	Ar+8%CO_2	182.0	3.6
	Ar+25%CO_2	181.3	2.9

Table 4. Microhardness values measured in the weld beads produced with the ER430Ti and ER430LNb wires.

Wire	Shielding gas	Microhardness (HV2.5)						
		1	2	3	4	5	Mean value	Standard deviation
ER430Ti	Ar+25%CO_2	245	269	304	304	364	292.3	20.2

Table 5. Microhardness values measured in the martensite formations in the fusion zone.

Figures 21 and 22 show graphically, for the ER430Ti wires ER430LNb respectively, the microhardness values observed in the fusion zones versus the shielding gas used. The microhardness values taken from the ferritic matrix of the weld beads produced with the ER430Ti wire and with argon and Ar+2%O_2 as shielding gases were very similar to those measured in the base metal ferritic matrix. With the addition of CO_2 in the shielding gas there was a significant rise in the microhardness levels, which was probably due to the large number of Ti precipitates present, as seen in Figures 16 to 20. With Ar+25%CO_2, martensite was formed along the grain boundaries, which is put in evidence by the elevated microhardness levels observed in this case.

Cardoso (2003) observed a microhardness value close to 350 $HV_{2.5}$ with the use of Ar+8%CO_2. According to him, such level of hardness would cause decrease in the union toughness. As microhardness values of magnitudes close to this level were found on the martensite region in the weld beads produced with the ER430Ti wire, loss of toughness in such weld beads is also expected to take place.

The microhardness values measured in the ferritic matrix of the weld beads produced with the ER430LNb wire were very close to the values found in the base metal ferritic matrix. In general, no major variation in the microhardness levels were noticed as the shielding gas was changed in this case.

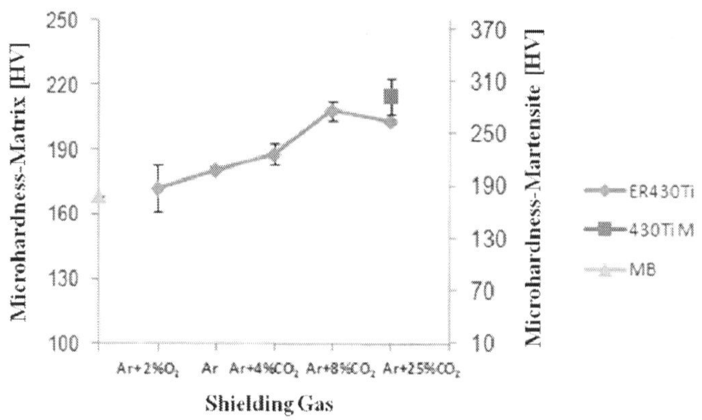

Fig. 21. Microhardness values in the fusion zone of the weld beads produced with the ER430Ti wire versus the shielding gas used.

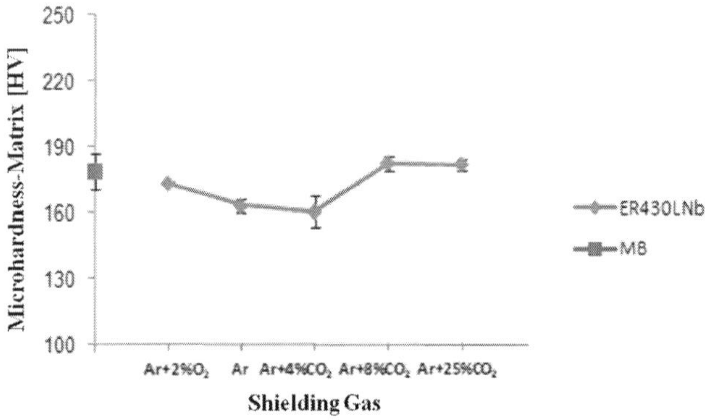

Fig. 22. Microhardness values in the fusion zone of the weld beads produced with the ER430LNb wire versus the shielding gas used.

3.5 Welded joints ductility

Figures 23 and 24 show the evolution of the loads supported by the welded samples during the stampability tests along with the aspect of the samples after the tests for, respectively, Ar+2%O_2 and Ar+25%CO_2 as the shielding gas.

For the sample welded using Ar+2%O_2 the failure (crack) of the weld bead took place along the welding direction. In contrast, for the sample welded with Ar+25%CO_2 the fracture took place transversally to the welding direction. Madeira (2007) also reported these two forms of fractures in stampability tests of welded joints. According to him, transversally cracks (failures) assure that the weld bead ductility is being evaluated.

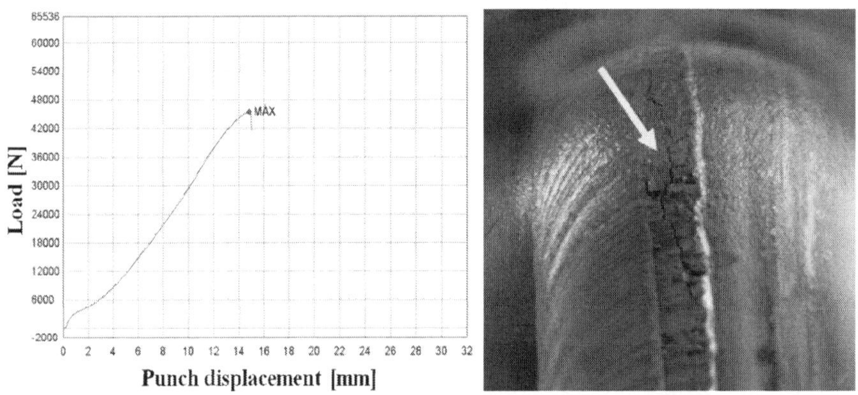

Fig. 23. Stampability test for the condition using the ER430Ti wire and Ar+2%O_2 as the shielding gas (loads supported by the sample (a) and aspect of the sample after the test (b)).

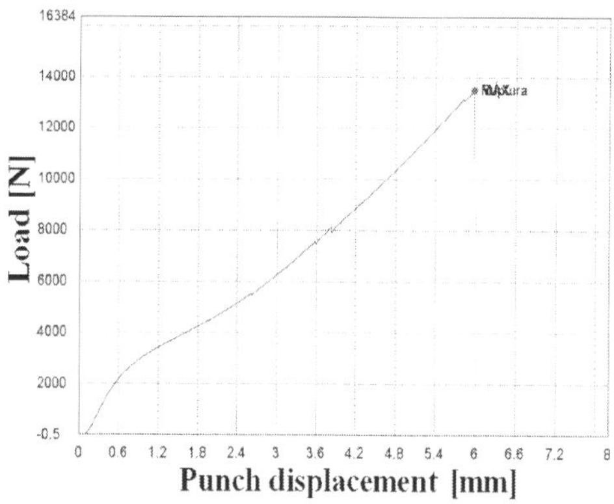

a

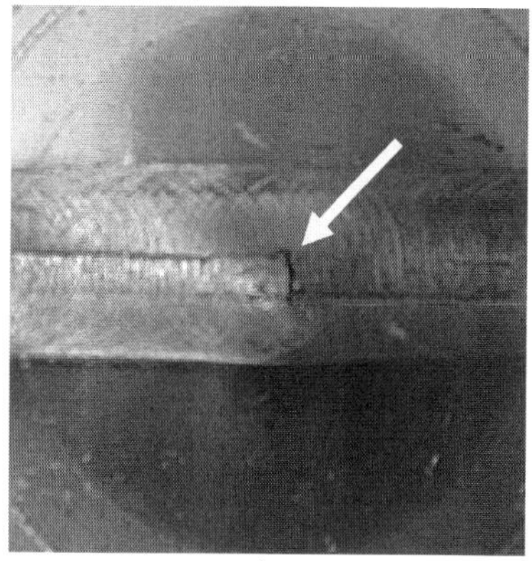

b

Fig. 24. Stampability test for the condition using the ER430Ti wire and Ar+25%CO_2 as the shielding gas (loads supported by the sample (a) and aspect of the sample after the test (b)).

Table 6 presents the values of the maximum loads supported by the samples and also the punch displacements and energies absorbed during the stampability tests of weld beads produced using the ER430Ti wire. Tests were carried out for loads applied both on the weld face and root.

Shielding gas	Loading side	Mean FMAX [N]	FMAX STD	Mean D 10-3 [m]	D STD.	E [J]	E STD
Ar	Root	41636	5370	13,5	1,9	265	85
Ar+2%O_2	Face	27883	13879	9,7	3,9	139	117
Ar+2%O_2	Root	26070	2448	9,9	0,7	149	35
Ar+4%CO_2	Face	22371	8412	8,1	2,3	98	42
Ar+4%CO_2	Root	23538	2662	8,7	0,9	92	21
Ar+8%CO_2	Face	18951	8370	7,2	2,2	66	44
Ar+8%CO_2	Root	29584	0	10	0,0	131	0
Ar+25%CO_2	Face	11326	3054	5,2	1,0	31	12
Ar+25%CO_2	Root	4909	375	2,4	0,3	7	2

FMAX = mean maximum load; D = punch displacement; E = energy absorbed; STD = standard deviations

Table 6. Values of the maximum loads supported by the samples, punch displacements and energies absorbed during the stampability tests of weld beads produced using the ER430Ti wire.

Figures 25 to 27 graphically present the trends found in the stampability tests of the samples welded with the ER430Ti wire. As seen, in general the evolution of the parameters analyzed was similar to each other and always lower than the value found for the base metal. This fact was also observed by Hunter & Eagar (1980), Sawhill & Bond (1976) and Redmond (1977). There was also a decreased in the values of the maximum loads supported by the samples, punch displacements and energies absorbed during the stampability tests as the CO_2 percentage in the shielding gas was increased, which was more evident for 25% of CO_2. When the CO_2 content in the shielding gas was increased in the welded samples produced with the ER430Ti wire there was an increase in the microhardness levels (Figure 21), decrease in the grain size and increase in the number of precipitates (Figures 16 to 20). For the case of 25% of CO_2 in particular, there was formation of martensite. All these factors justify the decrease in the levels of the stampability parameters assessed and so the decrease in the welded joints ductility.

Madeira & Modenesi (2010) also observed falls in the ductility of welded joints when they used the ER430Ti wire with an active shielding gas. They referred to the content of carbon, nitrogen and titanium in the fusion zone as the responsible for the fragilization of the joints as these elements form precipitates along the grains. However, the results found here are opposite to those presented by Washko & Grubb (1991), who state that the presence of titanium in the weld beads minimize the loss of ductility. Considering the high presence of carbon in the ER430Ti wire (Table 1) and the high levels of carbon getting into the fusion zone from the CO_2 added to the shielding gas, the titanium present in this filler metal does not offer adequate stabilization.

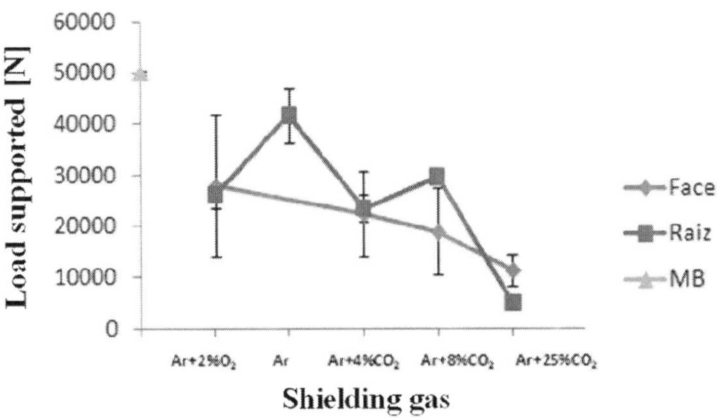

Fig. 25. Maximum loads supported by the samples produced with the ER430Ti wire versus the shielding gas used.

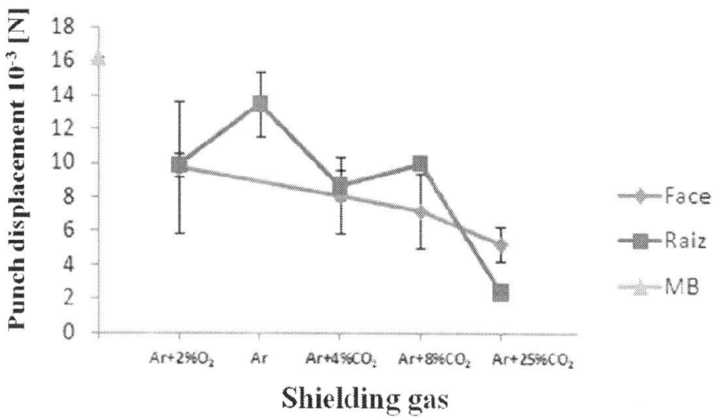

Fig. 26. Punch displacement in the samples produced with the ER430Ti wire versus the shielding gas used.

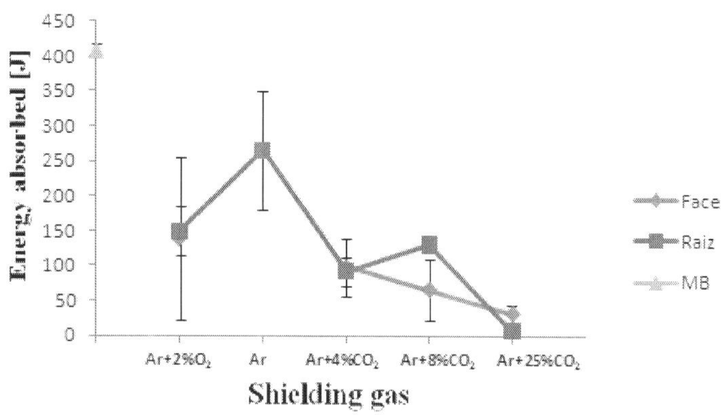

Fig. 27. Energy absorbed by the samples produced with the ER430Ti wire versus the shielding gas used.

Table 7 presents the values of the maximum loads supported by the samples and also the punch displacements and energies absorbed during the stampability tests of weld beads produced using the ER430LNb wire. As for the ER430Ti wire case, the tests were carried out for loads applied both on the face and root of the weld beads.

Shielding gas	Loading side	Mean FMAX [N]	FMAX STD	Mean D 10-3 [m]	D STD.	E [J]	E STD
Ar	Face	40065	11277	13.1	3.1	261	131
Ar	Root	17381	1514	7.0	0.6	56	10
Ar+2%O_2	Face	32792	10751	11.3	2.8	181	101
Ar+2%O_2	Root	24101	10286	8.9	3.2	108	84
Ar+4%CO_2	Face	36152	1413	11.7	0.6	184	18
Ar+4%CO_2	Root	32158	20096	11.2	5.4	210	211
Ar+8%CO_2	Root	21440	14611	8.8	3.0	102	85
Ar+25%CO_2	Face	34068	0	11.2	0	236	0
Ar+25%CO_2	Root	16845	8546	7.0	2.6	76	32

FMAX = mean maximum load; D = punch displacement; E = energy absorbed; STD = standard deviations

Table 7. Values of the maximum loads supported by the samples, punch displacements and energies absorbed during the stampability tests of weld beads produced using the ER430LNb wire.

Figures 25 to 27 graphically present the trends found in the stampability tests of the samples welded with the ER430LNb wire. With this wire no significant variations in the parameters assessed was recorded. The dispersion in the results for each shielding gas might have occurred due to possible fragilization in the weld beads that was not perceived during the visual analyses of the samples.

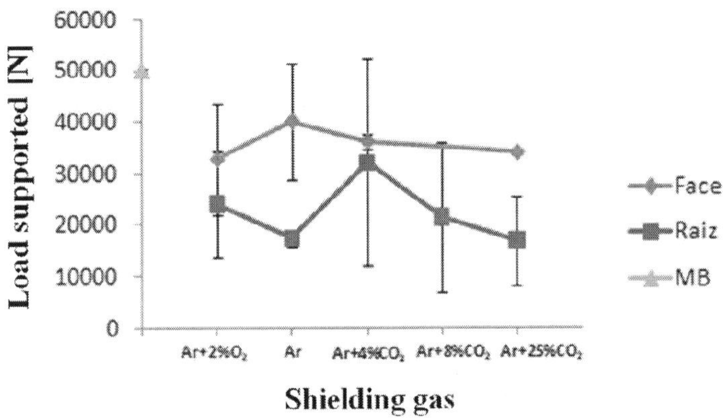

Fig. 28. Maximum loads supported by the samples produced with the ER430LNb wire versus the shielding gas used.

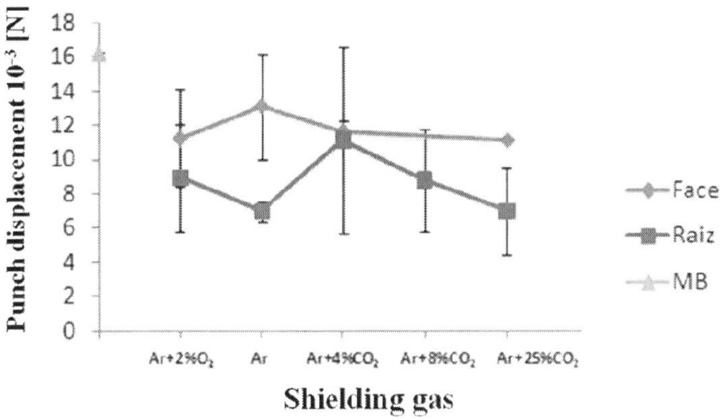

Fig. 29. Punch displacement in the samples produced with the ER430LNb wire versus the shielding gas used.

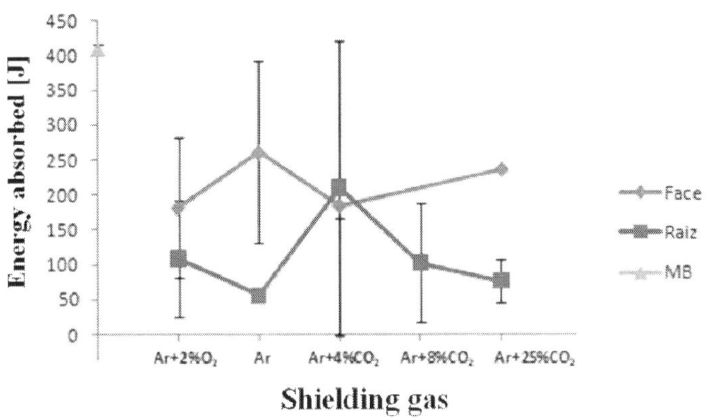

Fig. 30. Energy absorbed by the samples produced with the ER430LNb wire versus the shielding gas used.

Taking into account the results of the stampability tests, it is possible to consider that the increase in the CO_2 content in the shielding gas decreases the ductility of the welded joints if the ER430Ti wire is used. If the ER430LNb wire is utilized instead, it performs a better stabilization of the C present and the result is that no significant variations are recorded for the welded joints ductility even with the high levels of CO_2 added to the shielding gas.

4. Conclusions

Considering the conditions and results presented in this chapter, the conclusions can be summarized as:

- For the ER430Ti and ER430LNb wires, the addition of CO_2 in the shielding gas promotes an increase in the quantity of carbon and a decrease in the amount of manganese, silicon, and also in the stabilizing elements (titanium and niobium, respectively);
- In the welded layers (without dilution), the titanium present in the ER430Ti wire was insufficient to avoid the formation of martensite in the fusion zone with the use of levels of CO_2 higher than 4%. Also without dilution in the welded joint, but using the ER430LNb wire, martensite did not form with shielding gases with up to 8% of CO_2;
- In the weld beads produced in square butt joints using the ER430Ti wire, martensite was only noticed for the weld beads produced with 25% of CO_2. Also in square butt joints but using the with ER430LNb, the stabilization was effective and no martensite formation was verified even for such level of CO_2;
- An increase in hardness and therefore a fall in the ductility of the welded joints took place for the ER430Ti wire. This fact was not recorded for the weld beads produced with the ER430LNb wire.
- Therefore, the ER430LNb was the best wire utilized for the selected conditions.

In face of the conclusions, this manuscript shows the importance of correct stabilization of a filler metal in welding. Besides that, the shielding gas may play a decisive role in the ductility of welded joints, so as in the microstructures formed. As verified, it is possible to utilize ferritic stainless steel filler metals in welding approaches for ferritic stainless steel components by using low-cost shielding gases and at the same time preserve the joint properties. This shows that the tendency of using austenitic stainless steel filler metals with high-cost shielding gases for ferritic stainless steel welded components might be equivocated.

5. Acknowledgments

The authors express their gratitude to CNPq, CAPES, Fapemig, Fapergs, Federal University of Rio Grande, Federal University of Uberlândia and LAPROSOLDA/UFU for the infrastructure and, especially, to ACELORMITTAL and WhiteMartins for providing the materials used in the experiments.

6. References

Alves, H.J.B, Carvalho, J.N., Aquino, M.V., Mantel, M.J. (2002). Development of ferritic stainless steels for automotive exhaust systems. *Proceedings of 4th Stainless Steel Science and Market Congress*, Paris, France, June 2002.

Balmforth, M. C.; Lippold, J. C. (2000) A New Ferritic-Martensitic Stainless Steel Constitution Diagram. *Welding Journal*, Vol. 79, n. 12 (Dec. 2000), pp. 339s-345s, ISSN 0043-2296.

Cardoso, R. L.; Prado, E. M.; Okimoto, P. C.; Paredes, R. S. C., Procopiak, L. A. (2003). Avaliação da Influência de Gases Proteção Contendo Diferentes Teores de CO_2 nas Características dos Revestimentos Soldados Visando o Reparo de Turbinas

Erodidas por Cavitação. *Soldagem & Inspeção*, Ano 8, n. 2, (April-Jun. 2003), pp. 68-74, ISSN 0104-9224.

Chae, H. B., Kim, C. H., Kim, J. H., Rhee, s. (2008). The effect of shielding gas composition in CO2 laser-gas metal arc hybrid welding. *Proc. IMechE, Part B: J. Engineering Manufacture*, Vol. 222 (2008), pp. 1315-1324, ISSN 0954-4054.

Durgutlu, A. (2004). Experimental Investigation of the Effect of Hydrogen in Argon as a Shielding Gas on TIG Welding of Austenitic Stainless Steel, *Materials & Design*, Vol. 25, Issue 1 (Feb. 2004), pp.19-23, ISSN 0264-1275.

Faria, R. A. (2006). *Efeito dos elementos Ti e Nb no comportamento em fadiga em aços inoxidáveis ferríticos utilizados nos sistemas de exaustão de veículos automotores*, 245 f., (PhD Thesis) Universidade Federal de Ouro Preto, Ouro Preto, MG, Brazil..

Ferreira Filho, D. ; Ferraresi, V A ; Scotti, A. (2010). Shielding gas influence on the ferritic stainless steel weldability. *Proceedings of the Institution of Mechanical Engineers. Part B, Journal of Engineering Manufacture*, v. 224, (2010), pp. 951-961, ISSN 0954-4054.

Filho, Demostenes Ferreira ; Ferraresi, Valtair Antonio. (2010). The influence of gas shielding composition and contact tip to work distance in short circuit metal transfer of ferritic stainless steel. *Welding International*, v. 24, pp. 206-213, ISSN 0950-7116.

Ferreira Filho, D.; Ferraresi, V. A. . Influência do tipo de gás de proteção e da distância bico de contato-peça na transferência metálica do modo curto-circuito do aço inoxidável ferrítico. *Soldagem & Inspeção*, v. 13, n. 3, (Jul. 2010), pp. 173-180, ISSN 0104-9224.

Gülenç, B., Develi, K., Kahraman, N. , Durgutlu, A. (2005). Experimental Study of the Effect of Hydrogen in Argon as a Shielding Gas in MIG Welding of Austenitic Stainless Steel, *International Journal of Hydrogen Energy*, Vol. 30, Issues 13-14, (October-November 2005), pp. 1475-1481, ISSN 0360-3199.

Hiramatsu, N. (2001). *Niobium in ferritic and martensitic stainless steels*. Proceedings of the International Symposium Niobium, Orlando, Florida, USA, 2001.

Hunter, G. B., Eagar, T. W., (1980) Ductility of stabilized ferritic stainless steel welds. *Metallurgical Transactions A*, v. 11 A (Feb 1980), p. 213-218.

Inui, K., Noda, T., Shimizu, T. (2003). *Development of the Ferritic Stainless Steel Welding Wire Providing Fine Grain Microstructure Weld Metal for the Components of Automotive Exhaust System*, Proceedings of SAE International 2003, World Congress and Exhibition, Detroit USA, 2003.

Lee, C-H., Chang, K-H. And Lee, C-Y. (2008). Comparative study of welding residual stresses in carbon and stainless steel butt welds. *Proc. IMechE, Part B: J. Engineering Manufacture*, 222(B12) (2008), pp. 1685-1694, ISSN 0954-4054.

Liao, M. T., Chen, W. J. (1998). The effect of shielding-gas compositions on the microstructure and mechanical properties of stainless steel weldments, *Materials Chemistry and Physics*, Vol. 55 (1998), p. 145-155, ISSN 0254-0584.

Lundqvist, B. (1980). *Aspects of Gas-Metal Arc Welding of Stainless Steels*, Proceedings of Swedish. Sandvik AB, Sandviken, Sweden, 1980.

Madeira, R. P., Modenesi, P. J. (2007). *Estudo dos arames ferríticos 430Ti e 430LNb para a aplicação na parte fria de sistemas de exaustão automotivos*, Proceedings of XXXIII CONSOLDA, Caxias do Sul-RS Brasil, 2007.

Madeira, R. P., Modenesi, P. J. (2010). Utilização do Ensaio Erichsen para a Avaliação do Desempenho de Juntas Soldadas, *Soldagem & Inspeção*, v. 15, n. 1 (Jan/Mar 2010), p. 022-030, ISSN 0104-9224.

Madeira, R. P. (2007). *Influência do Uso de Arames Inoxidáveis Ferríticos nas Características da Zona Fundida de um Aço Inoxidável Ferritico com 17% de Cromo Bi-estabilizado,* 151 f., (Master's Dissertation). Universidade Federal de Minas Gerais, Belo Horizonte, MG, Brazil.

Modenesi, P. J. (2001). *Soldabilidade dos Aços Inoxidáveis,* Vol. 1, SENAI, ISBN 85-88746-02-6, Osasco, SP, Brazil..

Reddy, G. M.; Mohandas, T. (2001). Explorative studies on grain refinement of ferrítico stainless steel welds. *Journal of Materials Science Letters,* Vol. 20, pp. 721-723, *ISSN 0261-8028.*

Redmond, J. D. (1977). Climax Molybdenum Co. *Report RP.,* (Sept. 1977), p. 33-76, ISSN 0034-4885.

Renaudot, N.; Santacreu, P. O.; Ragot, J.; Moiron, J. L.; Cozar, R.; Pédarré, P.; Bruyère, A. (2000), *430LNb - A new ferritic wire for automotive exhaust applications.* Proceedings of SAE 2000 World Congress, Detroit, MI, USA, March 2000.

Resende, A. (2007) *Mapeamento paramétrico da soldagem GMAW com arames de aço inoxidável ferrítico e austenítico,* 126 f, (Master's Dissertation). Universidade Federal de Minas Gerais, Belo Horizonte, MG, Brazil.

Sawhill, J. M.; Bond, A. P. (1976). Ductility and Toushness.of Stainless Steel Welds, *Welding Journal.* v. 55, n. 2, p.33s. 1976, ISSN 0043-2296.

Schwarz; B.; Tessin, F. (2003). ESAB high-alloyed welding consumables for ferritic stainless steel exhaust systems, *Svetsaren The Esab Welding and Cutting Journal,* V. 58 N.2, p. 27, 2003.

Sekita, T., Kaneto, S., Hasuno, S., Sato, A., Ogawa, T., Ogura, K. (2004). Materials and Technologies for Automotive Use, *JFE GIHO* N. 2 (Nov. 2004), p. 1-16, ISSN 1348-0669.

Stenbacka, N., Persson, K (1992). Shielding gases for gas-metal arc welding of stainless steels, *AGA AB Inovation,* Suécia, 1992.

Strassburg F. W., Schweissen nichtrostender Stahle, DVS Band 67, DCS Gmbh, Dusselorf, FRG, 1976.

Tusek, J., Suban, M. (2000). Experimental Research of the Effect of Hydrogen in Argon as a Shielding Gas in Arc Welding of High-Alloy Stainless Steel, *International Journal of Hydrogen Energy,* Vol. 25, Issue 4 (April 2000), pp. 369-376 ISSN 0360-3199.

Wang, H. R., Wang, W. (2008). Precipitation of complex carbonitrides in a Nb-Ti microalloyed plate steel, *Journal of Material Science,* vol. 44, issue 2 (2008), pp. 591-600, ISSN 0022-2461.

Washko, S. D.; Grubb, J. F. (1991). *The Effect of Niobium and Titanium Dual Stabilization on the Weldability of 11% Chromium Ferritic Stainless Steels,* Proceedings of International Conference on Stainless Steels, Chiba, ISIJ, 1991.

Yasuda, K.; Jimma, T.; Onzawa, T, (1984). Formability of butt welded Stainless Steel Thin Sheet. Quartely *Journal of the Japan Welding* Society, v.2, n.3, p. 161-166, ISSN 02884771.

9

Corrosion Fatigue Behaviour of Aluminium 5083-H111 Welded Using Gas Metal Arc Welding Method

Kalenda Mutombo[1] and Madeleine du Toit[2]
[1]*CSIR/*
[2]*University of Pretoria*
South Africa

1. Introduction

Aluminium and its alloys are widely used as engineering materials on account of their low density, high strength-to-weight ratios, excellent formability and good corrosion resistance in many environments. This investigation focused on one popular wrought aluminium alloy, namely magnesium-alloyed 5083 (in the strain hardened -H111 temper state).

Aluminium alloy 5083 is one of the highest strength non-heat treatable aluminium alloys, with excellent corrosion resistance, good weldability and reduced sensitivity to hot cracking when welded with near-matching magnesium-alloyed filler metal. This alloy finds applications in ship building, automobile and aircraft structures, tank containers, unfired welded pressure vessels, cryogenic applications, transmission towers, drilling rigs, transportation equipment, missile components and armour plates. In many of these applications welded structures of aluminium are exposed to aqueous environments throughout their lifetimes.

Welding is known to introduce tensile residual stresses, to promote grain growth, recrystallization and softening in the heat-affected zone, and to cause weld defects that act as stress concentrations and preferential fatigue crack initiation sites. Fatigue studies also emphasised the role of precipitates, second phase particles and inclusions in initiating fatigue cracks. When simultaneously subjected to a corrosive environment and dynamic loading, the fatigue properties are often adversely affected and even alloys with good corrosion resistance may fail prematurely under conditions promoting fatigue failure.

The good corrosion resistance of the aluminium alloys is attributed to the spontaneous formation of a thin, compact and adherent aluminium oxide film on the surface on exposure to water or air. This hydrated aluminium oxide layer may, however, dissolve in some chemical solutions, such as strong acids or alkaline solutions. Damage to this passive layer in chloride-containing environments (such as sea water or NaCl solutions), may result in localised corrosive attack such as pitting corrosion. The presence of corrosion pits affects the fatigue properties of the aluminium alloys by creating sharp surface stress concentrations which promote fatigue crack initiation. In welded structures, pits are often associated with coarse second phase particles or welding defects [1-4].

A review of available literature on the corrosion fatigue properties of aluminium 5083 welds revealed limited information. Although the mechanical properties, corrosion behaviour and fatigue properties of this alloy have been studied in depth, the influence of filler wire composition and weld geometry on the fatigue behaviour of fully automatic and semi-automatic welds, and the behaviour of weld joints when simultaneously subjected to a chloride-containing corrosive environment and fatigue loading, have not been investigated in any detail.

This investigation therefore aimed at studying the mechanical properties and corrosion fatigue performance of 5083-H111 aluminium welded using semi-automatic and fully automatic pulsed gas metal arc welding, and ER4043, ER5183 and ER5356 filler wires. The influence of the weld metal and heat-affected zone, weld defects and the weld geometry on the mechanical properties and corrosion fatigue resistance was evaluated. The project also determined the fatigue damage ratio (the ratio of the fatigue life in a NaCl solution to the fatigue life in air) by comparing the S-N curves measured in NaCl and in air for 5083-H111 aluminium in the as-supplied and as-welded conditions.

The background section reviews the relevant literature on the welding of 5083 alloy, their corrosion behaviour in chloride-containing solution, mechanical properties and fatigue behaviour. The research methodology describes experimental procedure followed to characterise the microstructure, their mechanical properties, corrosion behaviour and fatigue properties (in air and in a 3.5% NaCl solution) of 5083-H111 in the as-supplied and as-welded conditions. The results obtained, including weld metal microstructures, hardness profiles, tensile properties, fatigue performance, corrosion behaviour and corrosion fatigue properties in NaCl, are also discussed. Finally, conclusions and recommendations regarding the corrosion fatigue performance of 5083-H111 aluminium alloy welds are provided.

2. Background

Aluminium and its alloys represent an important family of light-weight and corrosion resistant engineering materials. Pure aluminium has a density of only 2.70 g/cm^3, as a result, certain aluminium alloys have better strength-to-weight ratios than high-strength steels. One of the most important characteristics of aluminium is its good formability, machinability and workability. It displays excellent thermal and electrical conductivity, and is non-magnetic, non-sparking and non-toxic.

2.1 Aluminium alloy investigated

Aluminium alloys can be broadly divided into those that are hardenable through strain hardening only, and those that respond to precipitation hardening. Aluminium alloys with the number "5" as first digit in the alloy designation are alloyed with magnesium as primary alloying element. Most commercial wrought alloys in this group contain less than 5% magnesium. A typical chemical composition of such alloy is shown in Table 1.

Alloy	*Al*	*Mg*	*Mn*	*Fe*	*Si*	*Cr*	*Cu*	*Zn*	*Ti*
5083	Balance	4.0-4.9	0.4-1.0	0.4	0.4	0.25	0.1	0.25	0.15

Table 1. Typical chemical compositions of aluminium alloy 5083 (percentage by mass).

2.2 Welding of 5083 aluminium

2.2.1 Pulsed Gas Metal Arc Welding (P-GMAW)

Arc welding is the most widely used process in the shipbuilding, aerospace, pipeline, pressure vessel, automotive and structural industries. In gas metal arc welding (GMAW), the heat required to fuse the metals is generated by an electric arc established between a consumable electrode wire and the workpiece. The electric arc and the molten weld pool are shielded from atmospheric contamination by an externally supplied shielding gas or gas mixture. GMAW may be used in the semi-automatic mode (SA-GMAW), i.e. the filler wire is fed at a constant speed by a wire feeder, while the welder manipulates the welding torch manually, or in the fully-automatic mode (FA-GMAW), i.e. the filler wire is fed continuously at a constant speed, while the torch is manipulated automatically.

With a pulsed power supply, the metal transfer from the tip of the electrode wire to the workpiece during GMAW is controlled. Pulsed current transfer is a spray-type transfer that occurs in pulses at regularly spaced intervals rather than at random intervals. The current is pulsed between two current levels. The lower level serves as a background current to preheat the electrode (no metal transfer takes place), while the peak current forces the drop from the electrode tip to the weld pool. The size of the droplets is approximately equal to the wire diameter. Drops are transferred at a fixed frequency of approximately 60 to 120 per second. As a result, spray transfer can take place at lower average current levels than would normally be the case. Due to the lower average heat input, thinner plates can be welded, distortion is minimized and spatter is greatly reduced. The pulsed GMAW process is often preferred for welding aluminium. The lower average heat input reduces the grain size of the weld and adjacent material and reduces the width of the heat-affected zone (HAZ) [1-3].

The weld penetration, bead geometry, deposition rate and overall quality of the weld are also affected to a large extent by the welding current, arc voltage (as determined by the arc length), travel speed, electrode extension, electrode orientation (or gun angle) and the electrode diameter. Excessive arc voltages or high arc lengths promote porosity, undercut and spatter, whereas low voltages favour narrow weld beads with higher crowns. The travel speed affects the weld geometry, with lower travel speeds favouring increased penetration and deposition rates. Excessively high travel speeds reduce penetration and deposition rate, and may promote the occurrence of undercut at the weld toes [4].

The welding current, arc voltage and travel speed determine the heat input (HI) during welding. This relationship is shown in equation (1);

$$\mathrm{HI} = \eta \frac{VI}{v} \qquad \ldots(1)$$

where: V is the arc voltage (V), I is the welding current (A), v is the travel speed, and η is the arc efficiency factor (typically in the region of 0.7 to 0.8 for GMAW).

The mechanical properties of the welded joint, the weld geometry, occurrence of flaws and level of residual stress after welding depend mainly on the joining process, welding consumable and procedure employed.

2.2.2 Structure of the welds

The filler metal and the melted-back base metal form an admixture. The properties of the weld, such as strength, ductility, resistance to cracking and corrosion resistance, are strongly affected by the level of dilution. The dilution, in turn, depends on the joint design, welding process and parameters used. A more open joint preparation (for example a larger weld flank angle, ϕ , in Figure 1(a)) during welding increases the amount of filler metal used, reducing the effect of dilution. Joint preparations such as single or double V-grooves are often preferred to square edge joint preparations when welding crack susceptible material with non-matching filler metal [5].

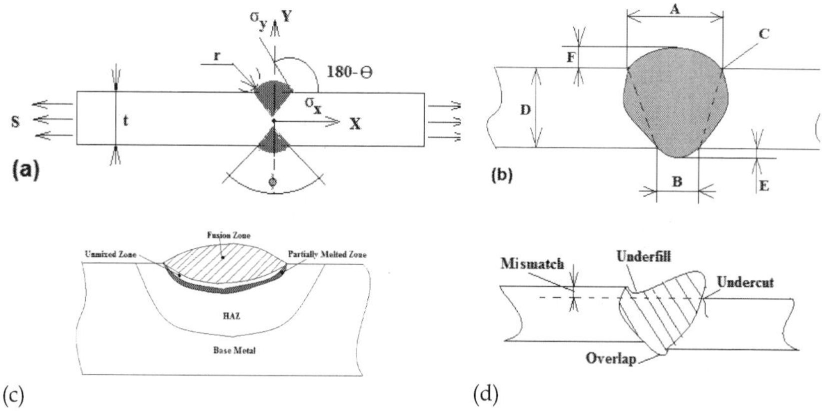

Fig. 1. Schematic illustration of (a) geometrical parameters of a typical butt weld with a double V edge preparation, where r is weld toe radius, ϕ weld flank angle and t plate thickness; (b) geometrical structure of a weld, where A is weld face, B the root of the weld, C weld toe, D the plate thickness or weld penetration, E root reinforcement, and F face reinforcement; (C) compositional structure of a typical weld; and (d) geometric weld discontinuities.

The thermal cycle experienced by the metal during welding results in various zones that display different microstructures and chemical compositions (Figure 1(c)). The fusion zone (composite zone or weld metal) melts during welding and experiences complete mixing to produce a weld with a composition intermediate between that of the melted-back base metal and the deposited filler metal. The unmixed zone cools too fast to allow mixing of the filler metal and molten base metal during welding, and displays a composition almost identical to that of the base metal. The partially melted zone experiences peak temperatures that fall between the liquidus and solidus temperatures of the base metal. HAZ represents the base metal heated to high enough temperatures to induce solid-state metallurgical transformations, without any melting [4].

Most welds contain discontinuities or flaws that may be design or weld related, with the latter category including defects such as undercut, slag or oxide inclusions, porosity, overlap, shrinkage voids, lack of fusion, lack of penetration, craters, spatter, arc strikes and underfill. Metallurgical imperfections such as cracks, fissures, chemical segregation and lamellar tearing may also be present. Geometrical discontinuities, mostly associated with imperfect shape or unacceptable bead contour, are often associated with the welding procedure and include features such as undercut, underfill, overlap, excessive reinforcement and mismatch (Figure 1(d)) [4].

Aluminium welds are also very susceptible to hydrogen-induced porosity. The weld pool may dissolve large amount of hydrogen from the arc atmosphere. On solidification, the solubility of hydrogen decreases and the entrapped hydrogen forms gas porosity. Typical sources of hydrogen contamination are lubricant residues, moisture and the hydrated surface oxide on the base metal or filler wire surface. These defects act as stress concentrations and may lead to rapid fatigue crack initiation if the weld is exposed to fluctuating stresses of sufficient magnitude [5-6].

Most weld flaws can be removed by grinding, machining and/or flush polishing, thereby improving the mechanical properties, corrosion resistance and fatigue properties of the joint. Subsurface flaws, which are more prevalent during SA-GMAW than FA-GMAW, are more difficult to detect and correct.

2.2.3 Weldability of aluminium 5083

Pure aluminium exhibits high electrical conductivity, about 62% of that of pure copper. Very little resistance heating occurs during welding, and high heat inputs are therefore necessitated when joining aluminium and its alloys to ensure complete fusion. Incomplete fusion may also result from the presence of a hydrated aluminium oxide layer that forms spontaneously on exposure to air or water due to the strong chemical affinity of aluminium for oxygen. This layer melts at about 2050°C, significantly above the melting range of aluminium. In order to prevent poor fusion, the aluminium oxide layer needs to be removed prior to or during welding. Suitable fluxes, chemical or mechanical cleaning methods, or the cleaning action of the welding arc in an inert atmosphere (cathodic cleaning) can be used to remove the oxide [4]. The high thermal expansion coefficient of aluminium (about twice that of steel) may result in distortion and high levels of residual stress in the welds, and precautions need to be taken to control distortion to within acceptable limits.

The weldability of aluminium, defined as the resistance of the material to the formation of cracks during welding, is affected by the physical properties (Table 2), chemical composition and prior temper state of the material [4-5]. As long as dilution is controlled to a minimum, these alloys can, however, be welded successfully using non-matching filler metal.

Base Metal	*Chemical Composition, wt %*						*Melting point, °C*	*TC at 25°C, W/m.K*	*EC, % IACS*	*GMAW W*
	Al	*Mg*	*Si*	*Mn*	*Cr*	*Cu*				
5083	Bal.	4.4	-	0.7	0.15	-	574-638	117	29	A

Where TC the thermal conductivity, EC the electrical conductivity, W the weldability and A indicates that the alloy is readily weldable.

Table 2. Typical chemical composition, physical properties and weldability of wrought aluminium alloy 5083 [4].

A dissimilar filler metal with a lower solidus temperature than the base metal is generally employed so that the hardenable base metal is allowed to completely solidify and develop some strength along the fusion line before weld solidification stresses develop. Filler wires containing approximately 5% silicon, such as ER4043, may be used. ER4043 filler wire solidifies and melts at temperatures lower than the solidification temperature of the base metal. Contraction stresses, which could cause cracking, are relieved by the plasticity of the still liquid filler metal, preventing the formation of cracks. The Al-Mg and Al-Mg-Mn filler alloys, such as ER5356 and ER5183, are employed more frequently as welding consumables since these materials provide an optimum combination of mechanical properties, corrosion resistance and crack resistance. The chemical compositions of these filler wires are shown in Table 3.

Aluminium alloys 5083 are normally welded with near-matching filler metal. Consumables, such as ER5356 and ER5183, increase the strength of the weld and reduce the crack sensitivity [5].

Filler Metal	*Chemical Composition, wt %*						*Melting range, °C*
	Al	*Mg*	*Si*	*Mn*	*Cr*	*Ti*	
ER4043	Bal.	-	5.25	-	-	-	574-632
ER5183	Bal.	4.75	-	0.75	0.15	-	579-638
ER5356	Bal.	0.12	-	0.12	0.12	0.13	571-635

Table 3. Chemical composition and melting point of filler metals typically used in joining aluminium alloys [4].

The mechanical properties, fatigue performance and corrosion resistance of the welded joint depend on the filler metal used, and an optimal filler wire for a specific application needs to be selected [5]. Tables 4 and 5 provide guidance on the selection of filler metals for welding 5083 aluminium.

Strength	*Ductility*	*Colour match*	*NaCl corrosion resistance*	*Least cracking tendency*
ER5183	ER5356	ER5183	ER5183	ER5356
	ER5556	ER5356		ER5183

Table 4. Recommended filler metals for welding 5083 [5].

Filler Alloys	*Filler Characteristics*				
	Ease of welding	*As welded strength*	*Ductility*	*Corrosion resistance*	*Colour match*
ER5183	A	A	B	A	A
ER5356	A	-	A	A	A

where: A, B, C and D are relative ratings, with A: best and D: worst.

Table 5. Filler metal selection for 5083 welds [4].

As shown in Tables 4 and 5, filler metal selection plays a major role in determining the corrosion resistance of the welded joint. The corrosion resistance is also affected by the cleanliness of the alloys, the chemical and physical environment and the welding process. In the as-welded condition, however, the weld metal and HAZ, and any welding defects, are most likely to become preferential corrosion sites.

2.3 Corrosion resistance of 5083 aluminium alloy

Aluminium and its alloys generally exhibit good corrosion resistance in a wide range of environments. The corrosion resistance of aluminium is derived from a thin, hard and compact film of adherent aluminium oxide that forms spontaneously on the surface. This thin oxide film, only about 5 nm (or 50 Å) in thickness, grows rapidly whenever a fresh aluminium surface is exposed to air or water. Aluminium oxide is dissolved in some chemical solutions, such as strong acids and alkalis, leading to rapid corrosion. The oxide film is usually stable over a range of pH values between 4.0 and 9.0, with water soluble species forming at low pH (Al^{3+}) and high pH (AlO_2^-) values (see Pourbaix diagram, Figure 2(a)) [7-8].

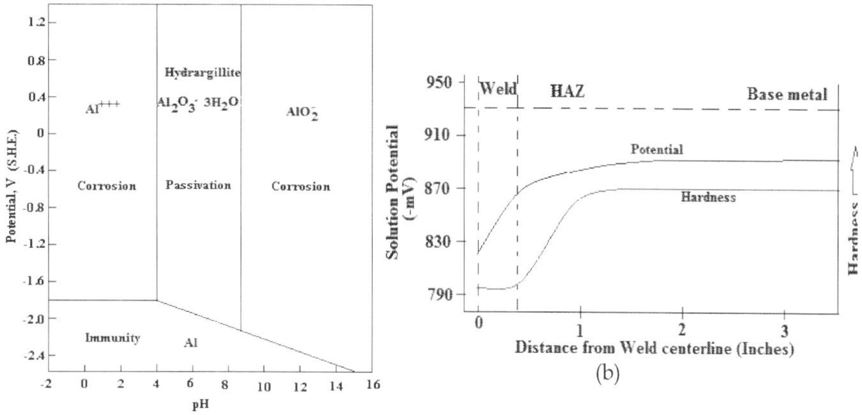

Fig. 2. (a) Pourbaix diagram for aluminium with areas representing the hydrated oxide film of hydrargillite($Al_2O_3.3H_2O$), and the dissolved species Al^{3+} and AlO_2^- at 25°C (potential values are given relative to the standard hydrogen electrode). (b) Schematic illustration of the change in solution potential and hardness in the weld metal and HAZ of 5083.

The corrosion resistance of 5083 aluminium is normally reduced by welding. An area within about 25 mm on either side of the weld tends to exhibit lower corrosion resistance [4].

2.3.1 Corrosion of 5083 welds

The passive oxide films, spontaneously formed on aluminium alloy surfaces, are however, susceptible to localised breakdown at the exposed surface or at discontinuities, which result in high dissolution rates of the underlying metal (pitting corrosion). The chemical composition of the weld metal and HAZ and the presence of inclusions, precipitates and second phases in the welded joint, however, produce slightly different electrode potentials in the presence of an electrolyte (Figure 2(b)). Selective or localized corrosion is therefore possible when the base metal and the weld metal or second phases possess significantly different electrode potentials. A galvanic effect can be created, with the more active region corroding preferentially to protect the more noble region with which it is in contact [4-5,8].
Welds produced by the GMAW process appear to be less resistant to pitting corrosion in salt water solutions than solid state friction stir welds, as reported by Moggiolino and Schmid [9]. Preferential attack occurs in the narrow interface between the weld bead and the HAZ, or between the HAZ and the base metal. As a result of the high peak temperatures experienced by the high temperature HAZ adjacent to the fusion line, grain coarsening, recrystallization and partial dissolution of intermetallic strengthening precipitates occurring during welding.

2.3.2 Mechanism of pitting corrosion in 5083 alloy

Pitting corrosion is a form of localized corrosion that occurs in environments in which a passive surface oxide film is stable. Pits initiate due to local rupture of the passive film or the presence of pre-existing defects, and then propagate in a self-sustaining manner. Localized corrosion can initiate as a result of the difference in corrosion potential within a localized galvanic cell at the alloy surface. These micro-galvanic cells can form at phase boundaries, inclusion/matrix interface regions, and at insoluble intermetallic compounds within the matrix.

The aluminium oxide passive film consists of two superimposed layers with a combined thickness between approximately 4 and 10 nm. The first compact and amorphous layer, in contact with the alloy, forms as soon as the material comes into contact with air or water, within a few milliseconds, according to the reaction shown in equation (2):

$$2Al + 3/2\,O_2 \rightarrow Al_2O_3 \quad (\Delta G = -1675\ kJ) \qquad ...(2)$$

The second layer grows over the initial film due to a reaction with the corrosive environment, likely by hydration (reaction with water or moisture). The second layer is less compact and more porous, and may react with the corrosive environment (Figure 3(a)) [8].

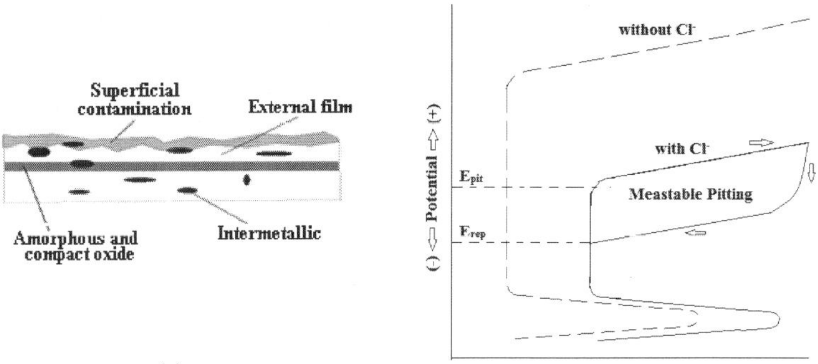

(a) (b)

Fig. 3. Schematic illustration of (a) typical structure of the aluminium oxide passive layer; and (b) polarization diagram, illustrating the position of the critical pitting potential, E_{pit}, and the repassivation potential (or protection potential), E_{rep}.

As shown in Figure 3(b), localized breakdown of the passive film initiates above the critical pitting potential (E_{pit}). The pitting potential is often stated to quantify the resistance of a material to pitting corrosion, and represents the potential in a particular solution above which stable pits may form. More noble pitting potential values (E_{pit}) signify increased resistant to pitting corrosion. The presence of aggressive anionic species, such as chloride ions (which increase the potentiostatic anodic current at all potentials) increase the likelihood of pitting corrosion.

The value of E_{pit} in NaCl solution remains unaffected by the dissolved oxygen concentration and moderate temperature variations (0°C to 30°C). At temperatures above 30°C, the pitting corrosion rate increases considerably. A rough surface finish increases susceptibility to pitting corrosion and reduces the pitting potential. The repassivation or protection potential (Figure 3(b)) represents the minimum potential at which existing pits can propagate, but new pits cannot form.

As described above, pitting corrosion typically develops in the presence of chloride ions (Cl^-). The chloride ions are adsorbed on the aluminium oxide layer, followed by rupture of the oxide film at weak points, with formation of micro-cracks that are a few nanometres wide. At the same time, oxygen is reduced on cathodic sites and rapidly oxidizes aluminium by forming an intermediate complex chloride, $AlCl^{4-}$, in areas associated with cracks in the oxide film [7-8].

As the pit deepens, the rate of transport of ions out of the pit decreases. The pit current density therefore tends to decrease with time, owing to an increase in the pit depth and ohmic potential drop. Repassivation may also occur if dissolution rate of the bottom of the pit is insufficient to replenish the loss of aggressive environment due to reaction, and the pit may stop growing after few days. Pitting can continue on fresh sites. Al^{3+} ions, highly concentrated in the bottom of the pit, diffuse towards the pit opening and react with the more alkaline solution on the plate surface, facilitating the formation of $Al(OH)_3$. Hydrogen micro-bubbles formed in the pit may transport the $Al(OH)_3$ to the pit opening where it forms an insoluble deposit that appears as white eruptions around the pit surface. The formation of positively charged Al^{3+} ions in the bottom of the pit may also attract Cl^- ions towards the underside of the pit, encouraging the formation of $AlCl^{4-}$ (through the reaction $Al^{3+}+Cl^-+6e^-\rightarrow AlCl^{4-}$), as shown in Figure 4. The accumulation of $Al(OH)_3$ forms a dome at the pit surface which progressively blocks the pit opening. This can hinder the exchange of Cl^- ions which may progressively retard or even arrest pit growth. A corrosion pit may therefore be considered as a local anode surrounded by a matrix cathode. Once pitting corrosion has initiated, pit growth becomes sustainable at lower potentials than the pitting potential [7-8,10].

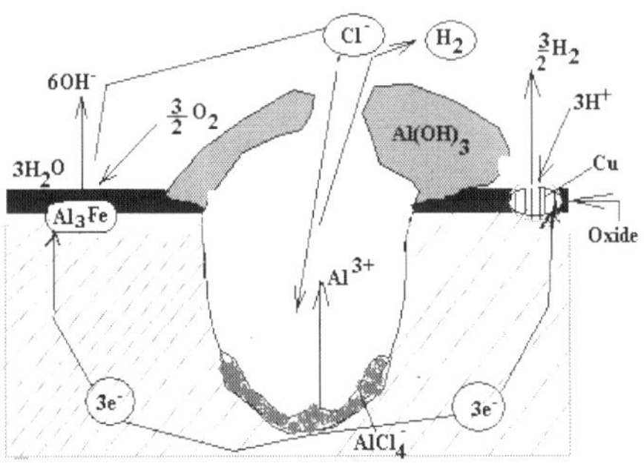

Fig. 4. Schematic illustration of the mechanism of pitting corrosion in aluminium.

Cold working generally reduces the corrosion resistance of the magnesium-alloyed grades, as the β-Al_3Mg_2 phase may precipitate on grain boundaries and dislocations, increasing susceptibility to stress corrosion cracking. Inclusions, impurities, pores, vacancies, dislocation walls and grain boundaries may generate galvanic cells in 5083 alloy. Cored structures promote galvanic interaction and point defects are usually more anodic than the surroundings [8,10].

Intermetallic phases, such as Al_3Mg_2, Al_3Mg_5 and Mg_2Si, are anodic with respect to the 5083 alloy matrix (Table 6), and promote rapid localized attack through galvanic interaction. Less electronegative intermetallic phases, such as Al_3Fe and Al_6Mn, are cathodic with respect to the 5083 aluminium matrix, leading to preferential dissolution of the alloy matrix [8,11-12].

Metal, alloy or intermetallic phase	*Potential, (V)*
Al_8Mg_5	-1.24
Mg_2Si (intermetallic phase)	-1.19
Al_3Mg_2 (intermetallic phase)	-1.15
Mg	-0.85
Al_6Mn (intermetallic phase)	-0.80
Al5083, Al5183	-0.78
Al_3Fe (intermetallic phase)	-0.51

Table 6. Relative electrochemical potentials for aluminium alloys and typical intermetallic phase in a NaCl solution (Potential versus standard calomel electrode) [8-13].

The pit density (spacing), the pit size and pit depth can be used to evaluate the pitting corrosion resistance of alloys. A pit depth measurement using an optical microscope is often the preferred way of evaluating pitting corrosion. The pitting factor *(p/d)* may also be used, where *p* is the maximum pit depth and *d* is the average pit depth. The pit depth increases, not only with time, but also with surface area, and can be estimated using equation (11):

$$d_1=Kt_1^{1/3} \qquad ...(3)$$

where: d_1 is the pit depth at time t_1 and k is a constant.

The presence of a weld often promotes corrosion due to changes in the local microstructure and precipitate distribution, and the increased likelihood of defects. Welding also affects the mechanical properties of the aluminium alloy in the vicinity of the joint. Any localized change in mechanical properties can, in turn, influence the corrosion behaviour and the fatigue properties of the material.

2.4 Mechanical properties of welded 5083 aluminium

Aluminium alloys 5083, produced via ingot casting, and cold working, contain precipitates that interact with moving dislocations, thus increasing strength at room temperature. When these alloys are welded, however, the precipitates dissolve and/or coarsen, reducing the mechanical strength significantly. This reduction in mechanical properties can be attributed to annealing and recrystallization, grain growth and precipitate dissolution and/or overageing; and uncontrolled grain boundary precipitation on cooling [14].

The HAZ of 5083 aluminium alloy is completely annealed and recrystallized during welding. The effect of any prior work hardening is lost when such an alloy is exposed to a temperature above 343°C for only few seconds. A reduction in hardness is therefore observed in the HAZ [14]. The degree of softening is mainly affected by the heat input, the welding technique, the size of the workpiece and the rate of cooling. Low heat input levels reduce the time at temperature and increase the cooling rate, thereby minimizing the degree of softening in the HAZ. Pulsed GMAW has the advantage of ensuring good penetration and adequate fusion at lower average heat input levels [4]. The amount of grain growth is reduced and the width of the HAZ minimized.

Welds in 5083 alloy display reasonable ductility and high strength when near-matching filler metals are used (Table 7).

Base Alloy	Filler Metal	Ultimate Tensile Strength, MPa	Minimum. Yield Stress, MPa	Tensile Elongation, % (50.8mm gauge)	Free Bending Elongation, %
5083	5183	276-296	165	16	34
5083	5356	262-241	117	17	38

Table 7. Mechanical properties of butt joints in aluminium 5083 welded using ER5183 and ER 5356 filler metals [4].

2.5 Fatigue behaviour of welds

Fatigue is a highly localised and permanent structural change involving the initiation and propagation of a crack under the influence of fluctuating stresses at levels well below the static yield stress required to produce plastic deformation. Under these conditions, fatigue cracks can initiate near or at discontinuities lying on or just below the free surface. These discontinuities may be present as a result of mechanical forming, or welding and cause stress concentrations in the form of inclusions, second phases, porosity, lack of fusion, lack of penetration, weld toe geometry, shape changes in cross section, corrosion pits and grain boundaries [4]. The fluctuating applied stress (amplitude stress (S_a)) leads to plastic deformation (long-range dislocation motion) which produces slip steps on the surface. The dislocations may concentrate around obstacles, such as inclusions or grains boundaries, promoting fracture of inclusions or second phase particles, decohesion between the particles and the matrix, or decohesion along grain boundaries. The lowest fatigue strength is usually associated with the highest stress concentration at the metal surface. The weld toe represents a sharp stress concentration in transversely loaded welds (Figure 5) and fatigue cracks often initiate at the weld toe, followed by propagation into the base metal. Uneven root profiles can cause crack initiation at the weld root, followed by propagation into the weld metal. Stop/start positions and weld ripples can act as stress concentrations in longitudinal welds. Lack of penetration and undercut are severe stress raisers and can accelerate fatigue crack initiation, whereas internal defects (such as porosity and slag inclusions) usually only initiate fatigue cracks if surface stress concentrations are removed. Geoffroy *et al.* [6] confirmed that poor weld quality causes a significant reduction in fatigue life.

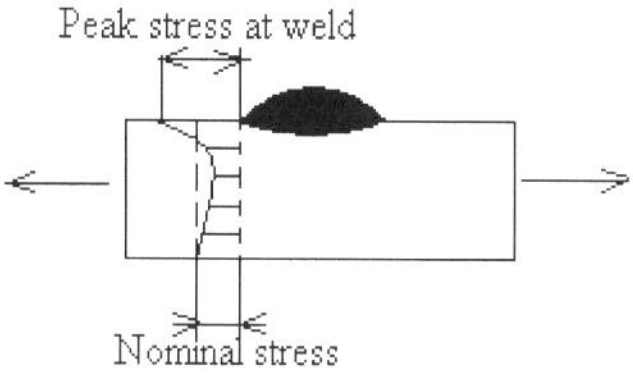

Fig. 5. Stress concentration caused by the weld toe geometry.

The formation of residual stresses in welds is a consequence of the expansion and contraction of the weld metal and base metal close to the heat source and the restraining effect of the adjacent base alloy at lower temperatures. On cooling, high tensile residual stresses in the weld metal and HAZ are balanced by compressive residual stresses in the adjacent plate material. The presence of high tensile residual stresses in the weld metal and HAZ has two important consequences. First, fatigue failure can occur under loading conditions that, nominally, introduce compressive stresses, and second, the fatigue strength of welded joints is often governed by the applied stress range regardless of the nominal applied stress ratio. Due to the lower tensile strength of welds in aluminium alloys, the applied stress ratio may influence the fatigue strength of the joint to a limited extent, but fatigue design is usually based on the stress range and a single *S-N* curve represents the performance of a given welded joint for any minimum/maximum ratio of load input [15].

2.6 Corrosion fatigue performance of 5083 aluminium

The fatigue behaviour of magnesium-alloyed aluminium after welding is determined by the weld microstructure and mechanical properties. Any stress concentration caused by a second phase particle of identifiable size and shape can nucleate a crack in a non-corrosive environment. This effect is enhanced in a corrosive environment where corrosion pits are often associated with second phase particles in the matrix. Such a combination of a pit and a second phase particle may present a greater stress concentration than a pit or particle alone. Precipitates, second-phase particles, pores and grain boundaries within the matrix facilitate the nucleation and growth of corrosion pits in aggressive media and promote fatigue crack initiation and growth.

Under corrosion fatigue conditions, the shape of the fatigue loading cycle, the frequency and any periods of rest have a considerable influence on the fatigue life. The growth rate of corrosion pits increases with increasing stress amplitude and cyclic stress frequency [6]. The corrosion fatigue fracture surface may be characterised as describe below.

2.6.1 Features of corrosion fatigue fracture surfaces

The characterization and understanding of the kinetics and mechanisms of corrosion fatigue are indispensable to the service life prediction, fracture control and development of fatigue resistant alloys. Corrosion fatigue is characterized as brittle failure caused by the combined effect of a fluctuating stress and a corrosive environment. The principal feature of this fracture mode is the presence of corrosion products and beach marks on the fracture surface [16].

In corrosion fatigue cracking (CFC), anodic dissolution at the root of the crack is facilitated by repeated rupture of the passive film at the crack tip by fatigue processes and the subsequent repassivation of the newly exposed metal surface. The mechanism of anodic dissolution may involve rupture of the brittle oxide layer, selective dissolution or dealloying, and/or corrosion tunnelling. The growth rate of a crack during environmentally-assisted corrosion fatigue is therefore controlled by the rate of anodic dissolution, the rupture of the oxide film, the rate of repassivation, the mass transport rate of the reactant to the dissolving surface and the flux of dissolved metal cations away from the surface. Anodic dissolution (commonly referred to as active path dissolution, slip dissolution, stress/strain enhanced dissolution or surface film rupture/metal dissolution) is the CFC mechanism by which the crack growth rate is enhanced by anodic dissolution along susceptible paths that are anodic to the surrounding matrix. Such susceptible paths can include grain boundaries, strained metal at the crack tip and the interface between second-phase particles and the matrix.

In this mechanism, a slip step is formed at the crack tip under fatigue loading conditions and fractures the protective surface oxide film. The freshly exposed metal surface at the crack tip reacts with the aggressive solution and partly dissolves until the crack tip repassivates and the oxide layer is restored. This process repeats during successive fatigue loading cycles as slip-steps break the oxide layer and fresh material is exposed to the corrosive environment. Factors affecting this process are mechanical variables (frequency, stress and waveform of the loading cycle), geometrical variables (crack size, weld geometry and specimen thickness), metallurgical variables (alloy chemical composition, microstructure, and the strength and toughness of the material), and environmental variables (electrolyte, corrosion species concentration and temperature) [8].

One of the most important parameters affecting the susceptibility to CFC is the loading frequency. The lower the frequency of the applied loading cycle, the higher the crack propagation rate per cycle (da/dN). Very high frequencies can completely eliminate the effect of the corrosive environment on fatigue by minimizing the interaction time between the environment and the crack tip [17].

As described earlier, the presence of corrosion pits induces stress concentrations responsible for promoting crack initiation. As the pit depth increases, the stress levels in the surrounding material increase. When the stress level reaches a threshold value (determined by the alloy microstructure and the corrosive environment), a crack is initiated. As reported by Pidaparti *et al.* [17], closely spaced pits and longer exposure times increase the stress levels in the material surrounding the corrosion pits, promoting crack initiation.

The influence of corrosion pits on fatigue crack initiation has been reported by a number of authors for various aluminium alloys. Chlistovsky *et al.* [18] showed that the fatigue life of 7075-T651 alloy is significantly reduced in 3.5% NaCl corrosive environment. This reduction is attributed to the initiation of cracks from stress concentrations caused by pit formation, and a combination of anodic dissolution at the crack tip and hydrogen embrittlement. A similar observation was made by Chen *et al.* [19] who studied the corrosion fatigue behaviour of aluminium 2024-T3. Their fractographic study revealed that fatigue cracks nucleated from one or two of the larger pits observed on the surface.

In order to study the corrosion fatigue behaviour of 5083 alloy in air and corrosive environment, fatigue testing has to be performed.

2.6.2 Corrosion fatigue testing

Laboratory fatigue testing can be categorized as crack initiation (fatigue life) testing or crack propagation testing. The loading mode can be direct or axial, plane bending, rotating beam, alternating torsion or a combination of these modes.

It is necessary for general design purposes to have fatigue data for positive and negative values of maximum stress (S_{max}), minimum stress (S_{min}) and S_m to cater for varying stress conditions. In this respect, a characterizing parameter known as the stress ratio, R, (given by, $R = S_{min} / S_{max}$) is often used in presenting fatigue data.

Fatigue test data is typically presented in the form of a S_a-log N_f curve, known as Wöhler's curve. Basquin [20] showed that a typical S_a-log N_f plot can be linearized with full log coordinates, thereby establishing the exponential law of fatigue shown in equation (4).

$$N (S_a)^p = C \quad (4)$$

where: N is the number of cycles, S_a is the stress amplitude, and C and p are empirical constants.

To perform corrosion fatigue testing, an environmental chamber of glass or plastic containing the electrolyte is introduced during testing. To minimize galvanic effects, the specimen must be gripped outside of the test solution. The solution is circulated through the corrosion chamber, which is sealed to the specimen. In this testing method, factors that influence the number of cycle to failure include the stress amplitude (S_a), the stress ratio (R), chemical concentration of dissolved species such H^+, O^{2-}, and other ions, the alloy properties (such as yield stress, hardness and microstructure), the waveform of the loading cycle, the test temperature and the electrolyte flow rate [17].

3. Research methodology

In order to evaluate the fatigue properties of aluminium 5083-H111 welds in the ambient atmosphere and in a 3.5% NaCl solution, the experimental procedure described below was followed during the course of this investigation.

3.1 Welding procedure

Flat aluminium sheets (with initial dimensions of 2000 mm long, 120 mm wide and 6.35mm thick) of aluminium alloy 5083-H111 (with chemical compositions given in Table 8) were supplied for examination. The -H111 temper designation in the case of the non-hardenable 5083 alloy refers to material strain-hardened to a level below that required for a controlled -H11 temper (corresponding to one eighth of the full-hard condition). These samples were joined using, pulsed SA-GMAW or FA-GMAW, as shown in Figure 6. Prior to welding, the plates for SA-GMAW were prepared with a double-V edge preparation, degreased with acetone and preheated to approximately 100°C. The FA-GMAW welds were performed with a square edge preparation. Three different aluminium filler wires were used, namely ER4043 (Al-Si), ER5183 (Al-Mg) and ER5356 (Al-Mg) (with typical chemical compositions given in Table 9).

Element %	*Al*	*Mg*	*Mn*	*Fe*	*Si*	*Cr*	*Cu*	*Zn*	*Ti*	*Others total*
5083-H111	Balance	3.66	0.39	0.40	0.22	0.14	0.04	0.03	0.02	<0.001

Table 8. Chemical compositions of the 5083-H111 aluminium plate material (percentage by mass).

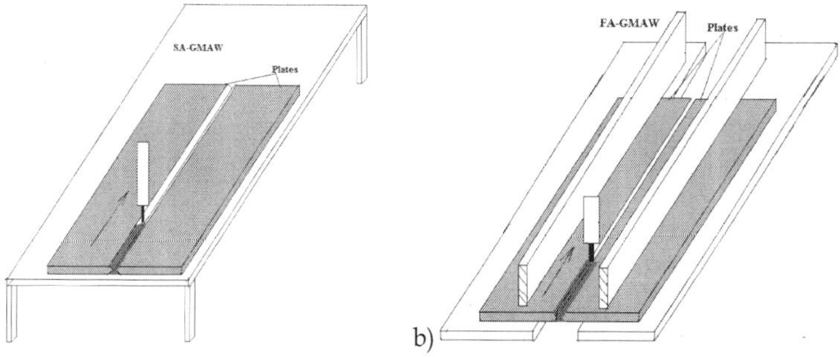

a) b)

Fig. 6. Schematic illustration of the pulsed GMAW process: (a) SA- GMAW; and (b) FA-GMAW.

Element %	*Al*	*Mg*	*Mn*	*Fe*	*Si*	*Cr*	*Cu*	*Zn*	*Ti*	*Others total*
ER4043	Balance	0.05	0.05	0.80	4.5-6.0	Not specified	0.30	0.10	0.20	Be 0.0008%
ER5183	Balance	4.3-5.2	0.50-1.0	0.40	0.40	0.05-0.25	0.10	0.25	0.15	Not specified
ER5356	Balance	4.5-5.5	0.05-0.2	0.40	0.25	0.05-0.20	0.10	0.10	0.06-0.20	Be 0.0008%

Table 9. Typical chemical compositions of filler wires (percentage by mass, single values represent minimum levels).

5083-H111 aluminium plates were welded in horizontal position using argon shielding gas. The welding parameters were selected to ensure a spray transfer mode for all the welds, and are given in Table 10.

Parameters	*Arc voltage*	*Welding current*	*Wire feed rate*	*Wire diameter*	*Nozzle to plate distance*	*Travel speed*	*Torch angle*	*Gas flow rate*
Unit	*V*	*A*	*m/min*	*mm*	*mm*	*m/min*	*Degrees*	*l/min*
SA-GMAW	24-29	133-148	6.1-7.6	1.2-1.6	15-20	0.8-1	60-80	18-33
FA-GMAW	20-23	133-148	6.1-7.6	1.2-1.6	15-20	0.4-0.6	60-80	19-28

Table 10. Measured pulsed gas metal arc welding process parameters.

3.2 Material characterization

In order to analyse the aluminium samples and to quantify the material properties that may influence fatigue resistance, the microstructures of the as-machined and as-welded specimens were analysed, the hardness was measured and tensile tests were performed. The corrosion resistance in 3.5% NaCl solution was evaluated using immersion testing.

3.2.1 Microstructural analysis

As-supplied and as-welded samples were sectioned and machined to produce rectangular fatigue and tensile specimens, with dimensions shown schematically in Figure 7. Samples were removed for microstructural examination in the long transverse (LT) direction, longitudinal (LD) direction and short transverse (ST) direction (Figure 7). Samples were prepared for microstructural analysis and etched using Keller's reagent as described in ASTM standard E340 [17]. The metallographic samples were examined with an inverted optical microscope (with Image-Pro PLUS 5.1™ or IMAGEJ™ image analysis software), and a scanning electron microscope (SEM) equipped with Energy Dispersive X-ray Spectroscopy (EDS) capabilities. The grain sizes were determined using the line intercept method.

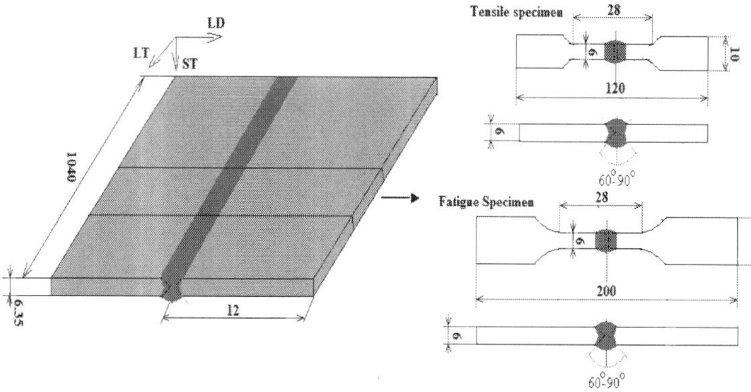

Fig. 7. Dimensions of the tensile and fatigue specimens machined from the welded plates.

3.2.2 Hardness measurements

In order to perform hardness measurements, machined specimens, in the as-supplied and welded condition, were wet-ground and polished using 1 µm diamond suspension, followed by final polishing using 50 nm colloidal silica, as described in ASTM standard E3 [21]. As-welded specimens were ground flush and polished to allow hardness measurements on the LT-LD plane (Figure 7).
Vickers hardness and Vickers micro-hardness tests were then performed according to the requirements of ASTM standards E92 and E384 [22]. An applied load of 100 grams and a holding time of 10 seconds were employed for the micro-hardness measurements. A hardness profile from the centreline of the weld, through HAZ, to the unaffected base metal was measured at 0.05 to 0.1 mm intervals.

3.2.3 Tensile testing

Tensile tests were performed according to ASTM standard E8-04 [13, 23], on unwelded, as-welded and dressed welded specimens. The machined specimens (Figure 7) were wet-ground flush in the longitudinal direction (LD) to remove all machining marks for unwelded and weld reinforcing for dressed weld specimens. Undressed welded specimens were wet-ground without changing the weld toe geometry. An INSTRON™ testing machine equipped with *FASTTRACK2*™ software was used to axially stress specimens at a cross head speed of 3.0 mm/min. The 0.2% offset proof stress, ultimate tensile strength and percentage elongation of unwelded and welded specimens were collected for comparison and evaluation.

3.2.4 Corrosion testing

Machined specimens for corrosion testing, in the as-supplied and as-welded condition, were wet-ground and polished. These specimens were cleaned and dried to remove dirt, oil and other residues from the surfaces. Immersion tests were then performed in NaCl solution using a Plexiglas corrosion cell (Figure 8) with a volume of 25 litres of salt water (3.5% NaCl by weight), according to the requirements of ASTM standards G31 [24] and G46 [25]. The 3.5% NaCl simulated sea water was prepared by dissolving 3.5 ± 0.1 parts by weight of

NaCl in 96.5 parts of distilled water. The pH of the salt solution, when freshly prepared, was within the range 6.9 to 7.2. Dilute hydrochloric acid (HCl) or sodium hydroxide (NaOH) was used to adjust the pH during testing. The ambient test temperature varied from 16°C to 27°C. Fresh solution was prepared weekly.

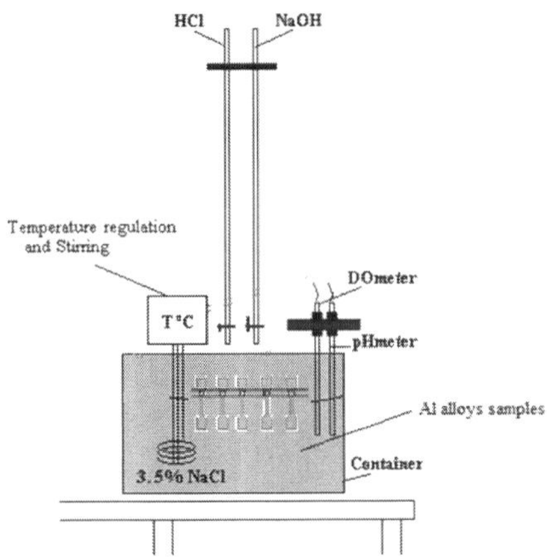

Fig. 8. Schematic illustration of the immersion test in 3.5% NaCl solution.

After a specific time of exposure the specimens were gently rinsed with distilled water and then cleaned immediately to prevent corrosion from the accumulated salt on the specimen surface. Loose products were removed by light brushing in alcohol. The specimens were then immersed in a 50% nitric acid solution for 2 to 4 minutes, followed by immersion in concentrated phosphoric acid for another 5 minutes, to remove bulky corrosion products without dislodging any of the underlying metal. The specimens were then cleaned ultrasonically and dried. The corroded specimens were examined after cleaning, to identify the type of corrosion and to determine the extent of pitting. The samples were inspected visually and microstructurally using Optical Microscope (OM) and SEM. One of the parameters used to quantify the pitting susceptibility was the pit depth, measured using the microscopic method described in ASTM standard G46 [26]. A single pit was located on the sample surface and centred under the objective lens of the microscope at low magnification. The magnification was increased until most of the viewing field was taken up by the pit. The focus was adjusted to bring the lip of the pit into sharp focus and the initial reading was recorded from the fine-focus adjustment. The focus was then readjusted to bring the bottom of the pit into sharp focus and the second reading taken. The difference between the final and the initial readings represents the pit depth. For comparison purposes, photographs and data on the pit sizes and depths were collected for pitting susceptibility evaluation.

3.2.5 Fatigue life assessment

Specimens were fatigue tested in air and in 3.5% NaCl simulated seawater environment using the crack initiation or fatigue life testing method. The specimen was subjected to number of stress cycles (stress controlled, *S-N*) required to initiate and subsequently grow the fatigue crack to failure for various stress amplitudes.

3.2.6 Fatigue testing in air

The axial fatigue life testing method was used to determine the fatigue properties of the specimens, as it takes into account the effect of variations in microstructure, weld geometry, residual stress and the presence of discontinuities.

The machined fatigue specimens (Figure 7) were ground flush and polished in the longitudinal direction to dress some of the welds. This negated the effect of the weld geometry on the fatigue resistance of the dressed welds. Undressed welded specimens were wet-ground in such a way that the weld toe geometry was not changed. The fatigue tests were performed using a symmetric tension-tension cycle (with a stress ratio $R = 0.125$) to keep the crack open during testing. A constant frequency of 1 Hz was used for all fatigue tests and the number of cycles to failure (N_f) was recorded for each specimen. To ensure repeatability, three to six tests were performed at each stress amplitude, depending on the quality of the weld. The number of cycles recorded to failure was then statistically. The fatigue tests in ambient air were performed at temperatures ranging between 17°C and 21°C and at relative humidity levels between 35.7 and 70.6% RH (relative humidity). INSTRON™ testing machines, equipped with calibrated load transducers, data recording systems and FASTTRACK™ software, were used to fatigue specimens to failure under amplitude stress control, as required by ASTM standard E466-02 [26]. Welded specimens were inspected before testing and any specimens with visual welding defects, such as large pores, underfill or excessive undercut, were discarded. The fatigue specimens were cleaned with ethyl alcohol prior to testing to remove any surface oil, grease and fingerprints. Care was taken to avoid scratching the finished specimen surfaces.

Following testing, the S-N_f curve (represented as stress amplitude-log N_f) was determined from the median number of cycles to failure at each stress level. The fracture surfaces were examined using a low magnification stereo microscope and a scanning electron microscope to reveal the primary crack initiation sites and mode of fracture.

3.2.7 Corrosion fatigue testing in 3.5% NaCl simulated seawater

A corrosion environment consisting of 3.5% NaCl in distilled water was used with the axial fatigue life testing method to investigate the effect of pitting corrosion on fatigue life. The corrosion chamber was designed and manufactured from Plexiglas (Figures 9(b) and 10) in such a way that the specimen was gripped outside the chamber (to prevent galvanic effects) and the chamber was sealed by rectangular rings away from the high-stress gauge section. The NaCl solution was re-circulated from 25 litre storage containers at a constant flow rate by means of a peristaltic pump.

The dissolved oxygen (DO) content, NaCl solution flow rate, pH, temperature, stress amplitude (maximum and minimum stress) and frequency were controlled, as shown in Figures 9(a) and 11. A frequency of 1 Hz was used to increase the interaction time between the specimen and the solution. The measured DO content varied between 7 and 8 ppm (parts per million) and the temperature between 17°C and 21°C during testing. The number of cycles to failure (N_f) was recorded for each stress amplitude (S) at the end of the test.

Following testing, the S-N_f curve was determined from the median number of cycles to failure at each stress level. In order to compare the fatigue resistance in air to that in NaCl, the damage ratio, which is the ratio of the fatigue life in the 3.5% NaCl solution to the fatigue life in air ($N_{f\ NaCl}/N_{f\ Air}$), was calculated and presented as a curve of stress amplitude against $N_{f\ NaCl}/N_{f\ Air}$.

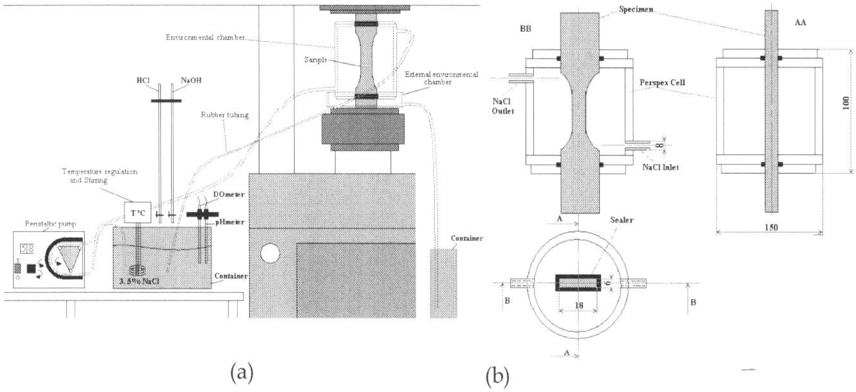

Fig. 9. Schematic illustration of the (a) experimental set-up used for corrosion fatigue testing; (b) corrosion chamber design.

Fig. 10. The experimental set-up used for corrosion fatigue testing in a NaCl solution.

4. Results and discussion

The major findings of this investigation are discussed below.

4.1 Metallographic investigation of 5083-H111 aluminium

The microstructure of the 5083-H111 in the as-supplied condition is shown in Figures 11. Microstructural analysis reveals coarse second-phase particles and fine grain boundary precipitates. The microstructure of plate material appears more equiaxed with an average grain diameter of 24.0 μm (standard deviation of 4.19 μm). Coarse second-phase particles and finer grain boundary precipitates are also evident.

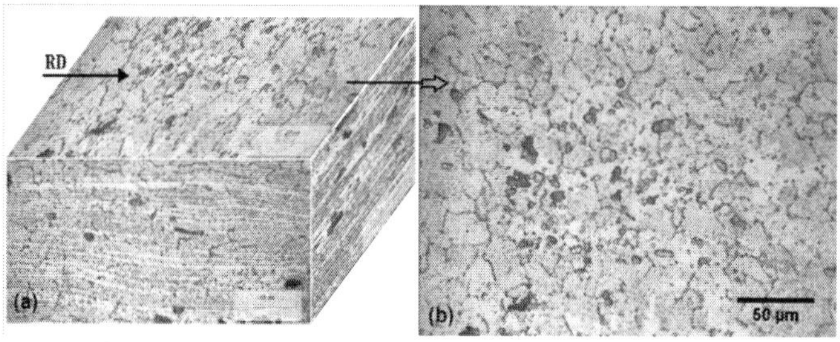

Fig. 11. Microstructures of 5083-H111 aluminium (in the as-supplied condition): (a) in three dimensions, (b) relative to the rolling direction (RD).

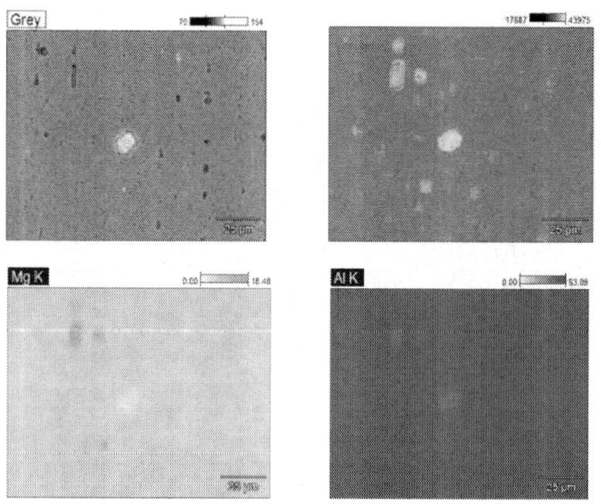

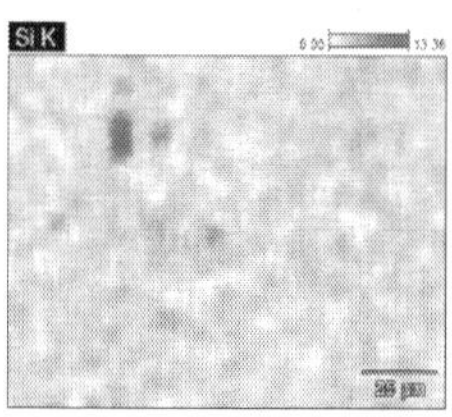

Element	*Wt.%*	*Error*
Mg K	4.14	+/-0.06
Al K	94.97	+/-0.22
Si K	0.26	+/-0.04
Mn K	0.63	+/-0.05
Total	100.00	

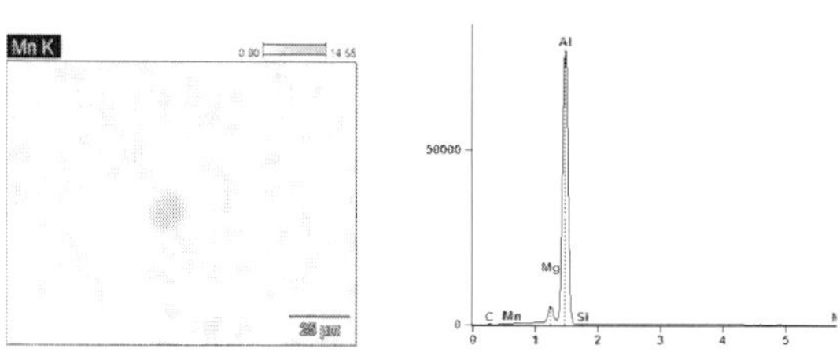

Fig. 12. SEM-EDS analysis of second phase particles observed in 5083-H111 in the as-supplied condition.

In order to identify the second phase particles observed in the microstructures, the EDS/SEM and elemental maps were constructed in the vicinity of a number of these particles. Typical elemental maps and EDS analyses of the second-phase particles are shown in Figures 12. The SEM-EDS elemental maps (voltage: 20kv and working distance: 10mm) suggest the presence of two types of particles. A coarse Mg-rich particle (depleted in Mg and Al and slightly enriched in Si) was identified as the Al_6Mn intermetallic phase, whereas smaller particles that appear to be enriched in Mg and Si were identified as an Al-Mg-Si intermetallic phase.
SA-GMAW weld, Figure 13(a), is full penetration joint welded from both sides. Considerable weld reinforcement and some porosity are evident on the macrograph. FA-GMAW welds, Figure 13(b), is full penetration joints welded from one side only, with a smooth profile and some evidence of misalignment and undercut at the weld root. The HAZ (Figure 13(c)) has coarser grain size than the base metal, with coarse second-phase particles, predominantly on grain boundaries. The HAZ grain structures of the SA-GMAW welds appear coarser than those of the FA-GMAW.

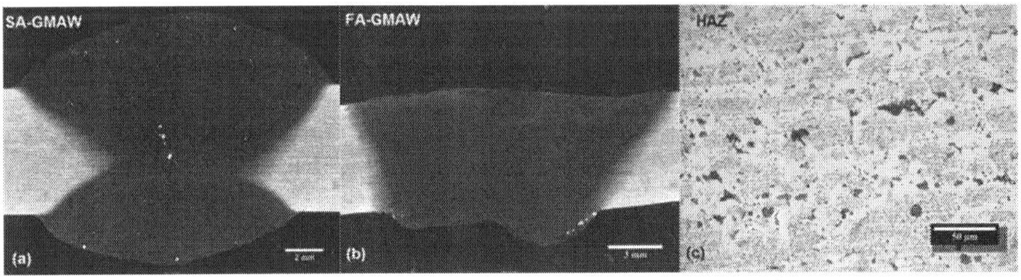

Fig. 13. Representative of 5083-H111 welds: (a) SA-GMAW weld; (b) FA-GMAW weld; and (c) HAZ.

Microstructural examination of the welds (Figure 14(a) to (d)) confirmed the presence of porosity in SA-GMAW welds, and also revealed some lack-of-fusion defects and microcracks in the weld metal. Although all samples with visual welding defects were omitted from mechanical testing, samples with internal flaws and defects were not excluded.

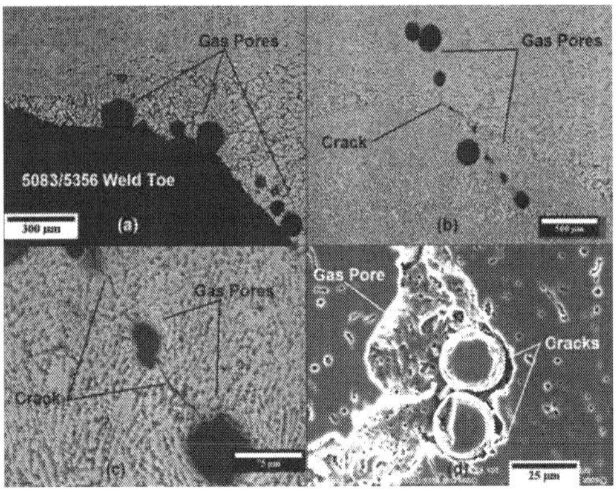

Fig. 14. Defects typically observed in 5083/ER5356 SA-GMAW weld: (a) gas pores, (b)-(d) gas pores and cracks.

Typical optical micrographs of the weld metal welded with ER5356, or ER5183 or ER4043 filler wire are shown in Figures 15(a) to (c). The weld microstructures appear dendritic in structure, characterized by an Al-rich matrix and second phases, present as interdendritic films in the case of ER4043, and as more spherical precipitates in the case of ER5356 and ER5183. Pulsed SA-GMAW welds generally displayed coarser grain structures than FA-GMAW welds.

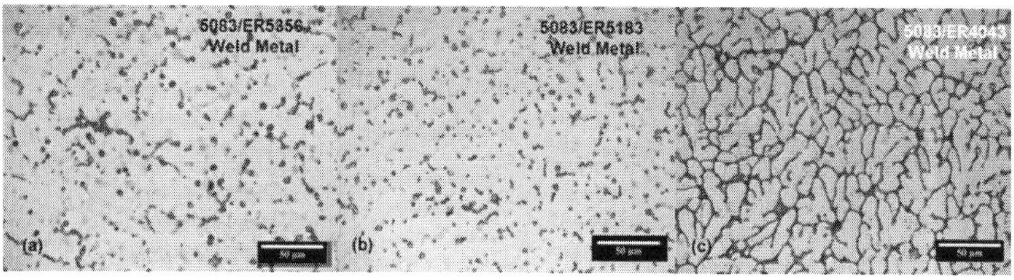

Fig. 15. Typical micrographs of (a) 5083/ER5356; (b) 5083/ER5183; and (c) 5083/ER4043 weld metal.

In order to identify the second phase particles observed in the weld metal, SEM-EDS elemental maps were constructed. A typical elemental map is shown in Figure 16 for a weld performed using ER5356 filler metal. ER5356 welds contain second phase particles and grain boundary regions enriched in Fe and Mg, and slightly depleted in Al. In welds deposited using ER5183 filler wire, second phase particles appear to be enriched mainly in Mg and Al (Figure 17).

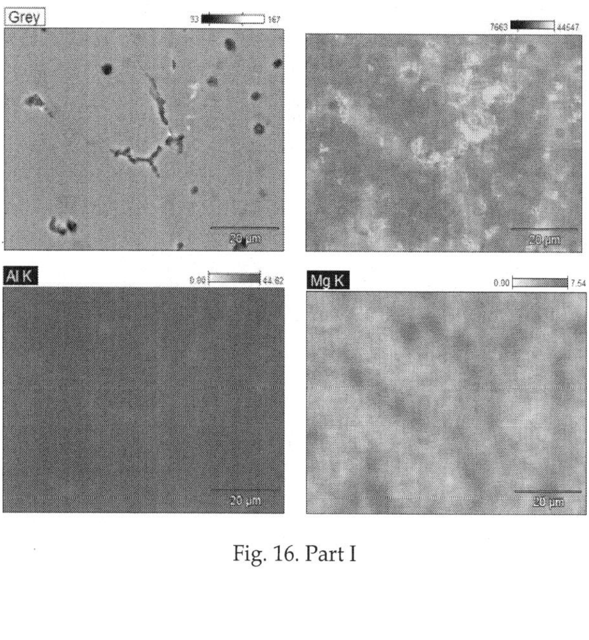

Fig. 16. Part I

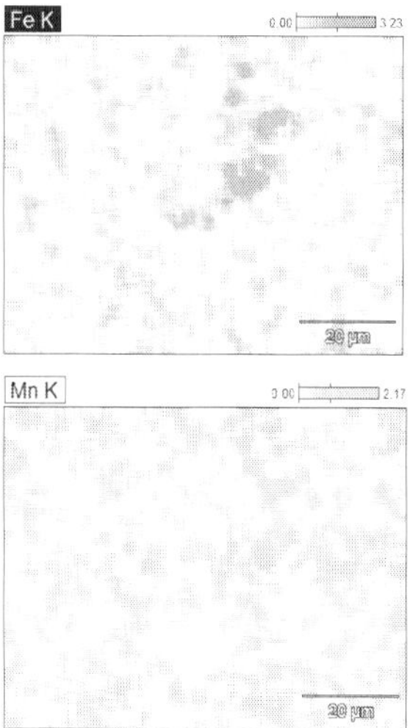

Elements	Weight %	%Error
C k	3.46	+/-0.43
O k	1.04	+/-0.11
Mg k	3.82	+/-0.05
Al k	90.66	+/-0.20
Si k	0.22	+/-0.04
Fe k	0.22	+/-0.04
Mn k	0.57	+/-0.04
Total	100.00	

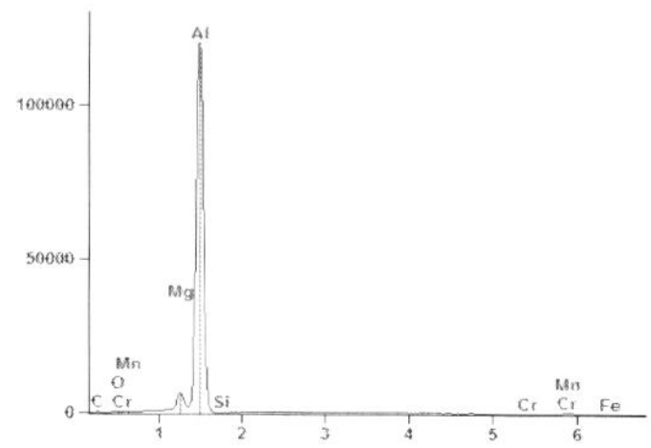

Part II

Fig. 16. Typical EM-EDS analysis of second phase particles observed in a weld performed using ER5356 filler wire.

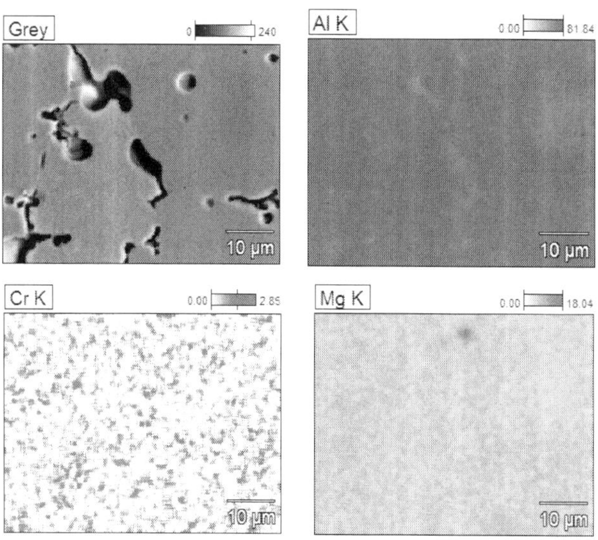

Fig. 17. Part I

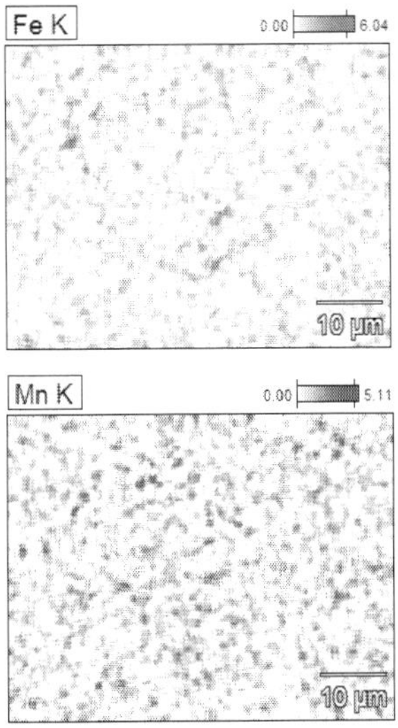

Elements	Weight%	%Error
C k	5.89	+/-0.16
O k	0.19	+/-0.07
Mg k	2.95	+/-0.03
Al k	90.45	+/-0.18
Cr k	0.12	+/-0.02
Mn k	0.33	+/-0.05
Fe k	0.08	+/-0.03
Total	100.00	

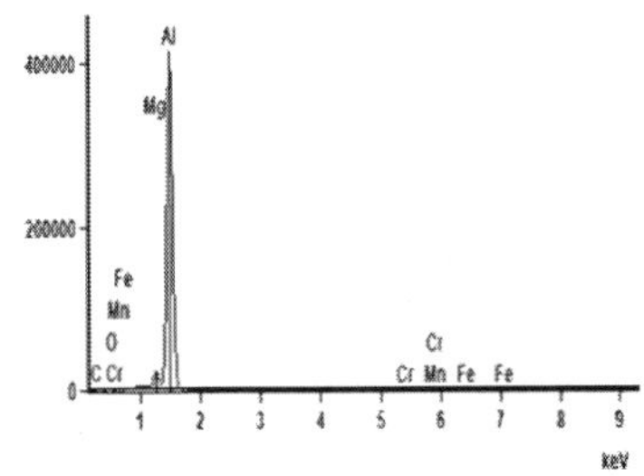

Part II

Fig. 17. Typical EM-EDS analysis of second phase particles observed in a weld performed using ER5183 filler wire.

The interdendritic component of weld metal deposited using ER4043 filler wire consisted of a fine Si-rich eutectic. Isolated Mg-rich particles are also evident (Figure 18).

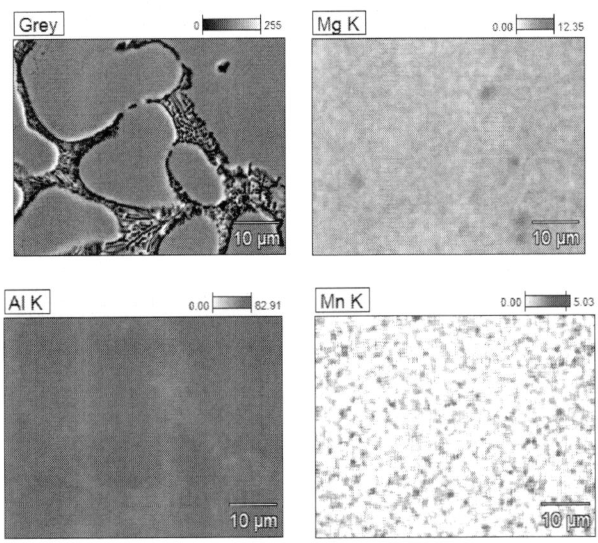

Fig. 18. Part I

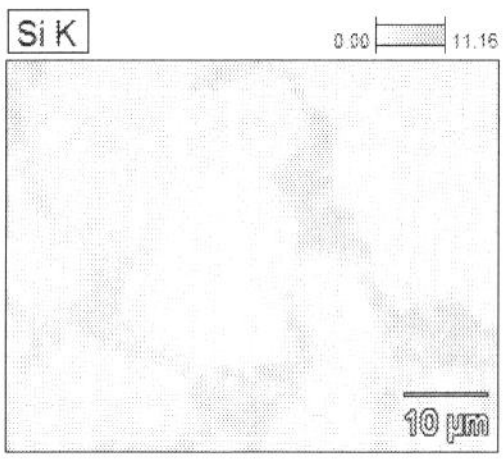

Element	Weight %	%Error
C k	3.05	+/-0.15
O k	1.21	+/-0.02
Mg k	1.46	+/-0.02
Al k	92.84	+/-0.19
Si k	1.00	+/-0.03
Cr k	0.13	+/-0.02
Mn k	0.32	+/-0.05
Total	100.00	

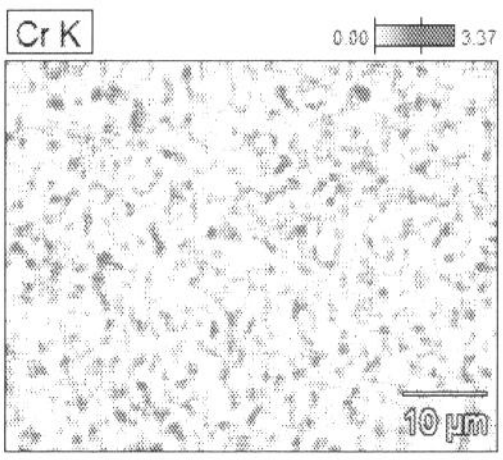

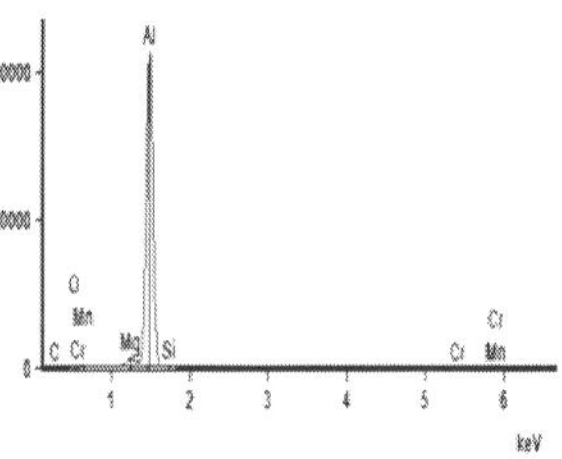

Fig. 18. Typical EM-EDS analysis of second phase particles observed in a weld performed using ER4043 filler wire.

4.2 Micro-hardness evaluation of 5083-H111 welds

The average hardness value of 5083-H111 alloy in the as-supplied condition is 91.81 HV (standard deviation of 13.20 HV) (Figure 19 and 20). Micro-hardness measurements revealed higher hardness values in the region of second-phase intermetallic particles (a maximum of about 794 HV). Figures 19 and 20 also display micro-hardness profiles across FA-GMAW welded joints performed using ER5356 and ER4043 filler wire, respectively. A similar trend was observed for the ER5183 filler metal. The lowest hardness values were observed within the weld metal of both the SA-GMAW and FA-GMAW welds produced using ER5356 or ER4043 filler wire. SA-GMAW welds displayed lower hardness values than FA-GMAW welds, regardless of the filler wire used.

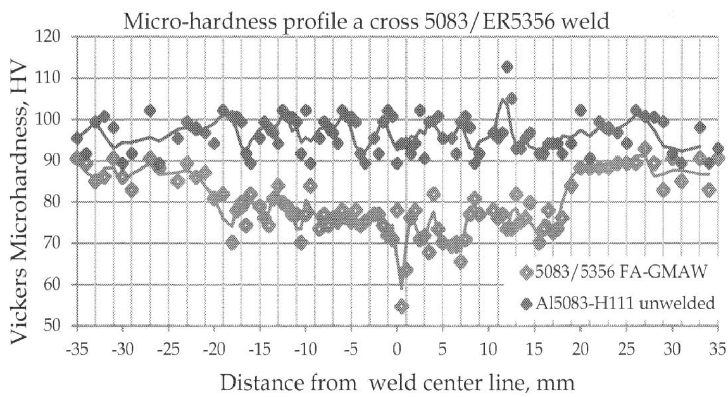

Fig. 19. Micro-hardness profile across a pulsed SA-GMAW joint welded with ER5356 filler wire.

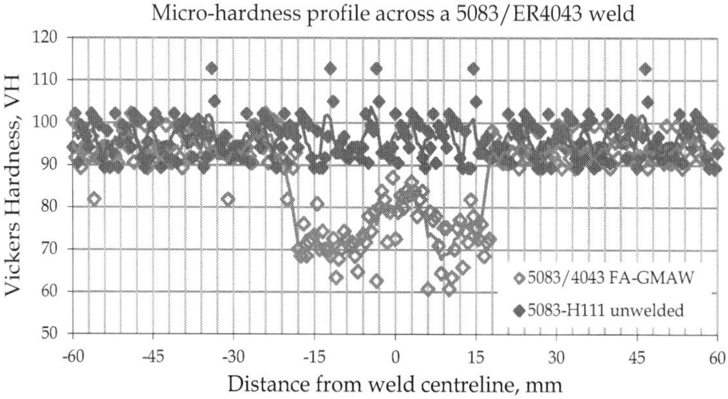

Fig. 20. Micro-hardness profile across a pulsed SA-GMAW joint welded with ER4043 filler wire.

4.3 Tensile properties of 5083-H111 aluminium

4.3.1 Tensile properties in the as-supplied condition

As shown in Figure 21, the unwelded 5083-H111 displayed moderate tensile and yield strength, and good ductility. During axial tensile testing, the crack path followed coarse second-phase particles within the matrix, as illustrated in Figure 21(a). This alloy fractured in a ductile manner, as evidenced by microvoid coalescence (dimples) observed around coarse precipitates or second-phase particles on the fracture surface (Figure 21 (b)).

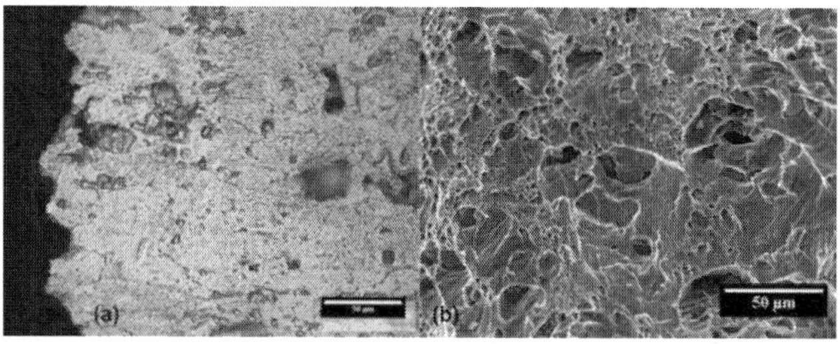

Fig. 21. Tensile fracture of the 5083-H111 as-supplied material: (a) fracture path; (b) microvoid coalescence.

4.3.2 Tensile properties of 5083-H111 welds

The transverse tensile properties of 5083 welded using ER5356, ER5183 and ER4043 filler wire are given in Figures 22 to 24, respectively. All fully dressed welds failed in the weld metal, as typically illustrated in Figures 25(a), 26 and 27. As shown in Figures 19 and 20, the hardness across welds is fairly uniform, with a moderate reduction in hardness in the weld metal. This reduction in hardness most likely prompted failure in this region during tensile testing. Any discontinuities in the weld metal, such as gas porosity or lack-of-fusion type defects, will also affect the measured tensile properties. The ultimate tensile strength (UTS) of FA-GMAW dressed welds performed using ER5356 filler wire was very similar to that of the base metal, with ER5183 and ER4043 generally yielding lower strength values due to the inherently lower strength levels of the consumables. Figures 23 to 24 also indicate that the ultimate tensile strength (UTS) values of FA-GMAW welds are consistently higher than those of SA-GMAW welds. This can most likely be attributed to the higher incidence of porosity and welding defects observed in SA-GMAW welds (Figure 14). The strength values of fully dressed welds using ER5356 filler wire, shown in Figure 23, are significantly higher than those of undressed welds, emphasising the detrimental effect of geometrical stress concentrations (at the weld toe and root) and weld defects (such as undercut) on the measured tensile properties. Undressed welds consistently failed in the HAZ at the weld toe or root (Figure 26(c)). The tensile fractures (Figure 26) and SEM micrographs of the weld metal fracture surfaces (Figure 26) confirm a predominantly ductile failure mode in ER5356 and ER5183 welds, with mixed mode failure along the interdendritic silicon-rich eutectic regions in the ER4043 weld metal.

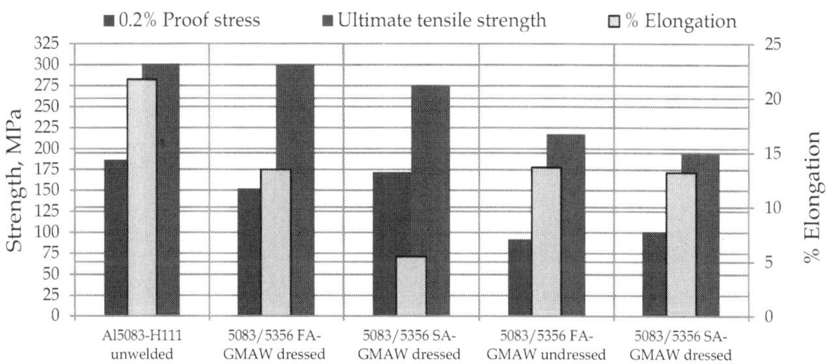

Fig. 22. Tensile Properties of 5083-H111/ER5356 welds.

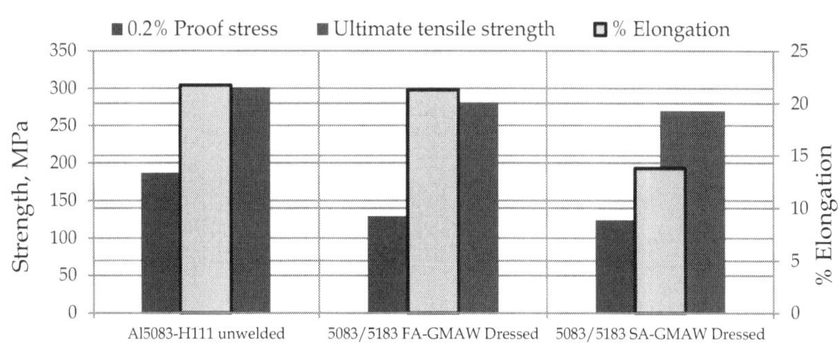

Fig. 23. Tensile Properties of 5083-H111/ER5183 welds.

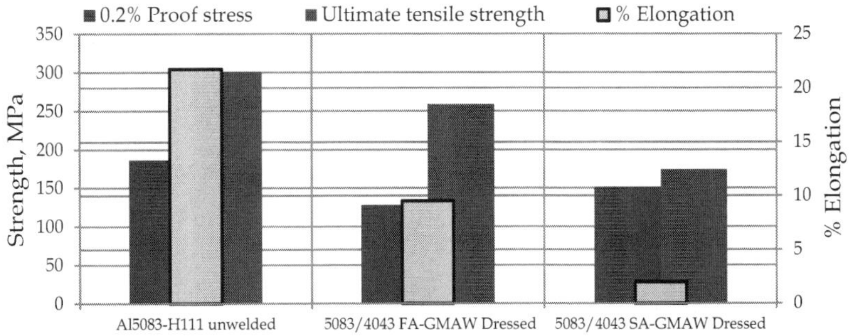

Fig. 24. Tensile Properties of 5083-H111/ER4043 welds.

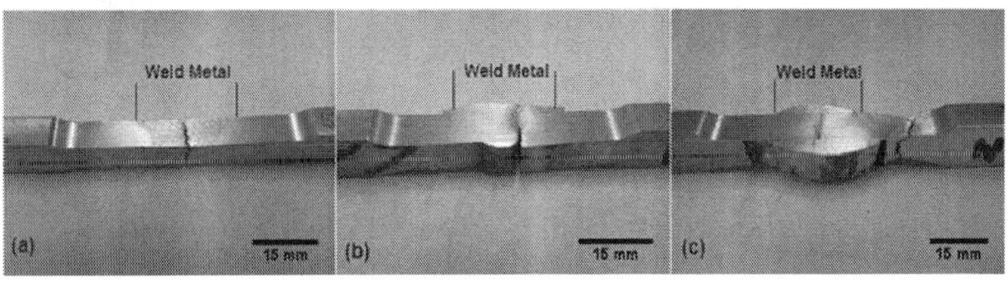

Fig. 25. Representative photographs of the tensile fractures observed in 5083 welded using ER5356, ER5183 and ER4043 filler wires: (a) fully dressed FA-GMAW weld; and (b) undressed FA-GMAW weld; and (c) undressed SA-GMAW weld.

Fig. 26. Tensile fractures of weld joint in 5083 alloy welded with (a) ER5356; (b) ER5183; and (c) 4043 filler wire.

Since failure occurs preferentially in the weld metal, filler metal selection plays a significant role in determining the transverse tensile properties of the welds. ER5356 welds display tensile properties very similar to those of the unwelded base metal, with ER5183 and ER4043 providing welds with lower strength.

In the case of undressed welds, failure at the fusion line was promoted by the presence of gas pores and lack of fusion type defects at the fusion line, as shown in Figure 28.

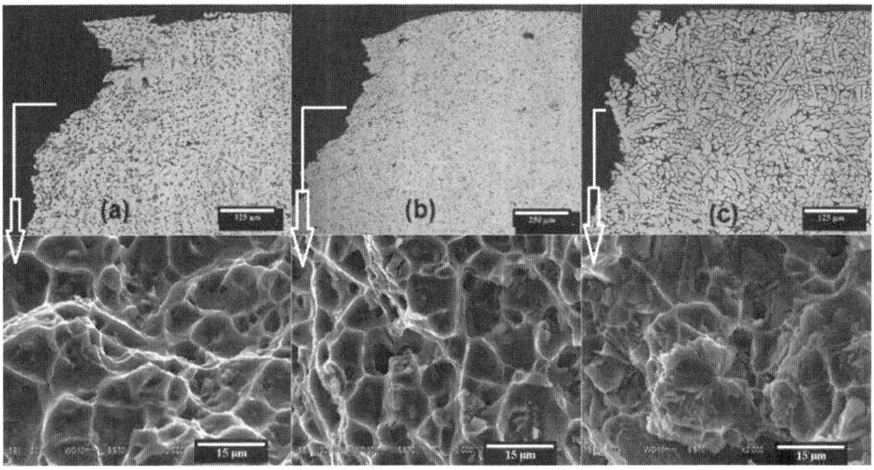

Fig. 27. Tensile fracture surfaces of fully dressed welds displaying predominantly ductile mixed-mode failure in the weld metal (a) 5083/ER5356, and (b) 5083/ER5183; and brittle failure in (c) 5083/ER4043 weld metal.

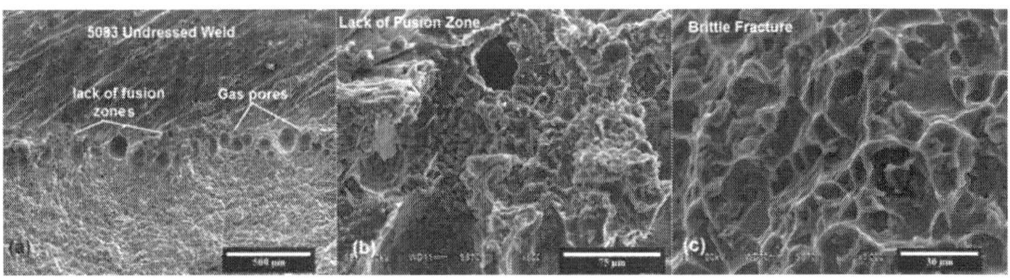

Fig. 28. Typical tensile fracture surfaces of undressed welds failing at weld/HAZ transition zone (a) lack of fusion and gas pores in weld/HAZ interface, (b) lack of fusion area; and (c) brittle mixed-mode failure in weld/HAZ interface.

4.4 Corrosion behaviour of 5083-H111 in a 3.5% NaCl solution

Aluminium 5083-H111 in the as-supplied condition exhibits pitting corrosion on immersion in 3.5% NaCl solution (maintained at a temperature of 16°C to 27°C, with a pH between 6.9 and 7.2). Typical corroded surfaces after various immersion times (with a total immersion time of 90 days) are given in Figures 29. Pitting attack is generally associated with second phase particles in the matrix (Figure 29(a)). This is consistent with the observation made on 6061-T651 aluminium alloy by K.Mutombo [27].

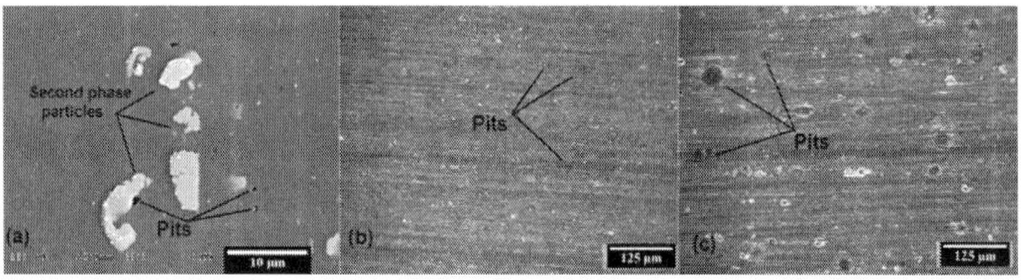

Fig. 29. Pitting corrosion observed after exposure to a 3.5% NaCl solution on the surfaces of 5083-H111 aluminium: (a) a polished surface immersed for 24 hours; (b) a ground surface immersed for 30 days; and (c) ground surface immersed for 90 days.

The pit dimensions observed on immersion of 5083-H111 in 3.5% NaCl solution are shown graphically as a function of exposure time in Figures 30. Longer exposure times increase the depth, length and width of the observed pits.

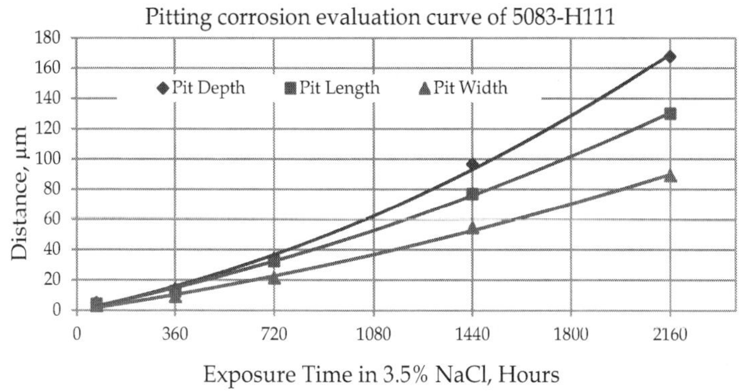

Fig. 30. Mean dimensions of pits observed in aluminium 5083-H111 exposed to 3.5% NaCl solution at temperatures between 25 and 27^{0}C and dissolved oxygen contents of 5.5 to 9 ppm.

Welding appeared to increase susceptibility to pitting corrosion. Welds suffered severe pitting attack on exposure to 3.5% NaCl solution. As shown in Figure 31, immersion in 3.5% NaCl solution of welds resulted in pitting of the weld metal and HAZ of ER5356 and ER5183 welds, with very severe corrosive attack of the ER4043 weld metal and HAZ.

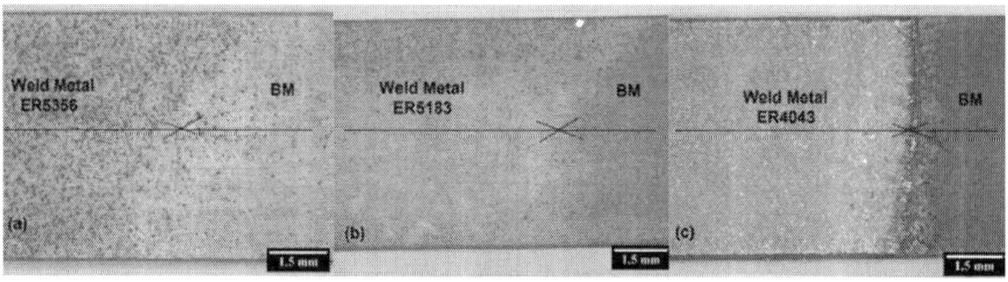

Fig. 31. Typical corrosion of welds after free immersion in 3.5% NaCl solution for 60 days: (a) ER5356; (b) ER5183; (c) ER4043 filler wire.

4.5 Fatigue properties of 5083-H111 in the as-supplied condition

4.5.1 Fatigue properties in the as-supplied condition

S-N_f curves of 5083-H111 in the as-supplied condition after testing in air and in 3.5% NaCl solution are shown in Figure 32. As expected, the number of cycles to failure increases with a decrease in the applied stress amplitude (Sa). The unwelded aluminium 5083-H111 displayed considerably longer fatigue life in air than in 3.5% NaCl solution, especially at higher stress amplitudes.

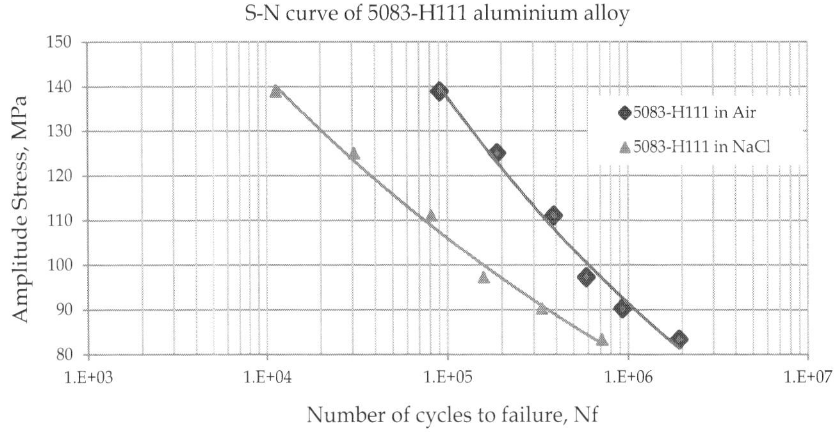

Fig. 32. S-N curves of 5083-H111 aluminium in the as-supplied condition.

During testing in air, fatigue cracks initiated preferentially at the free surfaces of the sample at discontinuities such as slip lines, polishing or machining marks and precipitates or inclusions. This is illustrated in Figure 33(a) to (c). Once initiated at the free surface (at inclusions (Figure 33(a) and (b)), the cracks propagated rapidly during testing, followed by final ductile failure when the remaining cross section of the sample could no longer sustain the applied stress.

Fig. 33. (a) Surface crack initiation at second phase particles; (b) crack initiation due to disbonding between precipitates and the matrix; and (c) crack propagation.

Immersion in 3.5% NaCl solution during fatigue testing shortened the fatigue life significantly. As shown in Figure 34, crack initiation was accelerated by the presence of corrosion pits at the surface of the samples. These pits formed preferentially at precipitates or inclusions due to the galvanic effect between the particle and the aluminium matrix.

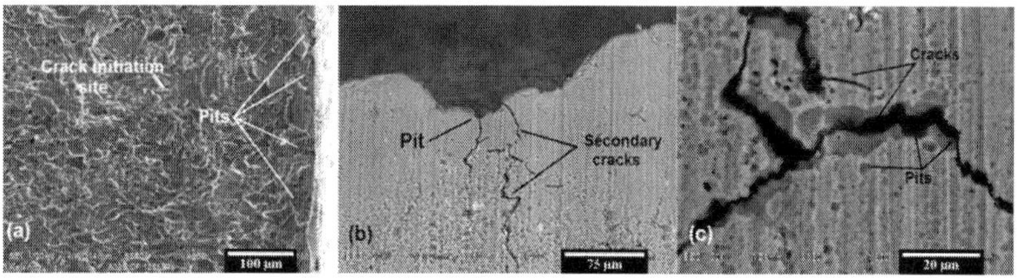

Fig. 34. (a) Multiple fatigue crack initiation at small corrosion pits; (b) crack propagation from corrosion pits; and (c) fatigue crack associated with small pits.

The fatigue damage ratio (DR) is the ratio between the number of cycles to failure in 3.5% NaCl solution ($N_{f\ NaCl}$) and the number of cycles to failure in air ($N_{f\ Air}$), as shown in equation (5).

$$DR = \frac{N_{fNaCl}}{N_{fAir}} \tag{5}$$

The limit values of the fatigue damage ration are zero (0) and one (1). The DR approaches zero only when $N_{f\ NaCl}$ approaches zero, i.e. pitting corrosion is the dominant process in determining the corrosion fatigue behaviour. Corrosion pits act as stress raisers by rapidly initiating fatigue cracks. When DR approaches one, or $N_{f\ NaCl}$ approaches $N_{f\ Air}$, the dominant process determining fatigue life is the fluctuating stress.

The fatigue damage ratios (DR) of 5083-H111 aluminium in the as-supplied condition are shown graphically in Figure 35. This graph indicates that the effect of pitting corrosion on fatigue properties in unwelded specimens is most pronounced at higher stress amplitudes. At high stress levels, corrosion pits act as sharp stress raisers, accelerating fatigue crack initiation. The effect of the corrosive environment on fatigue properties becomes less apparent at lower stress amplitudes, but is unlikely to approach a ratio of one (signifying that the corrosive environment has no influence on fatigue behaviour).

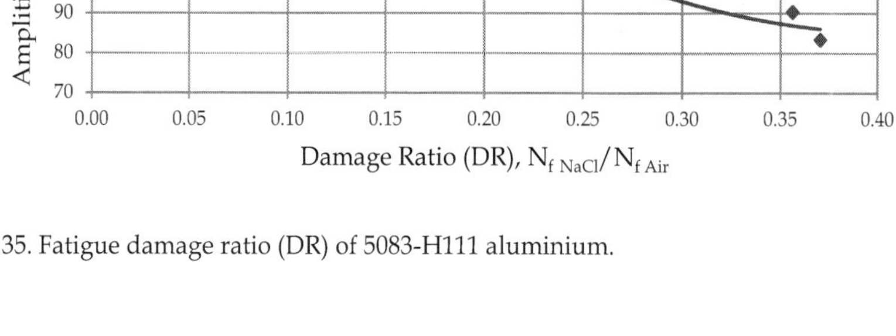

Fig. 35. Fatigue damage ratio (DR) of 5083-H111 aluminium.

4.5.2 Fatigue behaviour of 5083-H111 aluminium welds

4.5.2.1 Aluminium 5083-H11 welded using ER5356 filler wire

The results of fatigue tests in air of 5083 aluminium joined using ER5356 filler wire are shown in Figure 36. The data points shown represent median values, whereas the S-N_f curves were fit using the power law. It is evident that welding reduces the fatigue life significantly. SA-GMAW welds (fully dressed and as-welded) and as-welded (undressed) FA-GMAW welds display similar fatigue properties, with the fully dressed SA-GMAW welds performing marginally better than the undressed joints. The dressed FA-GMAW welds display much higher fatigue properties, which can be attributed to the absence of sharp stress concentrations at the weld toe and root, and the reduced incidence of welding defects such as porosity.

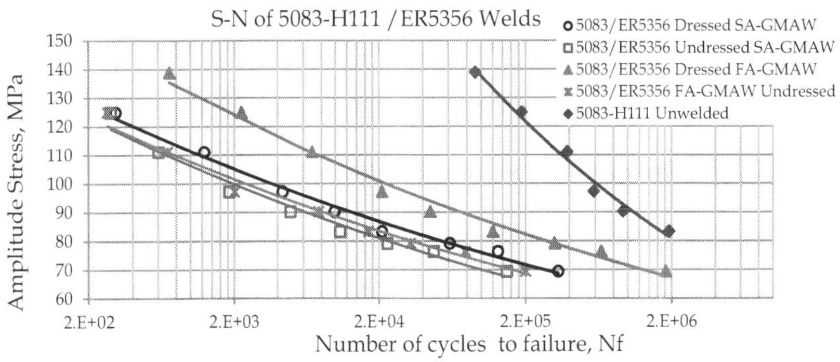

Fig. 36. Fatigue properties of 5083 welded with ER5356, tested in air.

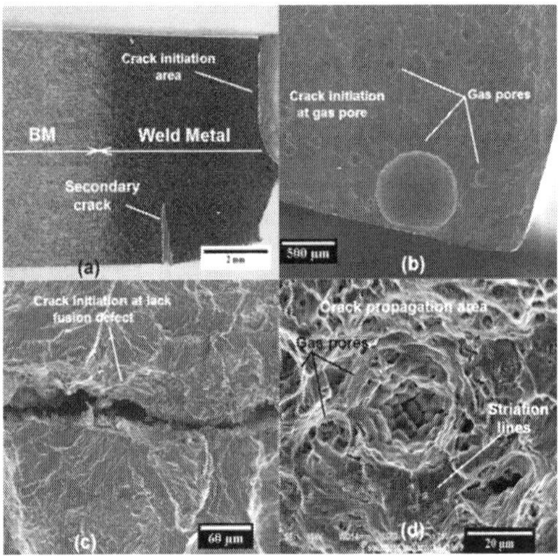

Fig. 37. Typical fatigue fractures in 5083 welds (a) crack initiation in the weld metal; (b) crack initiation associated with a large gas pore; (c) crack initiation at a lack-of-fusion type defect; (d) crack propagation associated with gas pores.

Fatigue cracks initiated preferentially at gas pores, lack-of-fusion type defects and incomplete weld penetration; and at the weld toes of undressed joints (Figure 37). Crack propagation occurred preferentially in the weld metal.
The results of fatigue tests of 5083 joined using ER5356 filler wire in 3.5% NaCl solution are shown in Figure 38. The results indicate that immersion in NaCl during fatigue testing reduces the fatigue properties of both the SA-GMAW and FA-GMAW welds. The advantage gained by FA-GMAW in reducing the number of welding defects are largely negated in the presence of a corrosive environment due to the introduction of corrosion pits as preferential crack initiation sites. The FA-GMAW weld therefore displays similar fatigue performance to the SA-GMAW weld under corrosion fatigue conditions.

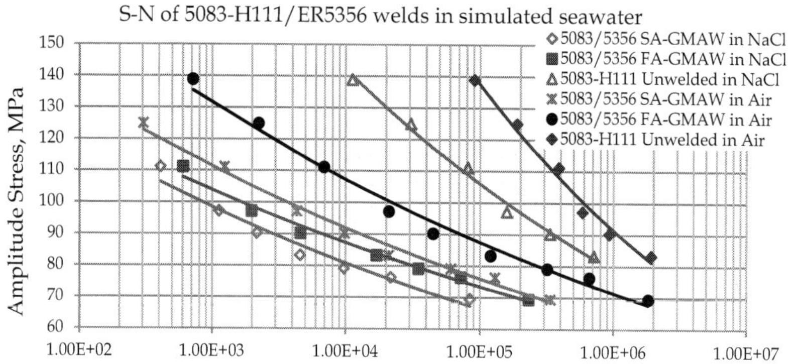

Fig. 38. Fatigue properties of 5083 welded with ER5356, tested in 3.5% NaCl solution.

The fatigue damage ratio (DR) curve of 5083/ER5356 welds shown in Figure 39 reveals that the effect of pitting corrosion on fatigue properties is most pronounced at stress amplitudes between 80 and 70 MPa.

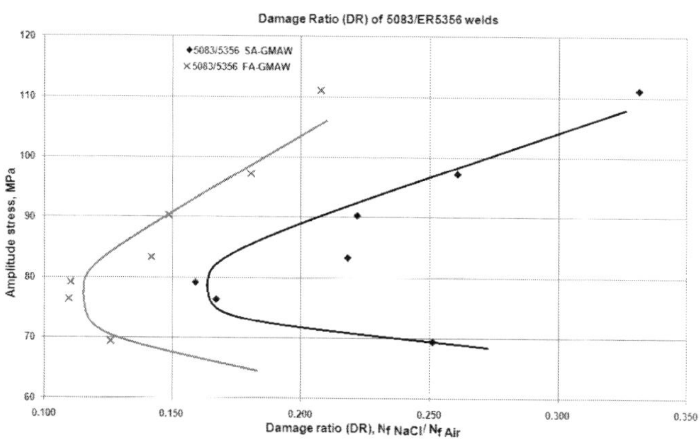

Fig. 39. Fatigue damage ratio of 5083 aluminium welded with ER5356 filler metal.

In this stress range, pits act as sharp stress raisers, accelerating fatigue crack initiation. Due to the presence of pre-existing weld defects, the effect of the corrosive environment on the fatigue properties of the welds becomes less apparent at higher stress amplitudes. At these higher stress amplitude values, the number of cycles to failure decreases and the effect of pitting corrosion becomes less apparent.

During fatigue testing in 3.5% NaCl solution, cracks initiate preferentially at pits in the weld metal (Figure 40(a) and (b)), or at discontinuities such as the lack-of-fusion type defect shown in Figure 40(d)).

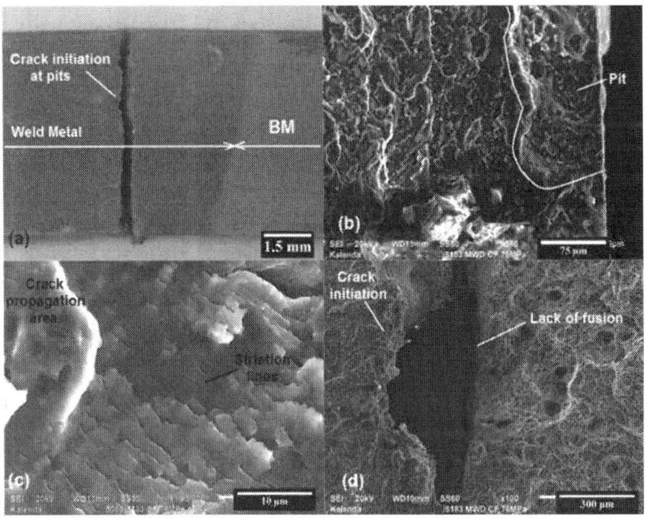

Fig. 40. Typical features of fatigue fracture in 5083/ER5356 welds tested in 3.5% NaCl: (a) and (b) crack initiation at pits in the weld metal; (c) crack propagation in the weld metal; and (d) crack initiation at a lack-of-fusion type defect.

4.5.2.2 Aluminium 5083-H111 welded using ER5356 filler wire

The fatigue properties of 5083-H111 aluminium welded with ER5183 filler wire using SA-GMAW or FA-GMAW are presented in Figure 41. Welding reduces the fatigue life tested in air considerably. Failure occurs in the weld metal, which has a lower hardness and strength than the base materials (as shown in Figures 20 to 21 and 23 to 25). Cracks initiate preferentially at weld defects such as gas pores, lack-of-fusion type defects, incomplete penetration and slag inclusions. The 3.5% NaCl corrosive environment reduces the number of cycles to failure for SA-GMAW and FA-GMAW welds. Cracking occurs preferentially in the softer weld metal which corrodes faster than the base material (Figure 40(a)). Crack initiation is accelerated by pits that formed prematurely in the weld metal at defects such as gas pores, lack-of-fusion defects and slag inclusions. Very little difference is evident between SA-GMAW and FA-GMAW welds, suggesting that the lower defect content of the automatic welds does not affect the fatigue behaviour significantly.

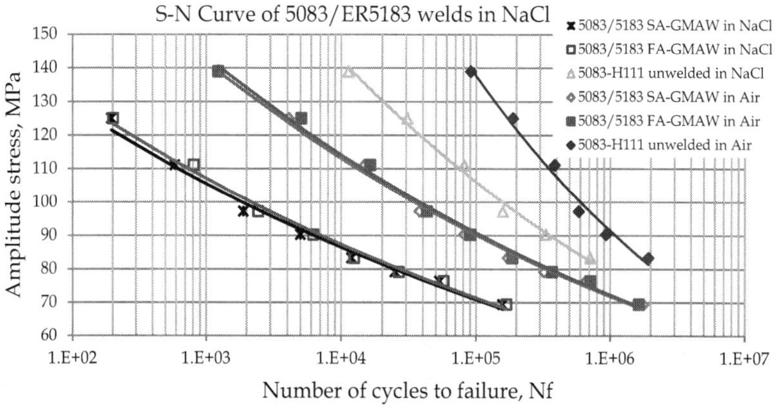

Fig. 41. Fatigue properties of Al5083/ER5183 welds tested in air and 3.5% NaCl solution.

The fatigue damage ratio of aluminium 5083 welded with ER5183 filler wire is shown in Figure 43 for SA-GMAW and FA-GMAW welds. Corrosion fatigue plays a more dominant role at higher stress amplitudes, where the presence of welding defects may accelerate pitting corrosion, leading to rapid fatigue crack initiation and propagation.

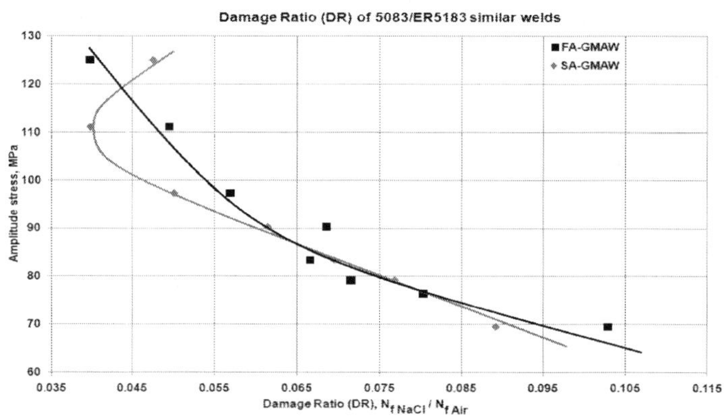

Fig. 42. Fatigue damage ratio of 5083-H111 welded with ER5183 filler wire.

4.5.2.3 Aluminium 5083-H111 welded using ER4043 filler wire

Figure 43 displays the fatigue properties of aluminium alloy 5083-H111 welded with ER4043 filler wire using SA-GMAW and FA-GMAW. Failure occurs in the weld metal and these welds exhibit reduced fatigue life compared to that of unwelded 5083-H111 tested in air. The 5083/ER4043 welds appear to be sensitive to the presence of weld defects, with the SA-GMAW welds displaying lower fatigue properties than the FA-GMAW welds. The SA-GMAW and FA-GMAW welds display similar corrosion fatigue life when tested in 3.5% NaCl salt water environment, suggesting that the presence of a higher percentage of welding defects in the SA-GMAW welds did not affect the fatigue behaviour in the corrosive environment to any significant extent. The short fatigue life of these welds can be attributed to the poor corrosion resistance of the ER4043 welds. Pit formation is rapid, with fatigue cracks initiating prematurely at corrosion pits.

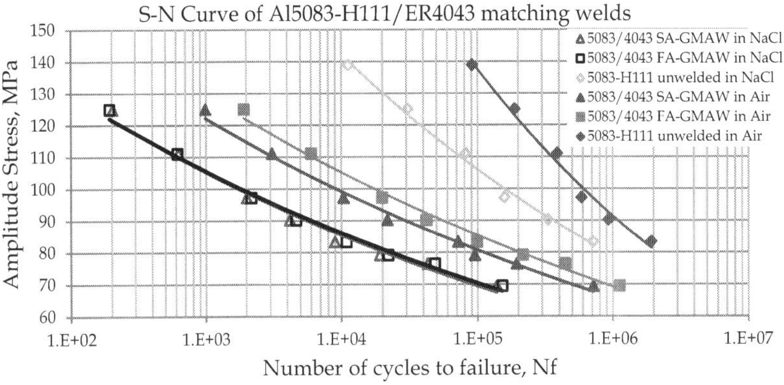

Fig. 43. Fatigue properties of 5083 aluminium welded with ER4043 filler metal.

The fatigue damage ratio (DR) values of 5083 aluminium, welded using ER4043 filler metal, are presented in Figure 44. The fatigue ratio values are low, suggesting that the fatigue properties of the welds are sensitive to pitting corrosion in the NaCl solution. The FA-GMAW welds appear to be affected to a greater extent by the presence of a corrosive environment. These welds displayed slightly better fatigue properties than the SA-GMAW welds in air, probably due to the lower number of weld defects. In the NaCl solution, however, both curves shift to lower numbers of cycles to failure, with the fully automatic curve shifting more than the semi-automatic curve. This suggests that the availability of welding defects ceases to dominate the fatigue behaviour, with crack initiation at corrosion pits becoming controlling. The rapid pitting rate of the ER4043 weld in NaCl therefore creates large numbers of potential pit initiation sites in both the FA-GMAW and SA-GMAW welds, resulting in a very low damage ratio, especially for the FA-GMAW welds.

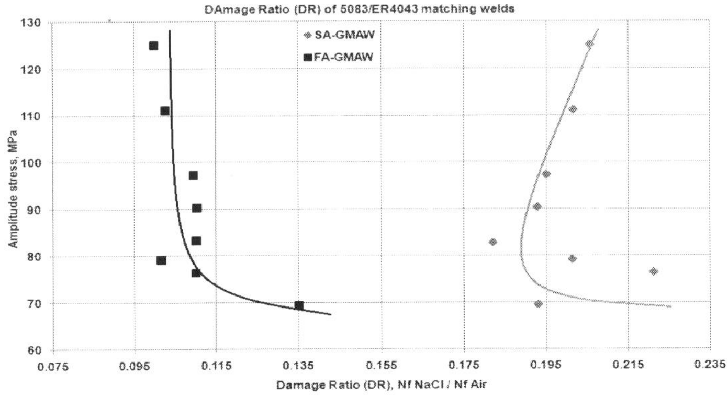

Fig. 44. Fatigue damage ratio of 5083/ ER4043 welds.

The results presented in this study confirm that the fatigue properties of 5083-H111 aluminium are adversely affected by welding. Fatigue cracks preferentially initiated at the stress concentration caused by the weld toes or the weld root (in as welded joints) or the weld metal at defects such as gas pores, lack-of-fusion type defects and incomplete weld penetration. Fatigue failure was accelerated by the presence of a corrosive environment due to the formation of corrosion pits.
The influence of filler metal selection on the mechanical properties of the welds is discussed below.

4.5.3 Effect of filler wire selection on the mechanical properties of 5083-H111 aluminium welds

The influence of filler metal selection (ER4043, ER5183 or ER5356) on the transverse tensile properties of welds in 5083-H111 aluminium is shown in Figure 45. The tensile strength of 5083 alloy welded with ER5356 is similar to that of the base metal, but the ductility is considerably lower. Welds performed using ER5183 filler wire display excellent ductility, good tensile strength, but low yield stress, whereas welds performed using ER4043 filler metal display poor strength and ductility.

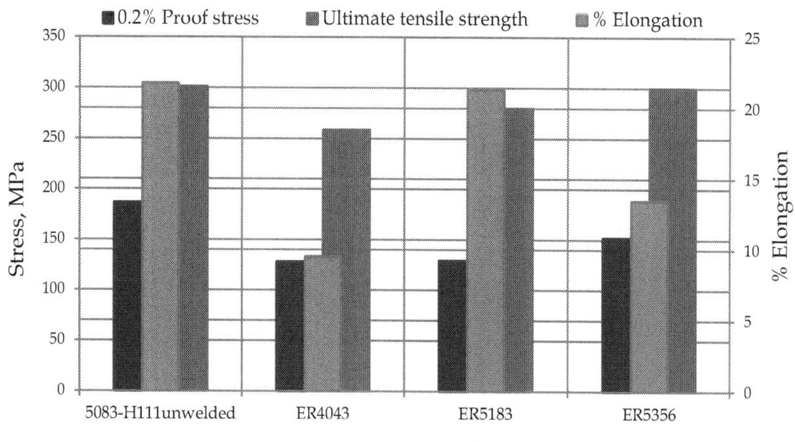

Fig. 45. Tensile properties of dressed welds in 5083-H111 aluminium joined using ER4043, ER5183 and ER5356 filler wires (pulsed GMAW).

Figure 46 presents the fatigue properties of 5083-H111 welded using ER4043, ER5183 and ER5356 filler metals. Although the fatigue properties are very similar, the weld performed using ER5183 filler wire displays slightly longer fatigue life in air. The good fatigue resistance of the ER5183 welds in air can probably be attributed to a good combination of high strength and excellent ductility. Exposure to a corrosive medium during fatigue testing reduced the fatigue life significantly, but despite the observed differences in mechanical properties, filler metal selection had very little effect on the corrosion fatigue properties. All three welds displayed similar corrosion fatigue properties during testing in 3.5% NaCl solution.

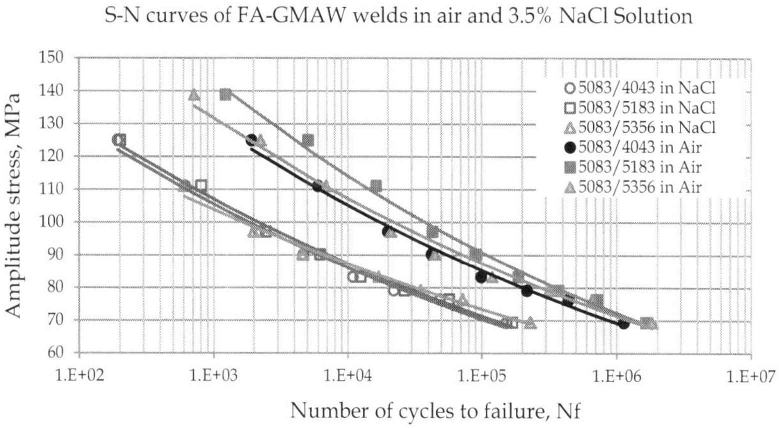

Fig. 46. Fatigue properties of FA-GMAW welds in 5083-H111 welded with ER4043, or ER5183, or ER5356 filler wire.

The results described above indicate that ER5356 or ER5183 filler metal is recommended for joining this alloy, with ER5356 yielding high strength welds and ER5183 providing a good combination of high strength and excellent ductility. Fully automatic welding, with its good control over weld dimensions and its lower defect content, ensures optimal resistance to fatigue failure.

5. Conclusions and recommendations

Both semi-automatic and fully-automatic welding reduces the strength and hardness of the Al5083-H111 welds produced with ER5356, ER5183 and ER4043 filler wires. This loss in strength and hardness was more pronounced in the semi-automatic gas metal arc welds. Lower hardness was revealed in the centre line of welds.

The fatigue lives were severely reduced in the undressed welds due to the severe stress concentration presented by the weld toes. Dressing of the welds improved the fatigue properties significantly. As a result of improved control over weld profile, fully-automatic welds consistently outperformed semi-automatic welds. Fatigue cracks initiated at gas pores, lack of fusion defects, incomplete weld penetration and weld toes. The presence of a corrosive environment accelerated the fatigue failure due to the formation of corrosion pits that initiated preferentially at weld defects or coarse second phase particles and facilitated rapid fatigue crack initiation.

ER5356 or ER5183 filler metal is recommended for joining 5083 aluminium alloys, with ER5356 yielding high strength welds and ER5183 providing a good combination of high strength and excellent ductility. Fully automatic welding, with ensures optimal resistance to fatigue failure.

6. Acknowledgements

My sincere thanks to the Light Metals Development Network and the THRIP programme for financial support during this investigation.

I would also like to express my appreciation to Professor Madeleine du Toit for her support and guidance. The contribution and technical advice of Professor Waldo Stumpf, Professor Chris Pistorius and Dr Tony Paterson are also gratefully acknowledged.

Finally the technical support of Willie Du Preez, Sagren Govender, Chris McDuling and Erich Guldenpfennig from the Council for Scientific and Industrial Research (CSIR) is gratefully acknowledged.

7. References

[1] Lean, P.P.; Gil, L. & Ureña, A. (2003). Dissimilar welds between unreinforced AA6082 and AA6092/SiC/25p composite by pulsed-MIG arc welding using unreinforced filler alloys (Al-5Mg and Al-5Si), *J.Mat.Pro.Tec.*, vol. 143-144, pp. 846-850

[2] Czechowski, M. (2004). Corrosion Fatigue of GMA Al-Mg Alloy, *Advances in Materials Science*, vol. 4, No. 2004, pp. 16-24.

[3] Praveen, P. & Yarlagadda, P.K.D.V. (2005). Meeting challenges in welding of aluminium alloys through pulse gas metal arc welding, *J.Mat.Proc.Tec.*, vol. 164-165, pp. 1106-1112

[4] Olson, D.L.; Siewert, T.A.; Liu, S.; Edwards, G.R. 2007, *Welding, Brazing, and Soldering*, ASM International, USA

[5] Wilcox, D.V.; Adkins, H.; Dickerson, P.B.; Hasemeyer, E.A. & Lockwood, L. (1972). Welding Aluminum, In: *Welding Aluminum*, 6th edn, American welding society, Florida, pp. 69.2-69.4-69.25

[6] Maddox, S.J. (1991). Fatigue Strength of Welded Structures, In: *Fatigue Strength of Welded Structures*, 2nd edn, Abington Publishing, Cambrige, England, pp. 19-27-30-36, 38-43, 66-70

[7] Jones, D.A. (1996). Principals and Prevention of CORROSION, In: *Principals and Prevention of CORROSION*, 2nd edn, Prentice-Hall, Inc., United States of America, pp. 200-200-220, 236-247, 262-263, 279-285, 307, 309-316, 322

[8] Vargel, C. (2004). Corrosion of Aluminum, In: *Corrosion of Aluminum* Elsevier, London, pp. 92-96-97, 124

[9] Maggiolino, S. & Schmid, C. (2008). Corrosion resistance in FSW and in MIG welding techniques of AA6XXX, *J.Mat.Proc.Tec.*, vol. 197, no. 1-3, pp. 237-240

[10] Cramer, S.D.; Covino, B.S. (2003). *Corrosion: Fundamentals, Testing, and Protection*, ASM International, USA

[11] Winkler, S.L.; Ryan, M.P. & Flower, H.M. (2004). Pitting corrosion in cast 7XXX aluminium alloys and fibre reinforced MMCs, *Corrosion Science,* vol. 46, no. 4, pp. 893-902

[12] Hatch, J.E. (1984) Aluminium: Properties and Physical Metallurgy In: *Aluminium: Properties and Physical Metallurgy* American Society for Metals, USA, pp. 242-242-248, 253.

[13] ASTM International. (2004). *Standard Test Method for Vickers Hardness of Metallic of Materials,* ASTM international, West Conshohocken

[14] Davis, J.R. (1998). *Aluminum and Aluminum Alloys,* 4th edn, ASM International, USA

[15] Geoffroy, N.; Vittecoq, E.; Birr, A.; de Mestral, F. & Martin, J. (2007). Fatigue behaviour of an arc welded Al–Si–Mg Alloy, *Scripta Materialia,* vol. 57, no. 4, pp. 349-352

[16] Ishihara, S.; Saka, S.; Nan, Z.Y.; Goshima, T. & Sunada, S. (2006). Prediction of corrosion fatigue lives of aluminium alloy on the basis of corrosion pit growth law, *Fatigue and Fracture of Engineering Materials and Structures,* vol. 29, no. 6, pp. 472-480

[17] ASM International. (2007). *Fatigue and Fracture,* ASM International, USA

[18] Pidaparti, R.M. & Rao, A.S. (2008). Analysis of pits induced stresses due to metal corrosion, *Corrosion Science,* vol. 50, no. 7, pp. 1932-1938

[19] Chlistovsky, R.M.; Heffernan, P.J. & DuQuesnay, D.L. (2007). Corrosion-fatigue behaviour of 7075-T651 aluminum alloy subjected to periodic overloads, *I.J.Fatigue,* vol. 29, no. 9-11, pp. 1941-1949

[20] Chen, G.S.; Wan, K.; Gao, M.; Wei, R.P. & Flournoy, T.H. (1996). Transition from pitting to fatigue crack growth—modeling of corrosion fatigue crack nucleation in a 2024-T3 aluminum alloy, *Materials Science and Engineering A,* vol. 219, no. 1-2, pp. 126-132

[21] ASTM International. (2002). Standard Test Method for Macroetching Metals and Alloys, ASTM International, West Conshohocken

[22] ASTM International. (2001). Standard Guide for Metallographic Specimens, ASTM International, West Conshohocken.

[23] ASTM International. (2002). Standard Test Method for Microindentation Hardness of Materials, ASTM International, West Conshohocken.

[24] ASTM International. (2004). Standard Test Methods for Tension Testing of Metallic Materials, ASTM International, West Conshohocken.

[25] ASTM International. (2004). Standard Practice for Laboratory Immersion Corrosion Testing of Metals, ASTM International, West Conshohocken.

[26] ASTM International. (1999). Standard Guide for Examination and Evaluation of Pitting Corrosion, ASTM International, West Conshohocken

[27] ASTM International. (2002). Standard Practice for Conducting Force Controlled Constant Amplitude Axial Fatigue Test of Metallic Materials, ASTM International, West Conshohocken

[28] Mutombo, K. (2011). Intermetallic Particles-Induced Pitting Corrosion in 6061-T651 Aluminium Alloy, *Materials Science Forum,* vol. 690, no. 2011, pp. 389-392

Part 4

Mechanisms, Models, and Measurements of Arc Welding

10

The Mechanism of Undercut Formation and High Speed Welding Technology

Zhenyang Lu and Pengfei Huang
Beijing University Of Technology
China

1. Introduction

With the economic integration and technological globalization, intensified competition in the manufacturing sector. Welding technology as an important means of processing, has become such as vehicles, shipbuilding, container, steel and many other fields in the core competitiveness of enterprises to improve the most important factor, especially as one of the assembly process, usually arranged in the manufacturing process the late or final stage, which has a decisive role in the quality of products.

Improve the welding speed is the subject of research for many years, but in recent years, with welding automation, especially in promoting the use of welding robots, along with the increasingly fierce market competition, making the welding process efficiency is particularly prominent, has become a constraint welding a key factor in productivity, a welder can not play a major potentials obstacle. Thus, how to improve the welding speed once again become a hot research topic.

While there arc welding speed to improve stability of the process parameters of the problem, but still the ultimate prerequisite is to ensure there are no welding seam welding quality defects. Thus, the case study high-speed welding of weld defects is to increase the welding speed of the mechanism of the important aspects of the process.

Welding speed is a key parameter of high speed welding when joining thin sheets. Undercut is one of the most important problems which restrict welding speed.

So this will be generated during welding undercut mechanism and high-speed welding technology to give detail.

2. Mathematical model of weld pool

2.1 Welding pool mechanics research situation

2.1.1 The force affected on pool

As the pool under the effect of the arc, it makes stress state complicated. The pool suffers not only gravity, surface tension, buoyancy, but also the arc force. In the melting pole gas shielded welding, it will suffer the droplet impacting when transfer.

Studies have shown that, among all of the forces, when the welding current under 200A, the arc force will be little shown in the figure 1. In relevant analysis, its effect is little, which can be ignored. Therefore, the force in the pool is the gravity, surface tension, buoyancy arc force and so on.

Studies have shown that, when the current is 150 A, welding speed is 3.33cm/s, the pool flow rate is 0.01m/s magnitude which caused by the buoyancy, electromagnetic force and the both showed the reverse flowing. The role of buoyancy promote the liquid of pool center upward movement from the bottom and push the liquid of the surface to edge from the centre, which make the melting width increase.

Electromagnetic force forces the surface liquid flow to centre from the edge and then to the bottom, which will impact the bottom so that the penetration increases. The vortex strength caused by the electromagnetic force is not only related with the current size, but also with material properties, the arc moving speed. Under the same current, moving faster, the greater the thermal conductivity, and the effect of electromagnetic force is more small.

Under the action of surface tension, the liquid flow at the rate of 1.0 m/s magnitude, the flow direction is related with the temperature coefficient of surface tension. The combined results of the pool owing to the forces , when the welding current is less than 200A, the size and distribution of the surface tension play a decisive role to moving speed and direction of the liquid.

2.1.2 The law that surface tension roles in liquid flowing in the pool

When small current (Ia < 200A) welding, electromagnetic force and buoyancy are small, which can be ignored, liquid metal flow of the pool is main related with surface tension and gravity. The gravity's law is sample, so the relationship between surface tension and temperature is the main influence on the liquid metal flowing. If the surface tension σ reduce with the increase of temperature T, that is $\partial\sigma / \partial T$ <0,the distribution of surface tension

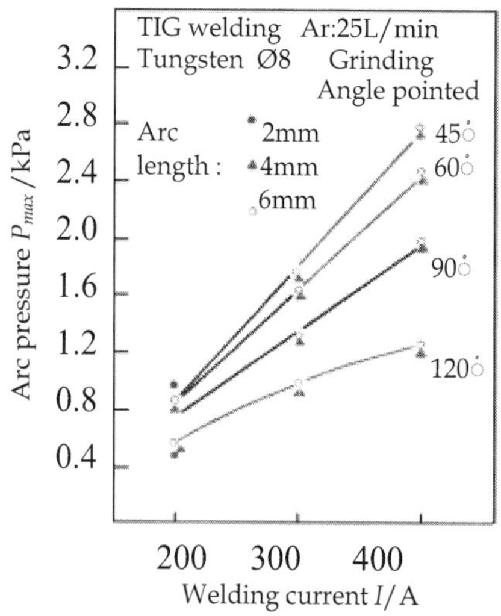

Fig. 1. Relation of pressure and current of arc.

on the pool surface is small near the center and large next to the edge, so the flow direction is from center to edge, that is the flow direction is from high temperature to low temperature. Shown in the figure 2 a 、b .At the same time, the flow of liquid metal will take the arc heat to pool edge from pool center which will form shallow and wide weld. If the surface tension increases with the temperature increased, that is $\partial\sigma / \partial T$ >0,the distribution of surface tension on the pool surface is large near the center and small next to the edge, therefore, the liquid metal will flow to center from pool edge. That is flow to high temperature from low temperature, shown in the figure 2 c, d. The heat of arc will be taken into pool bottom, which will form narrow and deep weld.

Generally believed that ， surface tension can be measured roughly with affinity strength of free surface atom. For most of substances, temperature coefficient of surface tension is

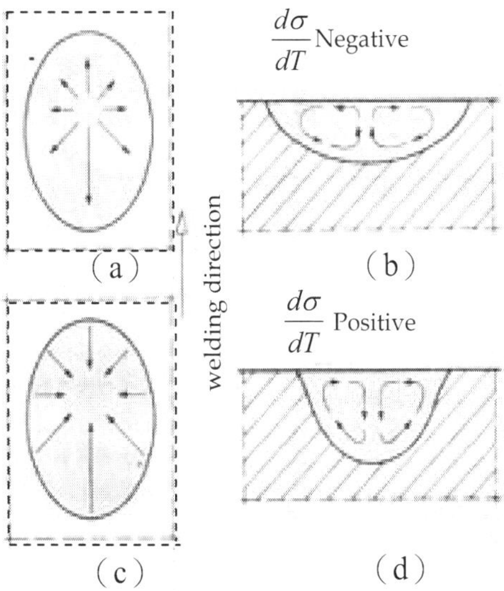

Fig. 2. Effect of surface tension on weld pool flow.

number, that is to say, surface tension reduces as the temperature increases. That is because high temperature led to affinity between atoms reduced. When it exists surfactant(such as oxygen, sulfur and so on), surfactant can enhance the affinity between the surface atoms as temperature increases, which will cause $\partial\sigma / \partial T$ in a positive state. That is increasing as the temperature rises. This means that because the temperature distribution is high in the middle and low close to the edge in the pool, which causes surface tension distribution large in the middle and small close to the edge, leading to the liquid metal flow to middle from edge.

The rules in A-TIG process has been confirmed.

2.1.3 Variation law of surface tension temperature coefficient

The variation of surface tension in the pool under the arc force is a complicated problem. This is not only because of the high arc temperature, the large temperature gradient distribution, the short welding time which results pool in a severe non-equilibrium, but also because non-uniform temperature distribution in the pool causes trace elements assemble, especially the surface active elements. All of these make distribution and variation of surface tension complicated.

Temperature distribution determined by the arc characteristics are high around the center and low close to the edge. It makes surface tension distribution of liquid metal show corresponding change. However, there is a more complicated problem. That is surface tension of liquid metal is not only related with temperature, but also related with its composition, especially related with the distribution and content of surface active elements. According to P.Sahoo and others findings, When the metal has surface-active substances (expressed by i) ,we can get the temperature coefficient by follows:

$$\frac{\partial\sigma}{\partial T} = -A - R\Gamma_s \ln(1+Ka_i) - \frac{Ka_i}{(1+Ka_i)} \frac{\Gamma_s(\Delta H^0 - \Delta\overline{H}_i^M)}{T} \tag{1}$$

In the formula: Γ_s-The super saturation of element i in the metal; K- absorption coefficient;ai-activity of element in the metal; ΔH^0 -Standard enthalpy; $\Delta\overline{H}_i^M$ -distribution enthalpy of element i in the metal. T-temperature.

From (1)we can see, $\partial\sigma / \partial T$ is a function of temperature but also a function of composition, and for pure metals, it is minus. When the surfactant concentration exists certain content, K a_i <<1, (1)will become:

$$\frac{\partial\sigma}{\partial T} = -A - \frac{Ka_i\Gamma_s(\Delta H^0 - \Delta\overline{H}_i^M)}{T} \tag{2}$$

When ΔH^0 is minus, $\partial\sigma / \partial T$ will become a positive number in (2). Figure 3、4 are the sulfur content in Fe-S system, oxygen content in Fe-O system and temperature on the influence of the surface tension.

We can see in the figure surface tension coefficient of Iron-based liquid metal increases and it curves gets upward with oxygen, sulfur content increases. Yet surface tension coefficient decreases with increasing temperature in each curve. More importantly, each curve with increasing temperature will come through the most important point: $\partial\sigma / \partial T$ =0. This is a turning point. When a point is above it, $\partial\sigma / \partial T$ >0,that is to say the surface tension of liquid metal will increase with temperature rising. While a point is under it, $\partial\sigma / \partial T$ <0, the surface tension of liquid metal will decrease with temperature rising. The turning point of the temperature range between roughly 2000K-2800K.

2.2 Liquid flowing in the pool

2.2.1 Liquid flow in pool when temperature coefficient of surface tension changes

As is mentioned earlier, surface tension is the main force on liquid flowing, yet the distribution and direction of surface tension are determined by temperature distribution of surface metal. In non-melting fixed-point arc welding conditions, if arc temperature distribution is in line with Gaussian distribution, because of it, the temperature next to the edge of pool is lowest(About the metal melting temperature),and the metal in the middle of pool is highest temperature(Up to the boiling temperature of the metal). When there is no

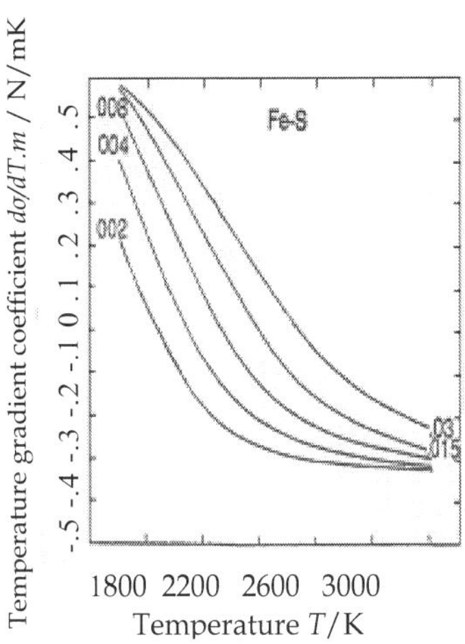

Fig. 3. Variation of temperature coefficient of surface tension as a function of temperature and S%.

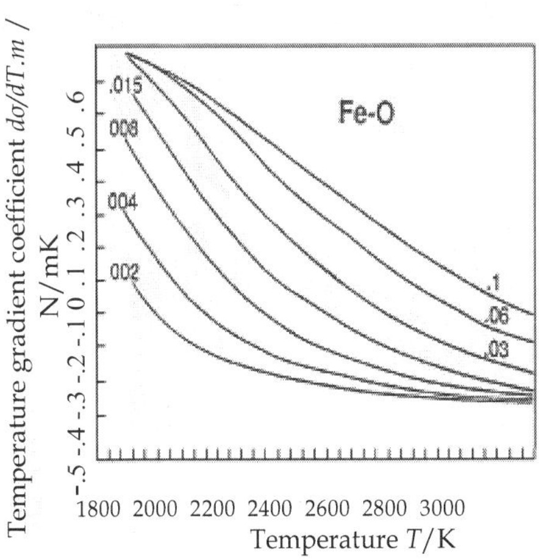

Fig. 4. Variation of temperature coefficient of surface tension as a function of temperature and O%.

surface active element, the temperature coefficient of surface tension $\partial\sigma / \partial T < 0$, just as figure 5 show us, the metal in pool will flow from the center of pool where the surface temperature is high and surface tension is small to the edge of pool where the surface temperature is low and surface tension is large, shown in the figure 5.

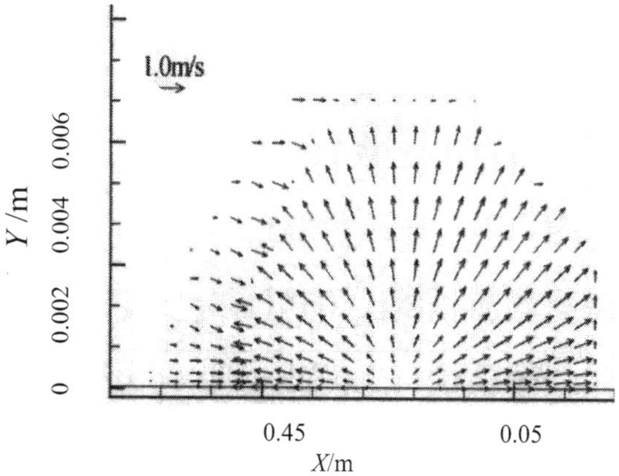

Fig. 5. Simulation of flow field as the change of $\partial\sigma / \partial T$ sign in welding pool.

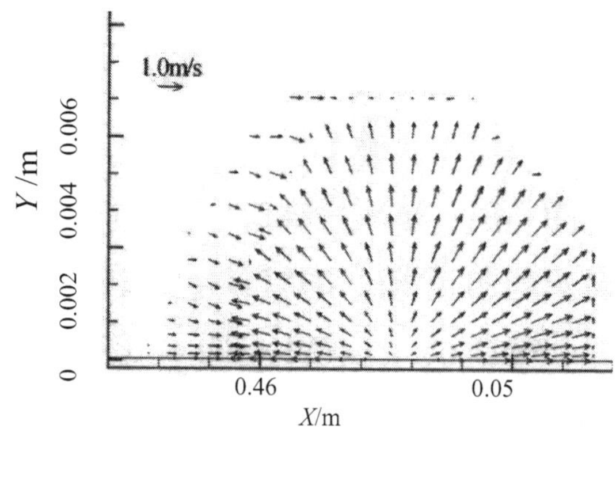

Fig. 6. Simulation of flow field as the change of $\partial\sigma / \partial T$ sign in welding pool.

However, when it exist surface active elements, the temperature of a certain part in pool surface is just the temperature in which the sign of $\partial\sigma / \partial T$ changes .It will form two areas, $\partial\sigma / \partial T$ is positive in one area, and negative in another. Therefore, it will produce two or more circulations in the internal pool. The figure 6 show us that it formed two circulations in the internal pool. The liquid in pool flows from the middle (where the temperature is high) to the edge in one area ,where the temperature is above the temperature that $\partial\sigma / \partial T$ changes its sign. Yet it flows from the edge (where the temperature is low) to the middle in another area, where the temperature is below the temperature that $\partial\sigma / \partial T$ changes. Literature is studying the corresponding law, which is the flowing of pool influence of penetration ,because temperature change lead to surface tension change. Thus in their simulation prerequisite, pool is assumed to be rigid surface. It is obviously a big gap between the actual situation, but it can be accepted to study penetration of pool. For this article, the main goal is study the influence that surface tension with temperature change impact to undercut. Obviously, the above assumption is unreasonable. But the simulation results, which is about flow state of liquid metal in pool affected by surface tension, to explain the cause of undercut has a theoretical significance.

2.2.2 Numerical simulation of pool free surface

M.S.Tsai in university of Wisconsin has studied the liquid convection in pool and pool shape of TIG under the assumption of the free surface. In order to describe the edge of pool surface, he used orthogonal curvilinear coordinates and determined morph of pool surface based the following formula:

$$P_a - \sigma_{22} - P + C = -\gamma\left[\frac{h''}{(1+h'^2)^{3/2}} + \frac{h''}{r(1+h'^2)^{1/2}}\right] \tag{3}$$

The above formula:

σ_{22} --Normal stress, N/ m^2 ;

P-- Environmental pressures，N/ m^2 ;

Pa-- The pressure from the equations of motion, N/ m^2 ;

C-- Undetermined coefficient；

γ --Surface tension coefficient, N/m；

h-- Surface deformation function.

Tsai who applied the Laplace equation, Ohm's law and Maxwell's equation calculated the flow velocity field and bath temperature in condition that the welding current is 150A and the distribution of arc is in line with Gaussian, just as the figure 7 shows us. In the picture, the surface next to pool center rises. The area near the outer part of the fusion line has a small concave. The explanation Tsai has given us is that the liquid metal near the surface flow inward, and the liquid metal near the pool center is forced flowing vertically downward, that caused the surface near the pool center being pushed up. The result is that the pool forms bump in the middle and concave near the edge. This is the first time to explain the reasons for the formation of undercut at the perspective of metal flowing in the pool.

3. Fixed-point arc welding results and analysis

3.1 TIG welding of fixed-point finite element analysis and experimental study

3.1.1 The same energy when the temperature and temperature gradient

In order to study the same energy, different welding current and welding time of pool behavior, we conducted a fixed-point arc TIG welding experiments, the following parameters.

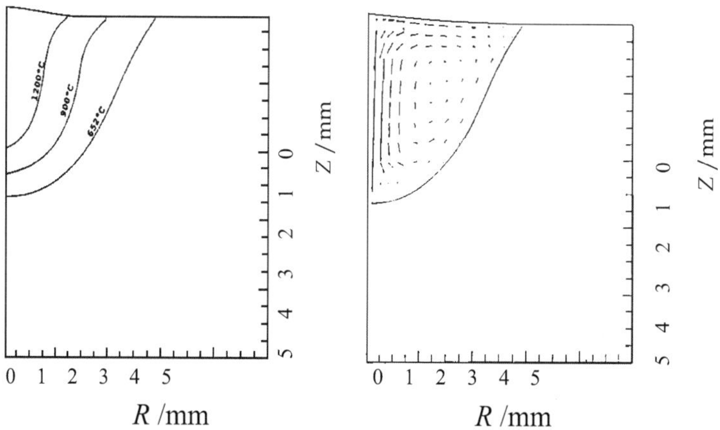

Fig. 7. Temperature, flow field and surface deformation under Gauss distribution of heat source (150A)

Shielding gas flow rate: 101/min, tungsten to the base metal distance: 5mm, length of tungsten out: 5mm, pre-aspirated time 1s, air time delay off: 4s, current rise and fall times of 0. Welding current and time relationship as show in the table 1.

number	1	2	3	4
current I （A）	190	150	110	70
time t （s）	2.4	3.4	5	9

Table 1. Parameters for equal energy standstill weld.

The same input energy, welding current and heating time is not calculated while the temperature gradient as shown in figure 8. It can be seen from the figure, the welding current large temperature gradient larger short time, while the small current for a long time heating the temperature gradient smaller.

In order to verify the results, this paper uses thermal imaging to measure the temperature field. Thermal imager which technical parameters are as follows:

Thermal imager model: HRX thermal imaging system

Manufacturer: Beijing Century Knight Technology Co, Ltd.

Detector spectral response range :8 - 14μm

Pixels: 256 × 256

Grayscale images: 256

Temperature measurement range :1550 - 2100K

Temperature measurement precision :2%

Data acquisition time:10ms

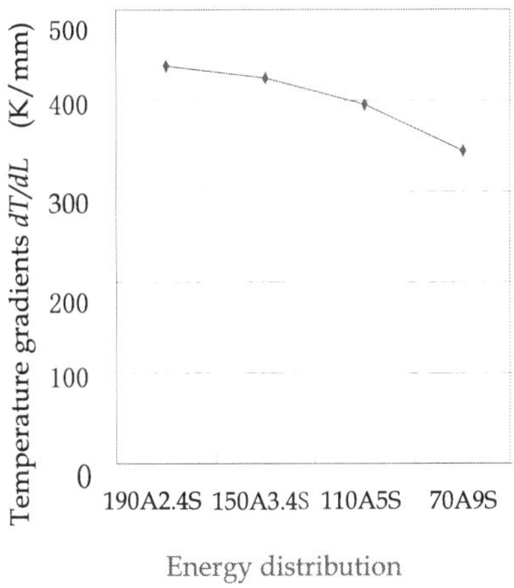

Fig. 8. Comparison of temperature gradients in same input energy

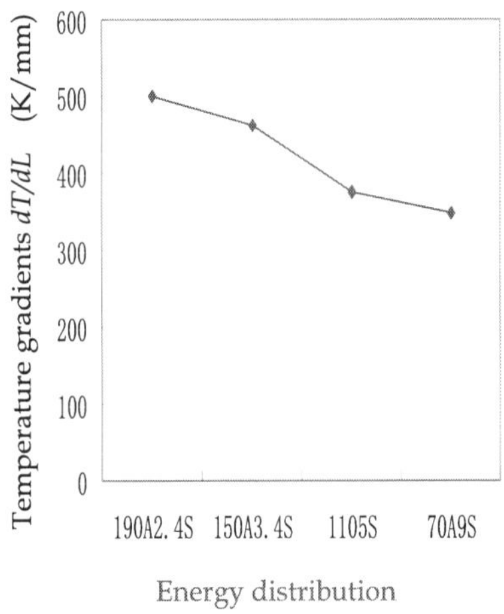

Fig. 9. Temperature distribution of welding pool surface in same input energy.

Experiments have shown temperature gradient in figure 9. From figure 9, we can see that although the same input energy there are, but when the way using small current had a long time heated, the fusion of the online temperature gradient is smaller. Instead, when changed large current for a short time heated, the temperature gradient is bigger.

As shown in figure 10 when it have the same energy input and different heating time, that is to say the calculated value depth-to-width ratio of the solder joints when energy distribution is not the same as the way. From the graph may safely draw the conclusion that, under the condition of the same energy input, the way using small current for a long time welding can get a more to big depth-to-width ratio value

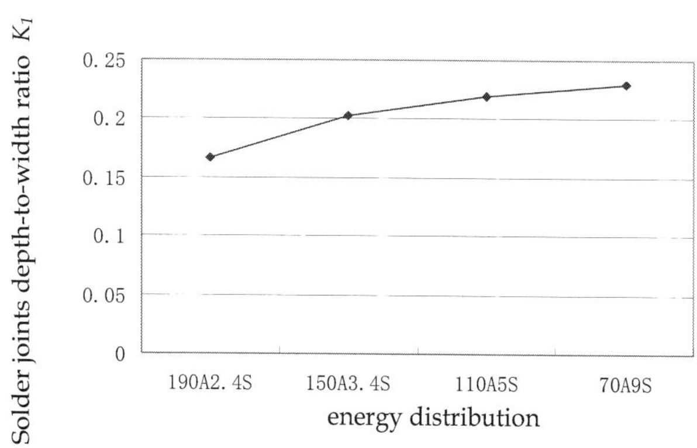

Fig. 10. Ration between depth and width of the bead.

3.1.2 The temperature and the temperature gradient of the same current

Figure 11 is the calculated temperature curve when the molten pool is cooling under the welding current 190 A, heating 2.4 seconds. Figure 12 is the calculated temperature curve when the molten pool is cooling under the welding current 190 A, heating 1.4 seconds.

From the figure 11 and 12, we can see when arc is burning in a long time, the molten pool is obviously bigger. But compared to two graph of the temperature change we can found the temperature change speed in the center parts of molten pool is far outweigh the pace of change in the surrounding. That is to say, in the cooling process of molten pool, though in the midst of the liquid metal finally solidified, but its temperature down than the temperature of the surrounding reduced much faster.

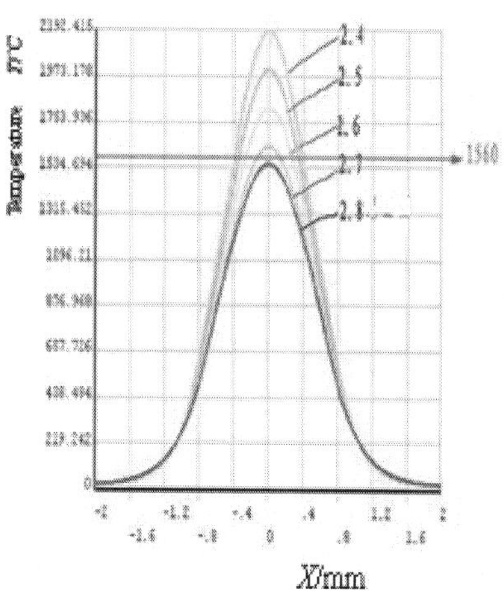

Fig. 11. Calculated result of temperature of welding pool surface as cooling (190A, 2.4S).

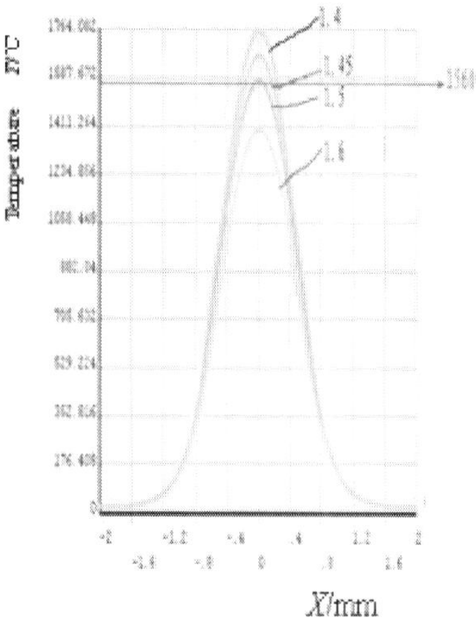

Fig. 12. Calculated result of temperature of welding pool surface as cooling (190A, 1.4S).

In figure 13 we can also see, with the time of heating extended, the fusion line near the temperature gradient decreased. It can be concluded, small standard slow welding can reduce the fusion line near the temperature gradient.

3.2 Designated welding experimental results

3.2.1 Experiment conditions and welding standard

Experiment mainly carried out from three aspects: one basic situation of contrast is the different welding current, different welding time, but heat input is the same; the second

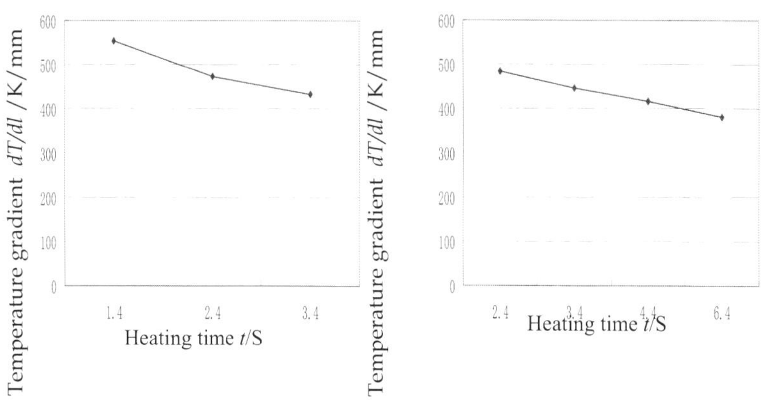

a) current is 190A

b) current is 150A

Fig. 13. Relationship between temperature gradient and arc time.

situation it is the same, but the welding arc burning time different changed; the third is under the same welding standard but welding materials is not the same as the sulphur content is different.

The specimens uses the Q235 steel plate. The actual weldedtestspecimen respectively in two kinds with sulphur content in the 0.02% and 0.05%. Table 1 is designated welding experimental parameters. The welding conditions: TIG welding dc electric reverse connect, Ar gas protection, gas flow 101/min, tungsten extreme view of 60 °, 6 mm distance to the workpiece.

The numbers in table 2 is the value of the welding current multiplied the welding time, because of the actual welding arc current difference can cause energy loss different and it can not stand of the truely welding heat input, but its can indirectly reflect the quantity of welding heat input. In the experiment, strictly control arc burning time and using thermovision and infrared thermometer measuring the temperature distribution of welding pool and some point of the heat cycle, especially near the fusion line of thermal cycle. After the specimen cooled, with precision surface measuring instrument measuring the surface shape of the solder joint we can get bite edge data.

The experiments in table 2 must be repeated at least three times, ×in the table stand for the test have not done. Some are due to welding current is too small, the time is short, basically the specimens does not melt to form a molten pool; some are due to welding current is too large, time is long, the measurement can not be carried out because of the specimen got burning through.

Time(s) / Current(A)	2	3	4	5	6	7	10	13	16
70	×	×	×	×	×	490	700	910	1120
110	×	330	440	550	660	×	×	×	×
150	330	450	600	750	×	×	×	×	×
190	380	570	760	×	×	×	×	×	×

Table 2. Parameter of standstill arc welding.

Note:

1. The × express the experiment have not done.
2. The corresponding numerical listed in the table express the product of multiplication of the current and the time.

Surface shape measuring instrument Hommel-Links PM2000 technical indexes are as follows: the vertical resolution 0.25 microns, level 1 micron, resolution for tip radius of 20 microns ± 5 microns.

3.2.2 Experiments results

(1) The same heat input experiment results. When the welding current is different and different welding time, but the same heat input, Welding undercutting depth will increase with the increased of the welding current and the reduce of the time. That is to say, large current and short time welding condition can produce the depth of undercutting more big, figure 14 is the actual measurement result. This is because large current and short time condition input can have a lager welding temperature gradient, surface tension is bigger, the

tendency of liquid metal near to the fusion line flow to the center of the molten pool increased, then undercutting depth increased.

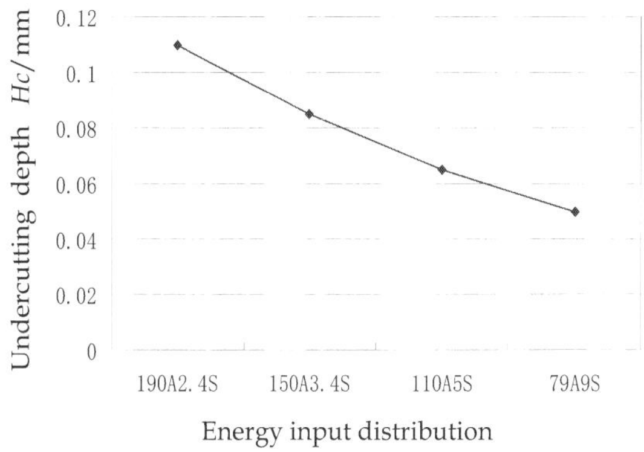

Fig. 14. Undercut depth of different welding current under the same input energy.

(2)The same welding current experimental results. When the welding current is the same, but the welding time is different, the welding undercutting depth will increas as the extension of welding time, as figure 15 shown. When the welding current is same, the molten pool size will increase as the extension of welding time, but the basic flow mode of molten pool is not change, still from the fusion line to pool center, so the welding undercutting depth increased.

(3)Different specimens in sulphur content experiment results. By the analysis above in this chapter shows that when the surface activity element proportion existing in the specimen is different, it will have an important impact on the flowing behavior of liquid metal. This experiment selected two kinds of Q235 steel specimens with the sulphur content in 0.02% and 0.05%. The actual welding test results in figure 16.

The figure 16 shown, with steel in different content was elected as the specimens, inthe same welding standard conditions the undercutting depth varies significantly. In 190A welding current conditions, the welding time is 2s, the specimens undercutting depth of 0.02% in sulfur content is almost three times of 0.05% . When the welding time for 4s, it still nearly twice. Therefore, the surface activity substance of the mother material has an important influence to the flowing of the welding molten pool, thus it can affecting welding results, lead to significantly change of the undercutting tendency.When the proportion of surface active substances rises, the absolute value of the surface tension get lower, but the regulation of the surface tension changing with temperature distributionare still the same.

In the experiment, the specimens with sulphur content in 0.02%, without exception are undercutting, and it is clearly visible with naked eye, as figure 17 shows. The background of the figure is under the currence is 70 A, 3 s TIG flat welding arc, the photo of looking down at the welding spot, the lower part is the cross section along the welding spot diameter and the measuring curve by the surface shape measuring instrument.The figure shows that, in the cooling process of the welding pool, the liquid metal have obviously tendency to gathering, formed the surface shape that among the middle convex and concave around.

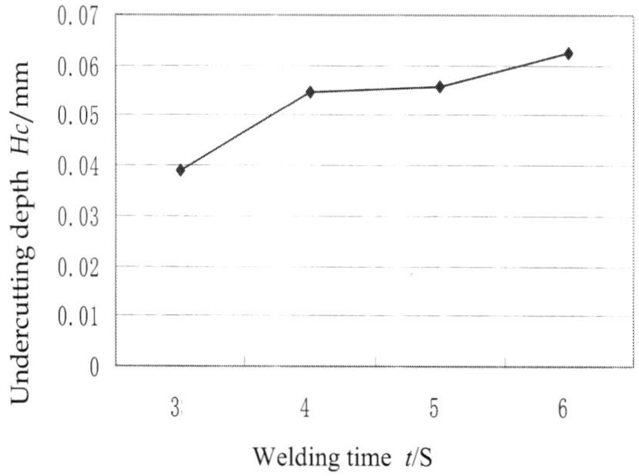

Fig. 15. Undercut depth of different time under the same current

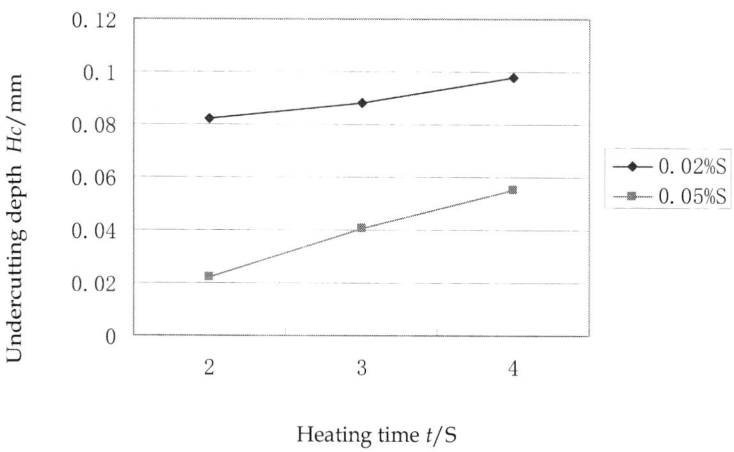

Fig. 16. Comparison of undercut as different S%.

Figure 18 shows the the welding spot under different welding parameters, including (a) photos for welding current 150 A, 4s welding results, (b) phptos for 190 A, 3s welding results, they have the same undercutting.

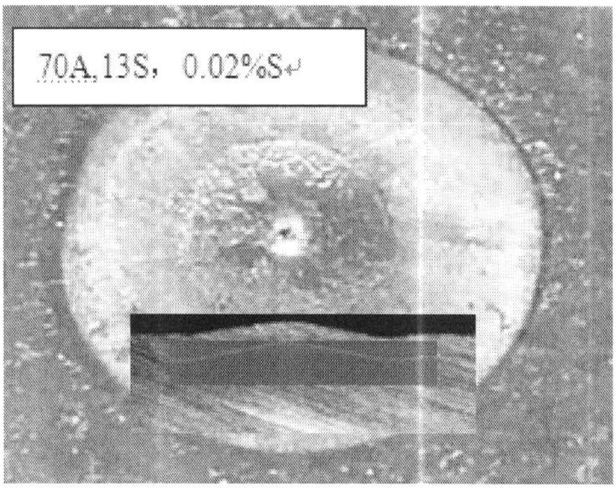

Fig. 17. Photo of welding spot.

3.2.3 Contrast experiments

For support the surface activity elements action mechanism and the conclusion above, the mother material lower in sulfur content (S content 0.01%, 0.004%) was chosen to do the contrast experiments, the experimental results shown in figure 19. We can see well spot surface didn't happened undercutting phenomenon. Observe carefully we will find, welding spot surface is tiny ripples, it is mainly because of the surface wave when molten pool

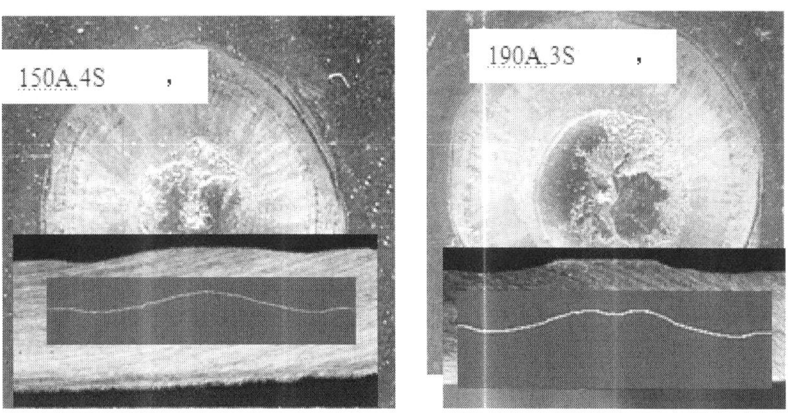

(a)150A、4s (b)190A、3s

Fig. 18. Photos of welding spots.

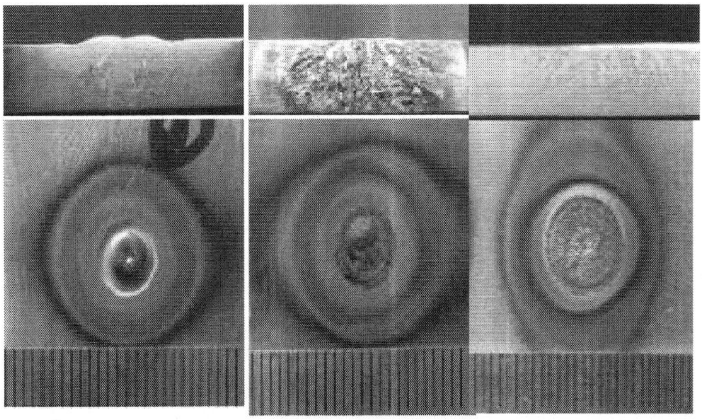

Fig. 19. Appearance of weld spot under the same welding condition.

freezes, but no middle convex phenomenon. This proved the fact that the metal surface activity elements can influence the direction of the metal flow, thus affects the undercutting when welding.

By the experiments above we can come to the conclusion: for the metal which contain very low surface active element content or no active surface elements, the metal surface tension temperature coefficient is less than zero in the range of above its melting point, liquid metal flow direction of molten pool is from the center point to fusion line, and it won't produce undercutting phenomenon in the designated welding.

To the metal contain active surface element, the degree of undercutting will get increase with the temperature gradient rise.

4. Experimental mobile welding results and analysis

4.1 Experimental study of a single TIG welding arc

To simplify the experimental conditions, comparison test has been done using the results which is got in different weld speeds in TIG welding process, shown in Figure 20. Experimental parameters are as follows: welding current is 190A, shielding gas flow is 10l/min, base metal thickness is 3mm, tungsten to the workpiece distance is 5mm, welding speed, respectively is 0.51m/min, 0.54m/minand 0.57m/min.

As can be seen from figure 20, when the welding speed is 0.51m/min (figure a), undercut doesn't exist; when the welding speed reaches to 0.54m/min time (figure b), undercut sometimes have sometimes no, in a transitional state; when the welding speed comes to 0.57m/min time (figure c), there is a clear continuous undercut phenomenon. Thus, the critical welding value of producing undercut is in 0.54m/min so in this welding conditions.

4.2 Study of single TIG transverse electromagnetic compression arc

To study the impact of the welding heat source shape on undercut, this passage uses magnetic control method to compress the single arc to make circular arc cross-sectional shape into the oval-shaped, as shown in figure 21.

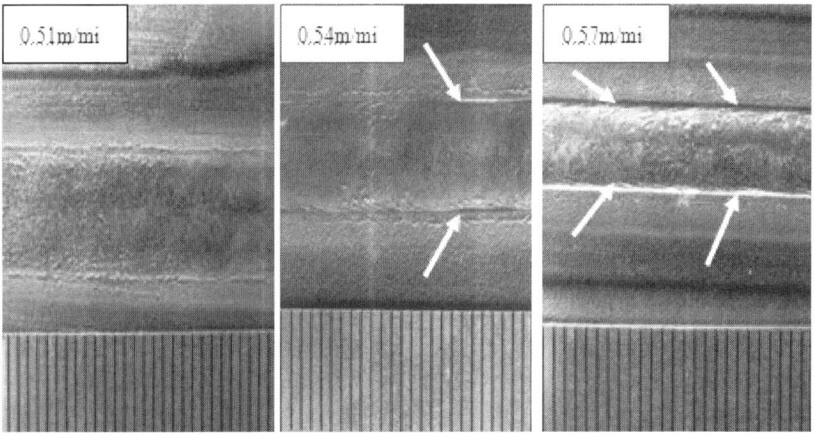

(a) (b) (c)

Fig. 20. TIG welding experiments.

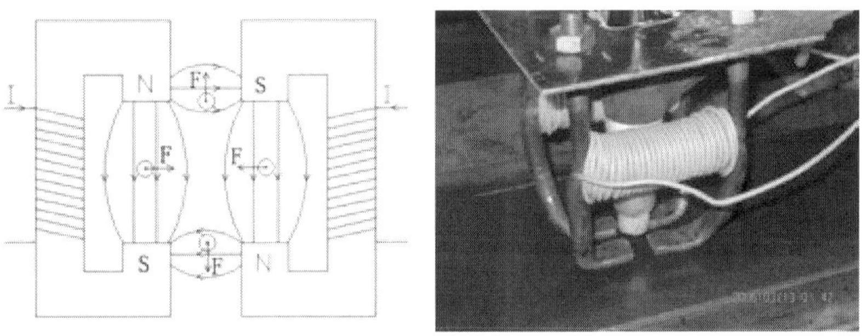

a, mechanism b, appearance

Fig. 21. The principle of electric magnetic field control and instrument photo.

4.2.1 Magnetic devices and experimental conditions

In this paper, arc electromagnetic control device has been designed using the principle of electromagnetic control, as shown in figure 21b. The coil turns is 100 turns, the control current is 3A, the magnetic pole spacing is 10mm, the magnetic pole area is 10mm×3mm, the distance between two pairs of poles is 10mm.

It can be seen from figure 21a, there is transverse magnetic field all around the arc, this magnetic field can compress the arc into an oval-shape.

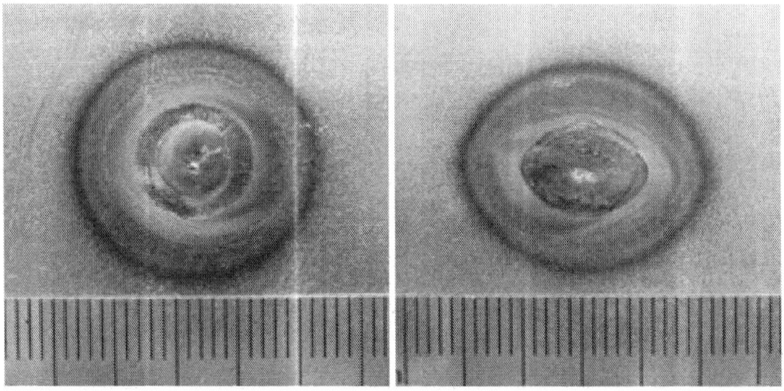

a. without magnetic press b. with magnetic press

Fig. 22. Welding spot of magnetic pressed arc or not pressed.

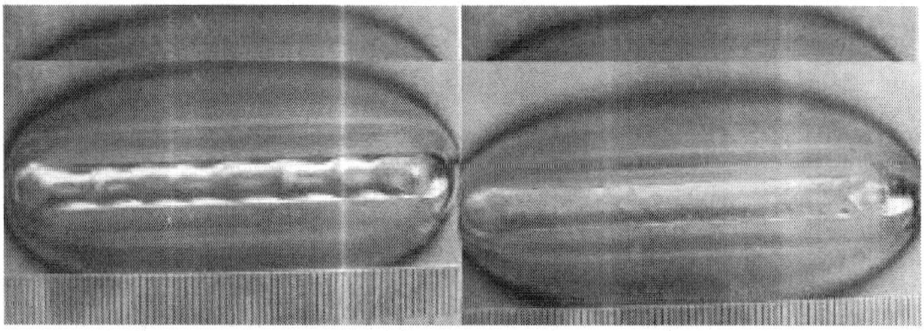

a. without magnetic press b. with magnetic

Fig. 23. Welding results of magnetic press arc or not press.

4.2.2 Magnetic compression fixed-point welding experiments

To verify the effect of magnetic arc compression, first fixed-point welding experiments have been done. Experimental conditions are as follows: welding current is 190A, welding time is 2.4s, shielding gas flow is 10l/min, and base metal thickness is 3mm. Fixed-point welding spot is shown in figure 22, a is the spot not applying magnetic compression, b is the spot applying magnetic compression. It can be seen that after magnetic compression fusion pool becomes oval-shape, magnetic compression effect is obvious.

4.2.3 Mobile welding magnetic compression arc experiments

Continuous welding experiments have been carried out along the long axis of oval in magnetic compression, welding current is 190A, welding speed is 0.63m/min, the results is shown in figure 23.

Figure 23, a is the result of no magnetic compression, weld undercut is scrious and there is a clear trend of bead hump. b is the result of applying magnetic compression, weldment is continuous and neat.

The temperature gradient is the most important factor of affecting the undercut extent. If setting the equivalent diameter d of circular welding arc, welding speed V_w, shown in the figure 24. The time that through cross-section A of welding arc is:

$$t = \frac{d}{V_w} \tag{4}$$

If the arc diameter d is a constant, and ignoring the deformation of the arc move, seen by the formula 4, the time t through cross-section A is inversely with the welding speed. In other words, actually, the welding speed increasing is achieved through the time t decreasing, which is similar to high current short time fixed-point welding, and it leads to temperature gradient increased, the tendency of undercut and extent arise.
If using oval-shaped heat to weld along the long axis of the b direction, shown in figure 24, when the welding speed is constant, because the heat distribution is larger along the welding direction, the time t1 is:

$$t_1 = \frac{b}{V_w} \tag{5}$$

Figure 24 shows, b>d, then, t1>t, it can be seen, when using elliptical arc to weld along with the long axis, the time which arc through the cross-section A increases and it equivalent to reducing the welding speed, and it can reduce the temperature gradient effectively and

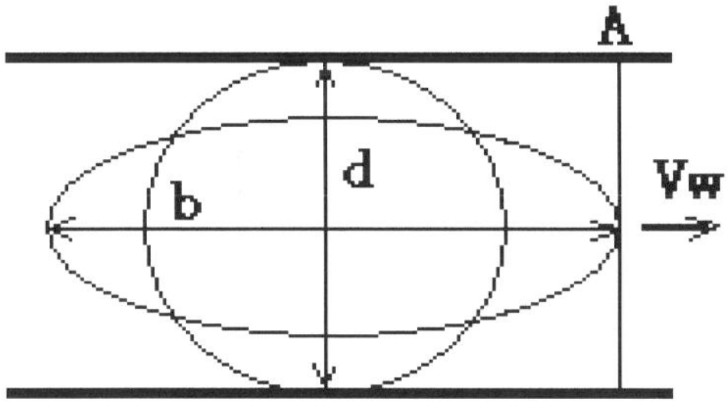

Fig. 24. Schematic diagram of arc.

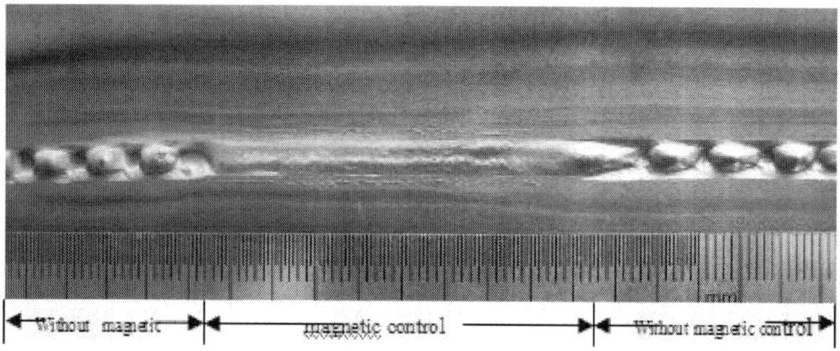

Fig. 25. Result of welding by magnetic control arc.

the tendency of undercut extent.

Secondly, when using the elliptical welding arc along the long axis, due to the width direction of the weld arc is compressed smaller, weld width will inevitably reduce, shown in figure 24. Thus, it can effectively improve the weld depth-with ratio, also help to inhibit the emergence of undercut.

Alternating magnetic arc welding results shown in figure 25. Thus, using of magnetic compression arc can effectively reduce weldment temperature gradient, inhabit undercut and increase welding speed.

5. Conclusion

The flowing direction of liquid metal is the main factor that depends on whether to produce undercut in the pool area near the fusion line. When the area's liquid metal flows from the pool center to the fusion line ,the undercut will not be produced; But when the area's liquid metal flows from the fusion line to t the pool center the undercut may be produced.

The temperature property and distribution of the pool liquid surface tension are one of the most important factors ,which affect the flowing direction of the liquid metal near fusion line area. When the existence of Surface Reactive Materials (sulfur or oxygen) makes the surface tension temperature coefficient become from negative to positive, the pool surface liquid will flow from the fusion line to the poll center, if the region can not be replenished by the metal in time, the undercut will be produced.

Temperature gradient is the most important factor that affects the degree of undercut of welding spot, the size of temperature gradient depends on the size of the surface tension gradient, and the latter directly affects a driving force of the pool metal . Under other unchanged conditions, undercut of the weld will increase with temperature gradient increasing.

Numerical simulation of the arc of movement shows that, with the welding speed increases, molten bath temperature gradients near the fusion line area increases, the dynamic aspect ratio decreases, both will increase the tendency of undercut. When increase the preheating temperature, reduce the thickness of welded parts, improve cooling conditions, can reduce the temperature gradient and improve the dynamic aspect ratio, help to curb the undercut.

When using electromagnetic compressed arc into an oval shape and welded along its long axis can effectively reduce the temperature gradient of direction perpendicular to the weld pool, increasing the dynamic aspect ratio, help to curb undercut generation, achieve higher speed welding.

6. References

T. Shinoda, J. Nakata, H. Miyauchi. Double Wire MIG Process and its Applications. IIW-Doc-XII-1543-98

E. Halmay The Pressure of the Arc Acting on the Weld Pool IIW DOC 212-368-76

J. Hedegard, J. Andersson, E. Tolf, K. Weman, M. Lundin. Enhanced Prospects for Tandem-MIG/MAG Welding. IIW-Xii-1808-04

Ken Michine, Stephen Blackman. Twin-Wire GMAW: Process Characteristics and Applications. Welding Journal. 1999,78(1):31~34

E. Lassaline, B. Zajaczkowski, T. H.North. Narrow Groove Twin-Wire GMAW of J. Hedegard, J. Andersson, E. Tolf, K. Weman, M. Lundin, Enhanced Prospects for Tandem-MIG/MAG Welding,IIW-DOC-XII-1808-04

Nomurh.H. Twin-Wire Gas Tungsten Arc Cladding Offers Increased Deposition Rates. Welding Journal. 1999,78(10):31~39

Amin.M. Pulse Current Parameters for Arc Stability and Controlled Metal Transfer in Arc Welding. Metal Construction. 1983,15(5):272~278

W.G. Essers, M.R.M.Van.Gompel. Arc Control with Pulsed GMA Welding. Welding Journal. 1984,63(6):26~32

Ghosh, P.K. Universality of Correlationships among Pulse Parameters for Different MIG Welding Power Sources. International Journal for the Joining of Materials. 2001, 13(2): 40~47

Mauro, Hiroshi, Hirata, Yoshinori; Noda, Yoshitaka. Effects of Welding Current Waveform on Metal Transfer and Bead Formation in Pulsed MIG Welding. Journal of the Japan Welding Society. 1984, 2(1):443~454

John Norrish, John Nixon. A history of pulsed MIG welding. Joining & Materials. 1989, 2(6):264~266, 268

Dzelnitzki, Dirk. Increasing the Deposition Volume or the Welding Speed?--Advantages of Heavy-Duty MAG Welding. Welding Research Abroad. 1999,45(3):10~17

G.M.Oreper,J.Szekely,Hear-and Fluid-flow Phenomena in Weld Pools,Journal of Fluid Mechanism,1984,147,pp53-79

Comparison with Experiments. Metallurgical Transactions B Volume 23B, 1992,(6):371-384
Geometry Mandal, Parmar, R.S. Effects of Pulse Parameters on Weld Bead Source. International Journal for the Joining of Materials. 1996, 8(2):69~75
M. S. Tsai et al. Numerical Heat Transfer. 1990, l(17): 73~89

11

Physical Mechanisms and Mathematical Models of Bead Defects Formation During Arc Welding

Wladislav Sudnik
R & E Center 'ComHi-Tech in Materials Joining'
Welding Department, Tula State University,
Russian Federation

1. Introduction

An increase in the productivity of arc welding is connected with an increase in both welding speed and welding current, which leads to the formation of welding defects, such as undercuts, humps, burn-through areas, etc. Fusion welding defects is classified due to the international standard ISO 6520-1.

In most cases mathematical models of weld formation describe the normal course of the process and establish the relation between welding parameters and weld pool sizes. It is much more difficult to construct the models which describe defective weld formation; the formation of the defects is described within the limits of the general model of the process at some combinations of welding parameters and weld pool sizes.

The arc pressure at the crater of a weld pool makes considerable impact on the basic processes occurring during the weld formation. Despite the arc pressure it is not a direct welding parameter, it has an important technological value as it defines the crater depth of a weld pool and essentially the lack of penetration and an incomplete fusion (Yamauchi et al., 1982). Free surface deformation of a molten weld pool is an important feature of fusion welding. For gas tungsten arc welding (GTAW), a significant weld pool deformation may take place at high current levels. In gas metal arc welding (GMAW), the free surface problem is more complicated due to the filler metal addition and droplet impact. Free surface deformation affects the fluid flow and heat transfer in the weld pool, which, in turn, affects the weld geometry. The first profound research on the formation of weld defects was made by Bradstreet, 1968, in which the basic mechanisms of the formation of weld defects were shown in fusion welding.

The first two decades of the 70s and 80s of the 20th century have been devoted to the research of the mechanisms of the formation of undercuts and the causes of the transition from a normal mode of the weld formation to that one with the appearance of defects, as well as to the creation of the correspondent mathematical models. Selected papers of that period are presented in Table 1.

Bradstreet, 1968	Effect of surface tension and melt flow on weld bead formation
Paton et al., 1971	Hypothesis of a weld pool hydraulic head and arc pressure balance, arc induced undercutting
Erokhin et al., 1972	Hypothesis of a level lowering of a weld pool and its fixation by solidification process
Yamamoto & Shimada, 1975	Hypothesis of a hydraulic jump and undercut bead, supercritical flow model
Nomura et al., 1982	Experimental confirmation of hypothesis validity Paton et al., 1971
Sudnik, 1985, 1991a, b	Mathematical modelling of solidified free surface profile in fusion welding. For the first time, undercutting had been modelled and simulated

Table 1. Selected papers of research on the mechanisms of the formation of undercuts.

For the last 20 years the mechanisms of the humps formation have been actively studied, and corresponding mathematical models have been developed, Table 2.

Tytkin, 1981	Theoretical study of coarse flaky surface. Kelvin-Helmholtz instability
Gratzke et al., 1992	Capillary instability model for humping
Lin & Eagar, 1983	Explained humping using vortex theory
Mendez & Eagar, 2003	Simple model for force balance between the gouged region and trailing region inside weld pool
Nguyen et al., 2005, 2006	Experimental study of hump formation by the instrumentality of LaserStrobe video imaging system; hump bead as a series of periodic fluctuation of swellings is one of dominating defect in high speed welding
Soderstrom & Mendez, 2006	Two types of humping formation: gouging region and beaded cylinder
Kumar & DebRoy, 2006	Unified mathematical model of humping in GTAW with Kelvin-Helmholtz instability
Cho & Farson, 2007	Thermohydrodynamic mathematical model and numeric analysis of hybrid process for prevention of weld bead hump formation
Chen & Wu, 2009	Thermohydrostatic mathematical model and numerical analysis of forming mechanism of hump bead in high speed GMAW

Table 2. Selected papers on the study of the humps formation and the creation of the correspondent models.

From a mathematical viewpoint, the modelling of the weld defects formation consists in simulation of the solidifying profile of the weld free surface. It demands the execution of numerical three-dimensional modelling of a weld pool with the deformable free surface, which for the first time had been executed by Sudnik, 1985, 1991 at predicting the quality of welds and simulation of the undercuts formation in GTAW. Then the first numerical thermohydrostatic mathematical models for GTAW with a deformable weld pool surface were created by Ohji *et al.*, 1983, and Zacharia et al., 1988.

In the 21st century there were developed fuller thermohydrodynamic models by Kumar and DebRoy, 2006; Cho and Farson, 2007, as well as the simplified thermohydrostatic models by Chen & Wu, 2009 for the simulation of the humps formation.

The purpose of the present work is the analysis of the physical mechanisms of the defects formation on the basis of the full and simplified mathematical models of the formation of undercuts and humps. The ways of thermohydrodynamic and thermohydrostatic modelling and simulation of the weld formation with correspondent equations and their boundary conditions will be presented. There will be stated the stready state mathematical models of the GTAW and GMAW processes, as well as the transient model for the GMAW processes that allows the reproduction of the formation of undercuts and burn-through areas. The task of the search for the welding parameters, providing faultless areas, will be illustrated on the examples of two-dimensional areas for GTAW- and three-dimensional areas for GMAW-processes.

2. Physical mechanisms of defects formation

Savage et al., 1979, showed that an increasing in welding speed beyond a critical limit results in weld-bead undercutting and/or humping, related to arc force and its distribution, Fig. 1.

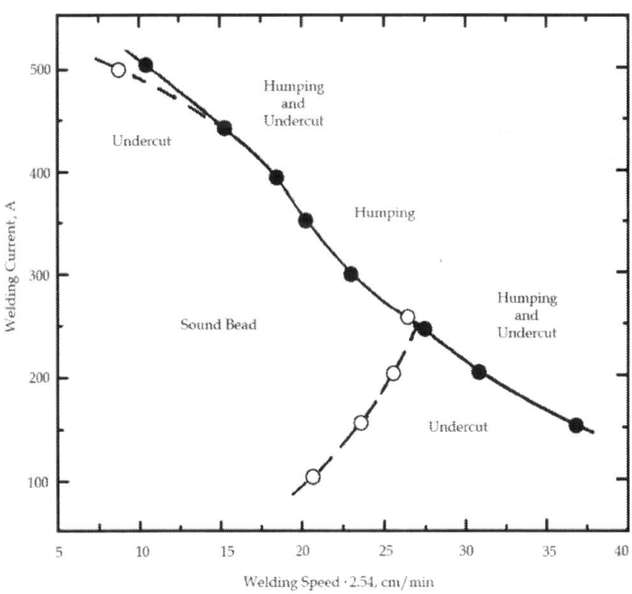

Fig. 1. Effect of welding current on welding speed limit, after Savage et al., 1979.

According to the results of this work there are some characteristic ranges of the weld current for the formation of undercuts and humps. For the currents below 250 A, critical welding speed corresponds to the beginning of formation of undercuts, and for the formation of humps higher speeds are required. In the range 250 - 400 A the speed limit corresponds to the beginning of the humps formation without undercuts. For the currents above 400-430 A, at first an undercut forms, whereas further increase in the welding speed leads to the humps.

2.1 Undercutting mechanisms

Bradstreet, 1968, used a high speed camera to observe undercutting and humping phenomena and showed that the undercutting is also strongly influenced by surface tension, which controls wetting at the edges of the weld pool. Paton et al., 1971, put forward a hypothesis about a weld pool hydraulic head and arc pressure balance. At an increase in the welding speed, the pressure at the weld pool bottom from a liquid column decreases at the constancy of the arc pressure, and at balance infringement probably formation of an undercut occurs. This hypothesis has been confirmed by Nomura et al., 1982 which have experimentally confirmed the justice of this hypothesis. Erokhin et al., 1972 put forward hypothesis of the level lowering of the weld pool and its fixation by the solidification process. Authors also considered that level of the liquid metal in the points of the maximal width of the weld pool where solidification at the weld edges begins, it appears below the surface of the basic metal owing to the big inclination of the weld pool mirror and the high speed of solidification. Chernyshov et al., 1979 considered that undercuts and other defects initiate difficult wave processes at the surface of the liquid metal and, in particular, hydraulic jumps of various forms. The hydraulic jump is known from hydraulics and is characterised by a spasmodic increase in height of the liquid level. The hypothesis (Yamamoto & Shimada, 1975; Shimada & Hoshinouchi, 1982) about the role of the hydraulic jump in the formation of welding defects, including the bead undercut has been undeservedly forgotten, Fig. 2.

They connect the occurrence of undercuts, first of all, with the arc pressure; therefore the aspiration to receive a smaller arc pressure has in their researches caused application of a welding process in the environment of inert gas of the lowered pressure (32 mm Hg). They have established that in comparison with welding at atmospheric pressure normal formation of the platen without undercuts can be received in wider range and at higher values of the weld current and the welding speed. At an increase in the welding speed instead of the normal wide platen there is a narrow deep platen, and on the front wall of a weld pool there is an area gouging. At excess of critical speed of formation of undercuts the area gouging extends on all a crater zone of a weld pool and reaches it midlength section. Thus liquid metal is displaced in a tail part of a weld pool with undercut formation at solidification of liquid metal.

Sudnik, 1985, 1991 developed a mathematical model of the weld formation in GTAW which, for the first time, allowed to model and to reproduce the undercut. The mathematical model of fusion welding process by Sudnik, 1991a, considered the energy equation in enthalpy statement and the equation of balance of pressure on deformable free surfaces of the weld pool at the full penetration with boundary conditions corresponding to the weld process. It has appeared that the undercut is one of natural forms of cross-section and arises at the fusion line at a level lowering of the surface of a weld pool on the sites of the beginning of the solidification.

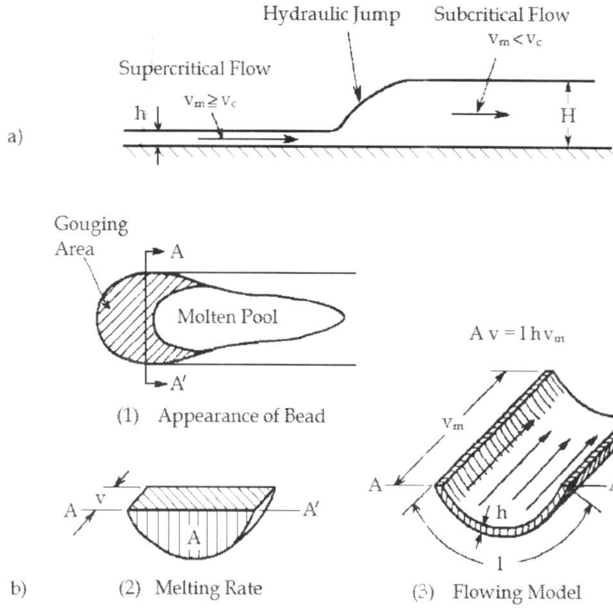

Fig. 2. Model of hydraulic jump and gouging region, after Shimada and Hoshinouchi, 1982

Ohji et al, 2004 presented a three-dimensional transient mathematical model of the GMAW-process for simulation of multiple pass welding and welding of fillet welds with cross-section fluctuations of the torch which also reproduces undercut formation.

2.2 Humping mechanisms

Bradstreet, 1968 is credited with publishing the first paper to directly recognise humping during GMAW. He has established that the hump defect was observed only at a high travel speed, that a leading ("push") weld gun travel angle suppressed hump formation, and that oxygen in the shielding gas exacerbated hump formation. He proposed a liquid instability theory that was later modified by Gratzke et al., 1992 explaining humping during GMAW. Yamamoto and Shimada, 1975 and Shimada and Hoshinouchi, 1982 used the theory of hydraulic jump to account not only for an undercut, but also for the humping phenomenon, Fig. 3.

Mendez and Eagar, 2003 offered an explanation of a penetration and occurrence of defects in a weld pool at high currents. Thus arc pressure pushes the fused metal to a back part of a weld pool, creating under an arc a thin layer of liquid metal. Premature solidification of this thin layer initiates formation of the humps, the split pool, parallel formation of humps, tunnel porosity and an undercut. The thin nature of a liquid layer is the reason of the increased penetration at a high current. They proposed a simple model for force balance between the gouged region and trailing region inside weld pool. Later, Soderstrom and Mendez, 2006 have offered two types of humping formation: gouging region and beaded cylinder. Nguyen et al., 2005, 2006 have lead an experimental study of hump formation by

the instrumentality of LaserStrobe video imaging system and have shown that hump bead as a series of periodic fluctuation of swellings is one of the dominating defects in high speed welding.
Cao et al., 2004 used a commercial software package, FLOW-3D, to simulate a transient moving weld pool under the impact of droplet impingement without the droplet generation. Choi et al., 2006 investigated hybrid processes of welding by the laser and GMAW for suppression of formation of humps and have shown that the heat input from the defocused laser beam applied in front of a GMAW pool suppresses formation of a weld bead hump defects. Kumar and DebRoy, 2006 have proposed the unified three-dimensional thermohydrodynamic mathematical model of humping defects in GTAW based on the Kelvin-Helmholtz instability of a free surface. This model can be used to help prevent humping when the effect of arc current, welding speed, shielding gas, electrode geometry, ambient pressure, torch angle, and external magnetic field are considered. Hu et al., 2007 studied weld pool dynamics under the periodical impingement of filler droplets and the formation of such defects as ripples. Cho and Farson, 2007 used a mathematical model of hybrid process laser + pulsed GMA welding for the prevention of weld bead hump formation and have also established how formation of bead humps in high-speed GMAW is prevented by additional laser heat input. Chen and Wu, 2009 have developed a thermohydrostatic mathematical model and have lead numerical analysis of forming mechanism of hump bead in high speed GMA welding.

3. Mathematical models of the formation of weld and bead defects

Generally anisothermic movement of a viscous compressed liquid in hydrodynamics is described by the system of the Navier-Stockes-Fourier equations that include.

- Three equations of Navier-Stokes (for a component of a velocity vector u, v, w).
- The continuity equation (for pressure p).
- The equation of convective heat conductivity (for calculation of temperature T).
- The equation of state which connects p, ρ, T (for ideal gases this equation is of the form $p = \rho RT$).

Thus, there are 6 equations with 6 unknown (u, v, w, ρ, T, p). This system is extremely difficult to solve even with the use of modern numerical methods on powerful personal computers. Therefore, often one should examine various simplifications of this system which in some cases adequately model a fluid flow.
The boundary problem for a concrete thermal and hydrodynamic process of fusion welding is formed by differential equations with initial and boundary conditions. The full formulation of this problem in scientific weld literature is absent. We will make an attempt to examine these questions with reference to arc welding and defects of a weld.
The weld pool model can be described by the three main conservation equations of energy, movement and continuity (Rykalin et al., 1985) with the boundary conditions defining interaction of heat and force flows with welded workpiece. The order of the record of the differential equations of impulse (movement), continuity and energy (a convective heat transfer) accepted in hydrodynamics, from the point of view of mathematical physics and the numerical iterative solution by the computer, does not play a special role, as the solution of the coupled problem is reached iteratively. For high-temperature weld processes the first-priority solution of the energy equation in most cases is defining and sets the melt volume and its form where hydrodynamic processes proceed.

3.1 Model of convective heat transfer and viscous fluid flow dynamics

3.1.1 Energy equation, or the equation of a convective heat transfer

The weld pool model according to Rykalin et al., 1985 is described by conservation equations of energy, movement after Navier-Stokes and continuity. The energy equation with a moving heat source can be presented as the equation of heat conductivity in enthalpy statement with a fluid flow (Kou & Sun, 1985)

$$\rho\left[\frac{\partial H}{\partial t}+\nabla\cdot(\vec{u}H)\right]+\rho v_{\mathrm{w}}\frac{\partial H}{\partial x}=\nabla\cdot(\lambda\nabla T) \tag{1}$$

where ρ is the density, is an enthalpy coupled with temperature T by relation , where , with c is thermal capacity, ψ_l is a fraction of a liquid phase in two-phase zone of solid-liquid melting / solidification area, ; T_l and T_s are liquidus and solidus temperatures, is the latent heat of melted alloy, is a flow velocity vector of a weld pool melt, v_W is the welding speed, and λ is heat conductivity coefficient.

3.1.2 Thermal boundary conditions

The boundary condition on the top surface describes the energy source heat flows and heat sinks on an air convection, radiation and evaporation

$$-\lambda\frac{\partial T}{\partial t}=\frac{\eta Q}{2\pi k_q^2}\exp\left(-\frac{r^2}{2k_q^2}\right)-\alpha(T-T_0)-\sigma\varepsilon(T^4-T_0^4)-q_{\mathrm{vap}} \tag{2}$$

where η is the process efficiency, Q is the heat source power, k_q is heat flow concentration coefficient. A is the heat exchange-coefficient, σ is the Stefan-Boltzmann constant, ε is the emissivity, q_{vap} is the heat sink on evaporation.

3.1.3 Movement conservation equation for a viscous liquid

The movement conservation equation for a viscous incompressible Newtonian liquid in Navier-Stokes approximation is presented according to Landau and Lifshiz, 1959

$$\rho\left[\frac{\partial\vec{u}}{\partial t}+(\vec{u}\cdot\nabla)\vec{u}\right]=-\nabla p+\mu\nabla^2\vec{u}+\rho\vec{g} \tag{3}$$

where p is the pressure, is the viscosity, is the acceleration vector of free fall. In this equation the left part describes inertial force, including local and convective acceleration. In the right part there is a pressure gradient, a friction force and elevating force of a free convection.

3.1.4 Force boundary conditions

Boundary conditions for a free surface may be of two types: kinematic, setting speeds on liquid boundaries, and dynamic, connected with pressure. Other boundary conditions for the walls of a weld pool are similar to hydrodynamic conditions on channel walls, i.e. to adhesion conditions according to which normal and tangential components of speed on a wall are equal to zero. The form of a free surface of a weld pool is defined from a condition of balance of internal and external forces on its surface. To avoid any formulations of this balance, we will make an attempt for its conclusion from the first principles.

Boundary conditions on a free interface a liquid - gas include:
- normal stresses from pressure balance of a liquid and forces of a viscous friction (Landau and Lifshiz, 1959; Batchelor, 1970), and also it is similar from balance of capillary forces and forces of an arc pressure in the arc welding conditions, and, the left part of expression is written down like expression taking into account a viscous stress tensor

$$-p+2\mu\frac{\partial u_{\mathrm{n}}}{\partial n}=-p_{\mathrm{arc}}+\sigma\left(\frac{1}{R_1}+\frac{1}{R_2}\right) \tag{4}$$

- and tangential stresses of Marangoni forces

$$\mu\frac{\partial u}{\partial n}=-\frac{\partial\gamma}{\partial T}\frac{\partial T}{\partial z} \tag{5}$$

where p is the pressure in a liquid on a free surface in a normal direction, μ is the dynamic viscosity of a melt, u_{n} is the velocity component of a fluid flow on a normal to a movement direction, γ is the surface tension, R_1 and R_2 are the main radiuses of curvature of the bent surface and p_{arc} is the arc pressure.
For the approximate calculation of curvature of surface Z = Z (x, y), set in an explicit form, in hydrodynamics the formula of a Landau and Lifshits, 1959 is used

$$\frac{1}{R_1}+\frac{1}{R_2}=-\left(\frac{\partial^2 Z}{\partial x^2}+\frac{\partial^2 Z}{\partial y^2}\right)$$

which is fair for poorly bent surfaces. At the big surface deflections the multiplier of formulas of the differential geometry, equal to a cosine of an inclination of a surface of a weld pool is in addition used.

3.1.5 The continuity equation, or the mass conservation law

The equation of continuity follows from the law of a mass conservation and registers

$$\frac{\partial\rho}{\partial t}+div(\rho\vec{u})=0 \tag{6}$$

This law of a mass conservation is important for modelling of a fusion welding process, including arc welding. At local heating there is a thermal expansion of metal and phase transformation during melting. Accepting an assumption about an immovability of solid boundary surfaces of a solidus and a constancy of mass of metal during heating and fusion (for example, during welding by a nonconsumable electrode), volume changes can be defined on the temperature change of density proportional to coefficient of expansion. The weld pool volume $V_{\mathrm{m},\,0}$ with density ρ_0 at reference temperature T_0 and with dependence of density on temperature ρ (T) is equal to weld pool volume V_{m} with convexity at the expense of thermal expansion and phase transformation at melting. The convexity form of a melt is fixed at the front solidifications that define the form of a profile surface of a weld. The subsequent volume shrinkage and return thermal compression at cooling of solid metal reduce the sizes, not changing of convexity and concavity forms of area, for example, a weld undercut a little. The increase in volume of a weld pool because of a filler addition or its reduction because of a technological gap is considered by addition or reduction of these volumes to volume V_m (Sudnik *et al.*, 1999). The account of these phenomena defines an additional geometrical condition of change of volume, or balance of masses, in a welding zone

$$\iiint\limits_{V_{m,0}} \frac{\rho_0}{\rho(T)} dV = \iint\limits_{Vm} Z(x,y)\, dxdy \tag{7}$$

The solution of the equations system (1) - (7) even on modern computers is connected with the big expenses of computer time. For example, for understanding of the humping formation during GMAW Cho & Farson, 2007 used software Flow3D which for reproduction of real-time 3 seconds has demanded expenses of almost 4 days of central processing unit time. Many simplified models are based on assumptions of weak changes of velocities and temperatures one direction under the relation with changes in other direction, for example, model of a boundary layer and other reduced Navier-Stokes equations.

3.2 Model of conductive heat transfer and inviscid fluid flow statics

For simplification of a dynamic problem (1) - (7) 2 assumptions are accepted:

- The fluid flow is accepted poorly convection, i.e. convection carrying over of heat to the fused metal is ignored,
- The fused metal is a nonviscous incompressible liquid.

It allows rejecting in the equations velocity terms who have less essential value for definition of behaviour of a weld pool which can be described by the theory of an ideal liquid of Euler.

3.2.1 Energy equation, or the equation of a conductive heat transfer

The energy equation from moving with a welding speed v_w a heat source in this case is represented by the usual equation of heat conductivity (without convection) in enthalpy statement

$$\rho \frac{\partial H}{\partial t} + \rho v_w \frac{\partial H}{\partial x} = \nabla \cdot (\lambda \nabla T) \tag{8}$$

with the thermal boundary condition (2).

At small velocities of movement of a source it is possible to neglect acceleration of a flow and viscosity of a melt, i.e. to sink inertial and viscous terms. From four basic driving forces of a fluid flow: gravity, pressure, friction and inertia remain only two: gravity and pressure which allow describing the phenomena in a weld pool under the hydrostatic law as the pressure created by a column of a liquid. Thus, three known conditions of a hydrostatics should be satisfied according Myshkis *et al.*, 1987:

- the Euler's condition in a weld pool volume,
- the Laplace law (condition) on a melted free surface
- the Dupre-Yuong condition on a contact line of three environments: a liquid - gas - solid body.

At balance the equation of continuity (6) becomes simpler (), that means, that a density field is permanent.

The equation of Navier-Stokes (2) of fluid dynamics passes in Euler's condition, or the first condition of a hydrostatics

$$\nabla p = \rho g, \tag{9}$$

and the hydrodynamic boundary condition of normal stresses (4) becomes simpler and takes the form the Laplace law, or the second condition of a hydrostatics

$$p = \sigma\left(\frac{1}{R_1} + \frac{1}{R_2}\right) + p_{arc} \tag{10}$$

Integrating Euler's condition (7) where z is the height of a liquid column, we receive where C is the arbitrary constant. This equality does not impose any restrictions on melt position, and only defines pressure distribution in it. The Laplace's law thus modified for weld conditions (9) will play a role of the differential equation of a weld pool surface, and a geometrical condition (7) - a role of a corresponding boundary condition.

3.2.2 Momentum conservation equation for a nonviscous liquid

The modified equation of Laplace for conditions of fusion welding taking into account an arc pressure is of the form as

$$\sigma\left(\frac{1}{R_1} + \frac{1}{R_2}\right) = \sigma\nabla \cdot \frac{\nabla Z}{\sqrt{1 + |\nabla Z|^2}} = \sigma\left[\frac{Z_{xx}^2\left(1 + Z_y^2\right) - 2ZxZyZxy + Z_{yy}^2\left(1 + Z_x^2\right)}{\left(1 + Z_x^2 + Z_y^2\right)^{3/2}}\right] = \rho g z + \frac{F}{2\pi k_p^2}\exp\left(-\frac{r^2}{2k_p^2}\right) + C \tag{11}$$

where Z = Z (x, y) is the free surface equation of the weld pool set in an explicit form in the Cartesian system of co-ordinates x, y, z; the z is vertical coordinate of a deformable free weld pool surface, Z_x, Z_y, Z_{xx}, Z_{yy}, Z_{xy} are private derivative functions Z=Z (x, y) on corresponding co-ordinates; *F* is force action of a source, the is concentration coefficient of pressure and *r* is distance from an axis of a pressure source.

For calculation of curvature of a surface in hydrodynamics the formula for the first time presented by Landau and Lifshits, 1959 is used. The constant C in the modified Laplace's equation (9) pays off from a condition satisfaction of an additional condition the continuity equations (7). Mathematically, this constant C is the Lagrange multiplier in extreme statement (Landau and Lifshiz, 1959; Kim and Na, 1999), and physically, the constant C designates caused by surface deformation average change of pressure in a melt (Sudnik & Erofeev, 1986; Radaj, 1999). Its value is defined from a boundary condition of a mass conservation

3.2.3 Boundary condition

The boundary condition (7) dynamic problems (1) - (7) does not change

$$\iint_{V_m} Z(x,y)\,dxdy = \iiint_{V_{m,0}} \frac{\rho_0}{\rho(T)} dxdydz \tag{7'}$$

Thus, the equations of a weld pool surface in hydrostatic approximation, and taking into account the equation of a conductive heat transfer (the energy equation) are formulated; such model can be named a thermohydrostatic model of a weld pool.

3.3 Models of weld bead defects

Defects of welded structures can be divided into two groups:

- having the quantitative characteristic (width and depth of a penetration of a weld, depth and undercut radius, height of convexity or depth of concavity, etc.);
- having qualitative character (presence of cracks, of pores, of burn-through etc.).

Current values of the first defects pay off from the corresponding mathematical models, the second defects are found out by calculation of the physical sizes influencing occurrence of defects, and comparison of these sizes and their relations with some critical values.

For the calculation of welding processes on weight it is necessary to reveal possibility of occurrence of a burn-through that is fixed by quantity of excesses on a profile of cross-section section of the bottom surface of a pool. Sudnik, 1991a has established that occurrence of the first bending point is a necessary condition of safe loss of stability of a surface, and the second bending point a sufficient condition of occurrence of a burn-through. The condition of loss of stability of a surface of an underside of a weld pool, or a burn-through, mathematically registers as follows:

$$\frac{\partial^2 Z(x,y)}{\partial y^2} = 0 + \xi_{dz}, \qquad (12)$$

where y is the cross-section coordinate of the fused metal surface, is the calculation error.

3.4 Steady state mathematical models for arc welding

3.4.1 Model for GTAW

The three-dimensional model for the GTAW-process which predicts undercuts and burn-through areas in butt welding is described by the system of three main equations:

1) Energy conservation

$$\rho v_{\mathrm{w}} \frac{\partial H}{\partial x} = \mathrm{div}\left[\lambda_{\mathrm{eff}}(T)\mathrm{grad}T\right] \qquad (13)$$

where $\lambda_{\mathrm{eff}}(T)$ is the effective heat conductivity coefficient, depending on temperature T and considering a fluid flow in a weld pool, , λ_{L} is the heat conductivity coefficient at liquidus temperature T_{L} and T_0 is the ambient temperature.

Thermal boundary conditions are usually represented as

$$-\lambda\frac{\partial T}{\partial z} = \frac{\eta Q}{2\pi\sigma_{\mathrm{T}}^2}\exp\left(-\frac{r^2}{2\sigma_{\mathrm{T}}^2}\right) - \alpha(T-T_0) - \varepsilon\sigma_0(T^4 - T_0^4) \qquad (14)$$

where ηQ is the effective heat source power, σ_{T} is the distribution parameter of the heat flow of anode power, other designations standard.

2) Movement conservations of a weld pool free surface $Z = Z(x, y)$ with adaptation for welding conditions

$$\pm\sigma(T)\nabla\cdot\frac{\nabla Z}{\sqrt{1+|\nabla Z|^2}} = \rho g Z + p_{\mathrm{arc}} + C \qquad (15)$$

where σ(*T*) is the melt surface tension, is the Nabla-symbol, g - acceleration of free fall, p_{arc} is the arc pressure, *C* is the constant designate caused by surfaces deformation average change of pressure in a melt, or the Lagrange multiplier in extreme statement.
The distributed arc pressure is defined as

$$p_{arc} = \frac{F_{arc}}{2\pi\sigma_p^2}\exp\left(-\frac{r^2}{2\sigma_p^2}\right) \tag{16}$$

where F_{arc} is the full force of an arc pressure, σ_p is the parameter of its distribution, *r* is the distance from an arc axis.
3) Mass conservation

$$\iiint_{V_{M0}} \rho_{T_0}\, dxdydz = \iiint_{V_{M0}} \frac{\rho_{T_0}}{\rho(T)}\, dxdydz \tag{17}$$

where the weld pool volume at a room temperature in left side of equation is equal to volume of the fused pool taking into account convexity at volume thermal expansion and phase transformation «solid - liquid» in right side of equation, that is considered through temperature change of density ρ (*T*).
Demonstration examples of GTAW process calculation of an austenitic steel, such as 304, by sheet thickness 2,2 mm a tungsten electrode with a sharpening corner 30 ° by means of the over formulated model and comparison calculated and experimental geometry are shown in Fig. 3.
The macrograph illustrates the longitudinal section of a weld pool on a mode: I = 265 A, l_{arc} = 2 mm and v_w = 1,1 sm/s which is compared with a corresponding calculated profile of the same weld pool.

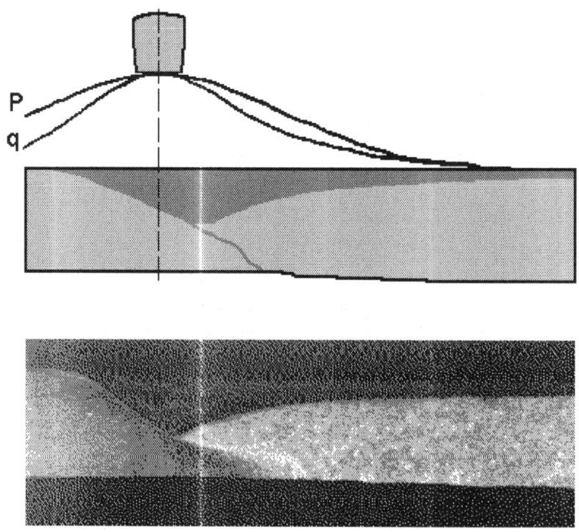

Fig. 3. Comparison calculated (top) and experimental (bottom) longitudinal sections of a seam (welding modes are specified in the text); *p* and *q* are the curve distributions of an arc pressure and its heat flow, after Sudnik, 1991.

Cross-section sections of the hardened seam with an undercut and a three-dimensional weld pool at GTAW are presented in Fig. 4.

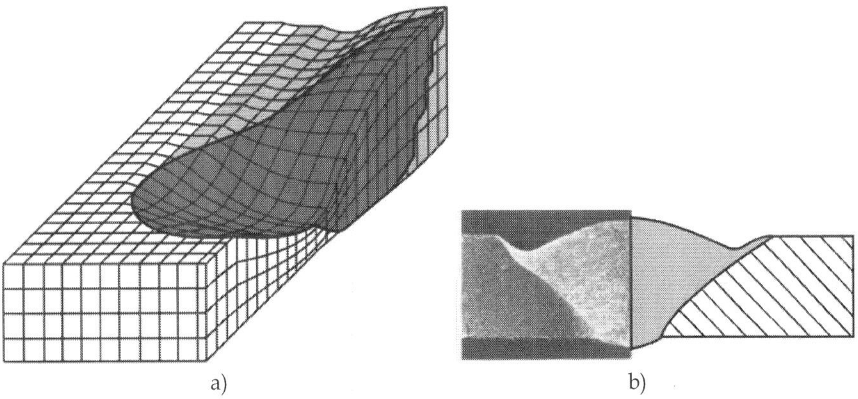

Fig. 4. Computed three-dimensional weld pool (a) and comparison between the macrograph and calculated two- dimensional finished weld shapes with undercutting (b) during GTAW at welding current 430 A, travel speed 30 mm/s and arc length 1 mm of CrNi-alloyed steel, plate thickness 2,2 mm; after Sudnik, 1991b

The equations were solved numerically by control volume method. The numerical approximation of nonlinear mathematical model was realised in 1988 by programming the language FORTRAN with operational system OS DVK on a computer DVK-3 with an operative memory volume of 64K, manufactured in the former USSR. Visualisation of a three-dimensional weld pool and its free surface with an undercut of the solidified weld was executed and, for the first time in the world, was published by the author in 1991.

Undercutting as quantitative weld bead defects

The numerical analysis of formation of an undercut of a fusion line is executed in the thesis for a doctor's degree by the author. It is shown, that two major factors defining the form of an undercut are the level of liquid metal before front of solidification and the position of the last. The first factor depends on balance of the distributed forces in a weld pool and pool hydrodynamics, and the second solidification on thermal conditions. Ways of prevention of undercuts are: 1) redistribution of an arc pressure by a cathode deviation forward or use of the hollow cathode for reduction of an arc pressure and an exception of a gouging and 2) use of heating or decrease in temperature of a pool (for reduction of a gradient of temperatures at the front solidification) and 3) transition to two-and to the multiarc processes effectively realising both above-mentioned ways.

Burn-through as quality weld bead defects

Nishiguchi et al., 1984 were the first who theoretically and experimentally have proved a prediction method of a burn-through in fusion unsupported welding, Fig. 5.

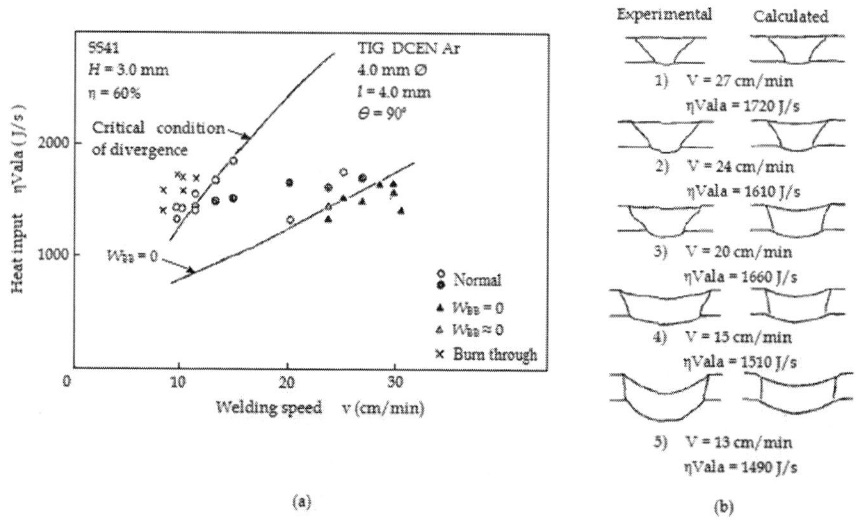

Fig. 5. Comparison between computed and experimental tolerance zones for mild steel, sheet thickness 3 mm during GTA unsupported welding, after Ohji et al., 1992

In the doctoral thesis, Sudnik, 1991a, it is shown that ways of prevention of burn-through areas are: 1) reduction of weld pool weight, 2) decrease in temperature and recoil vapour pressure, for example, at the expense of introduction of a filler wire or use of electromagnetic stirring and 3) imposing behind an arc of the external cross-section magnetic field creating at interaction with a current in a weld pool, vertical volume forces.

Two-dimensional area of defectless welds

The two-dimensional area of defectless formation of a weld and weld defects formation such as lack of penetration, and also burn-through and the continuous undercut, depending on a welding speed and a current, is shown in Fig. 6.

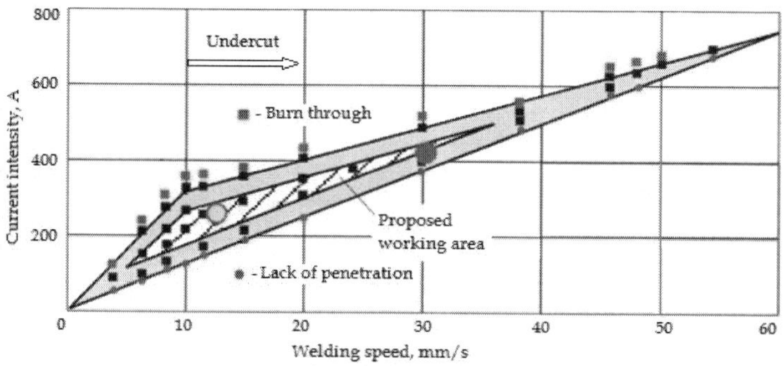

Fig. 6. Defectless area and defects of a GTA weld depending on welding speed and arc current for austenitic steel received by modelling of heat transfer and hydrostatics; after Sudnik, 1991b.

3.4.2 Model for GMAW

Three-dimensional numerical model GMA welding process is described by system of the energy, masses and movement equations (Sudnik et al., 1999a). In the energy equation, arc electric power is the sum of powers selected in anode and cathode areas, and also in arc column plasma ΔQ_{col}. Anode power is divided into two components - volume q_{vol}, and surface q_{surf}.

In this case, the system of three main equations is given as 1) energy conservation

$$\rho v_{w} \frac{\partial H}{\partial x} = \mathrm{div}\left[\lambda(T)\mathrm{grad}T\right] + q_{vol} \tag{18}$$

with boundary conditions

$$-\lambda \frac{\partial T}{\partial z} = \frac{q_{surf,a}}{2\pi\sigma_a^2}\exp\left(-\frac{r^2}{2\sigma_a^2}\right) + q_{surf,c} + q_{surf,col} - \alpha_s(T - T_0) \tag{19}$$

where , v_w is the wire feeding rate, A_w is the area of its cross-section, T_s is the solidus temperature, H_m is the melting enthalpy; is the anode power, ΔH_w is the enthalpy of drops overheat, ; T_{vap} and T_L are the evaporation and wire melting temperatures, σ_a is the distribution parameter of anode power of drops.

2) Movement conservations of a weld pool free surface

$$\pm\sigma(T)\nabla \cdot \frac{\nabla Z}{\sqrt{1+|\nabla Z|^2}} = \rho g(h - Z) + p_{arc} + p_v + C \tag{20}$$

where h is the height of a column of a melt, p_v is the recoil vapour pressure.

3) Mass conservation weld connection with the account of its local increase from receipt of drops of electrode metal is given as

$$\iiint_{V_{M0}} \rho_{T_0} dxdydz = \iiint_{V_{M0}} \frac{\rho_{T_0}}{\rho(T)} dxdydz + \frac{\pi d_w^2 v_f L_{wp}}{4 v_w} \tag{21}$$

where L_{wp} is the average length of a weld pool.

Continuous weld bead defects

A typical continuous weld bead defect, such as the weak undercut of a fillet weld, is shown in 7.

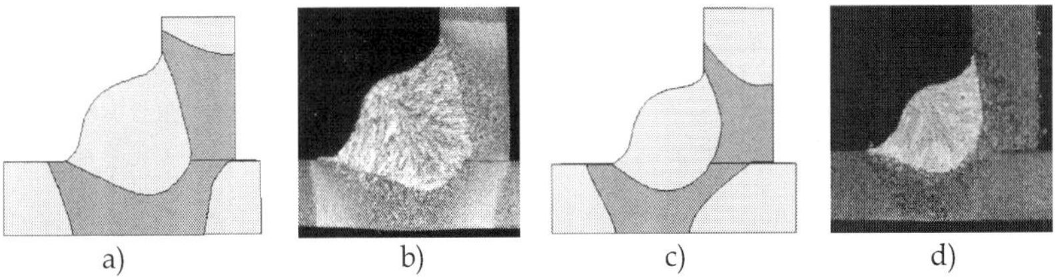

Fig. 7. Comparison calculated (a, c) and experimental (b, d) macrographs; sheet thickness 2 + 2 mm, cross-section, torch inclination 45 °; travel speed v = 0,7 m/min, wire speed v_w = 4 m/min, current I = 185 A and arc voltage U = 19 V for sections (a) and (b) and v = 1,6 m/min, v_w = 6 m/min, I = 250 A and U = 20 V for sections (c) and (d); after Sudnik, 1999.

Defectless welds with optimization of the process parameters

The choice of the best value is based on the solution of an optimizing problem taking into account welding parameters. The algorithm of search of welding parameters I, U and v_w in the field of the current-voltage characteristic of an arc includes following steps:

- the choice of the maximum weld current;
- calculation of a corresponding arc voltage;
- the task of some initial welding speed and search by a method of gold section of the maximum welding speed v_{max} at which the bottom run is provided;
- search of the minimum welding speed v_{min} at which the full penetration of one of details is provided;
- the estimation of probability of defects of a run-out of a pool and an undercut with updating if necessary v_{min};
- calculation of i- coefficients of variation P_i;
- repetition of procedures of calculation, since new value of current and calculation of new coefficient of a variation;
- the choice of the greatest value P_i as optimum and storing of optimum values of t I_{so}, U_s and v_{so}.

The screen copy of results of finding an optimum point in admissible area of change of a current and voltage, and also GMA welding speed in mixture CO_2 + 18 % Ar of butt welds of a low-alloyed steel is depicted in Fig. 8.

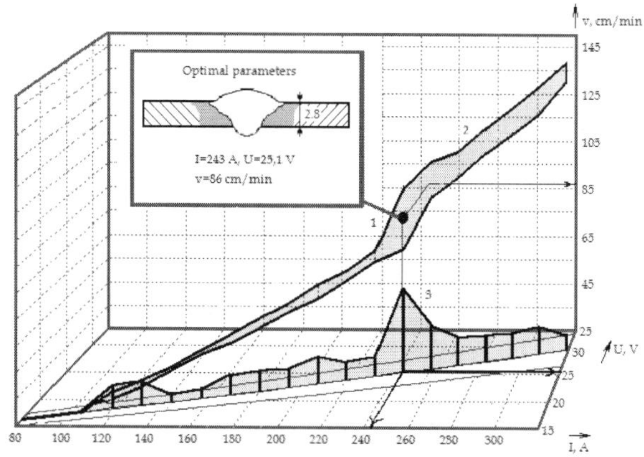

Fig. 8. Results of the process optimal parameters (operating point 3) in region (1) at plate thickness 2.8 mm, wire diameter 1 mm, electrode extension 16mm; after Sudnik et al., 1997.

3.5 Transient mathematical models for arc welding

Non-stationary model of GTAW or GMAW processes differs a dynamic term in the energy equation (Sudnik et al., 1999b)

$$c\rho \frac{\partial H}{\partial t} + \rho v_{\mathrm{w}} \frac{\partial H}{\partial x} = \mathrm{div}\left[\lambda(T)\mathrm{grad}T\right] + q_{\mathrm{vol}} \tag{22}$$

with boundary conditions

$$-\lambda\frac{\partial T}{\partial z}=\frac{Q_{\mathrm{an}}}{2\pi\sigma_{\mathrm{an}}^2}\exp\left(-\frac{r^2}{2\sigma_{\mathrm{an}}^2}\right)+\frac{Q_{\mathrm{cat}}+nQ_{\mathrm{col}}}{2\pi\sigma_{\mathrm{cat}}^2}\exp\left(-\frac{r^2}{2\sigma_{\mathrm{cat}}^2}\right)-\alpha_{\mathrm{s}}(T-T_0) \qquad (23)$$

where n is the fraction of power of a column of the arc, spent for heating of crater walls and surfaces of fillet welds.

The mass and movement conservation equations for a mode of welding without current programming and wire feeding rate do not change.

3.5.1 Beginning and end of weld

In the transient process of pulsed arc welding without additional programmed control parameters in the seam start there are excess of convexity of a welded seam (humps), and in the seam end - depressions. A comparison of experimental and calculated cross-sections of transient process of pulsed arc welding of aluminium sheets (fig. 9) shows their good conformity.

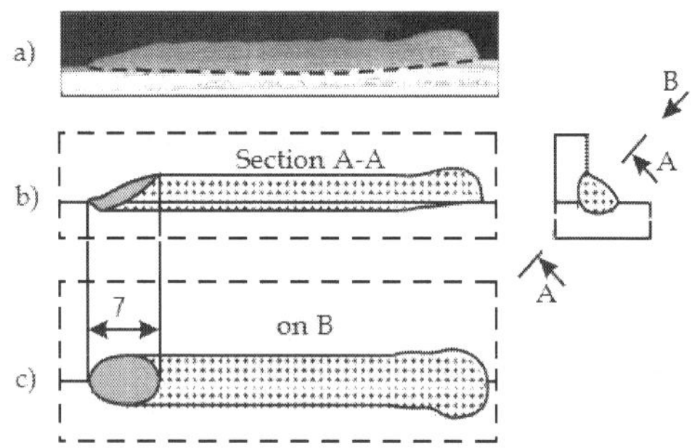

Fig. 9. Experimental longitudinal section (a), calculated cross sections of the welded joint (b, c) in pulsed metal inert gas welding of aluminium alloy AlMg2,7Mn, sheet thickness 3.4 mm, welding speed 0.78 m/min; after Sudnik et al., 2002.

3.5.2 Discontinuous weld bead defect or humping

Chen & Wu, 2009 have offered the simplified thermohydrostatic mathematical model and have conducted numerical analysis of forming mechanism of hump bead in high speed GMA welding. Authors have taken into account both the kinetic energy and heat content of backward flowing molten jet They have entered into equation (20) the kinetic term describing an impulse of the melt, flowing back

$$\sigma\left[\frac{Z_{xx}^2\left(1+Z_y^2\right)-2ZxZyZxy+Z_{yy}^2\left(1+Z_x^2\right)}{\left(1+Z_x^2+Z_y^2\right)^{3/2}}\right]=\rho g z+p_{\mathrm{arc}}+p_{\mathrm{d}}+\frac{1}{2}\rho u_{\mathrm{h}}^2+C. \qquad (24)$$

In this study Chen & Wu, 2009, a presumed distribution of fluid flow velocity is employed, and emphasis is put on its effect on the hump formation. The experimental observations (Hu & Wu, 2008) show that the gouging region is a very thin layer of liquid that transports molten metal to the trailing region, and the backward flowing molten metal is the main driving force toward the rear of the weld pool. Thus, only the fluid velocity in rearward-direction U is taken into account in determining the momentum of backward flowing. At any transverse cross-section of weld pool, the fluid velocity u_h takes its maximum at the pool centre, and decreases along y-direction as described.

$$u_h(x,y)=\begin{cases}\sqrt{(k_v v_0)^2+2a_1(x-x_w)}\cdot\dfrac{\xi(x)-y}{\xi(x)} & \text{for gouged region}\\ \sqrt{(k_v v_0)^2+a_1L+2\dfrac{(k_v v_0)^2+a_1L}{L}(x+\dfrac{L}{2}-x_w)}\cdot\dfrac{\xi(x)-y}{\xi(x)} & \text{for humped region}\end{cases} \tag{25}$$

where v_0 is the droplet velocity when it impinges on the pool surface, is the x-coordinate of wire centre line, $\xi(x)$ is the half width of weld pool at different x-coordinates, L is the distance from wire centre line to the rear edge of weld pool, k_v and a_1 are coefficients depending on the process parameters.

In high speed GMAW, backward flowing molten metal delivers most of the droplet heat content to the rear of weld pool. The thickness of molten metal layer varies along the pool length direction. It is very thin at the pool front, while it is thicker at the pool rear. It is assumed that the distribution depth of overheated droplet heat content is related to the molten layer thickness. Then, the source term S_V describing the distribution of heat content of transferred droplets in Eq.(?) may be expressed as

$$S_V(x,y,z*)=\frac{Q_d}{\iint\limits_{\Omega} k_v h_l(x,y)\,dxdy} \tag{26}$$

where Q_d is the heat content of droplets transferred into the weld pool, $h_l(x,y)$ is the molten layer thickness inside the pool, and Ω is the domain with boundary of melt-line at the top surface of workpiece.

From Fig. 10 it is visible, as the hump arises and develops: at t=1.6 s middle part of the pool begins to solidify, at t=1.7 s middle parts solidified and the first humping formed, and at t=1.8 s the first humping solidified and the second humping appeared.

According to Wu et al., 2007 at high-speed welding it is possible to avoid defects of a welded seam, such as an undercut and formation of humps if the value of deposited metal is a constant on unit of length of a seam. It means, that wire melting velocity should be high enough but in the meantime arc thermal energy should be divided between a wire and the basic metal. The requirement of higher current of a wire and lower heat input in the basic metal becomes the contradiction. The modified arc weld process named double-electrode gas metal arc welding has been developed by Zhang et al., 2004 at University of Kentucky to uncouple a current of the basic metal from a wire current in GMAW so that the high current could be used to fuse a wire and to reach high speed of melting, to fill cutting in one pass while the heat input to the basic metal is lowered.

In Fig. 11 and 12 the results of virtual reproduction of the humping received by means of thermohydrodynamic and thermohydrostatic models are shown.

It is visible that the simplified model has advantages on speed of reproduction of process and prospect of application for process control in high-speed arc welding.

Double-electrode gas metal arc welding process can increase a critical welding speed and suppress defects of a welded seam for two reasons. The first is a scope of an arc on the basic metal in double-electrode gas metal arc welding more than it is in usual GMAW, and

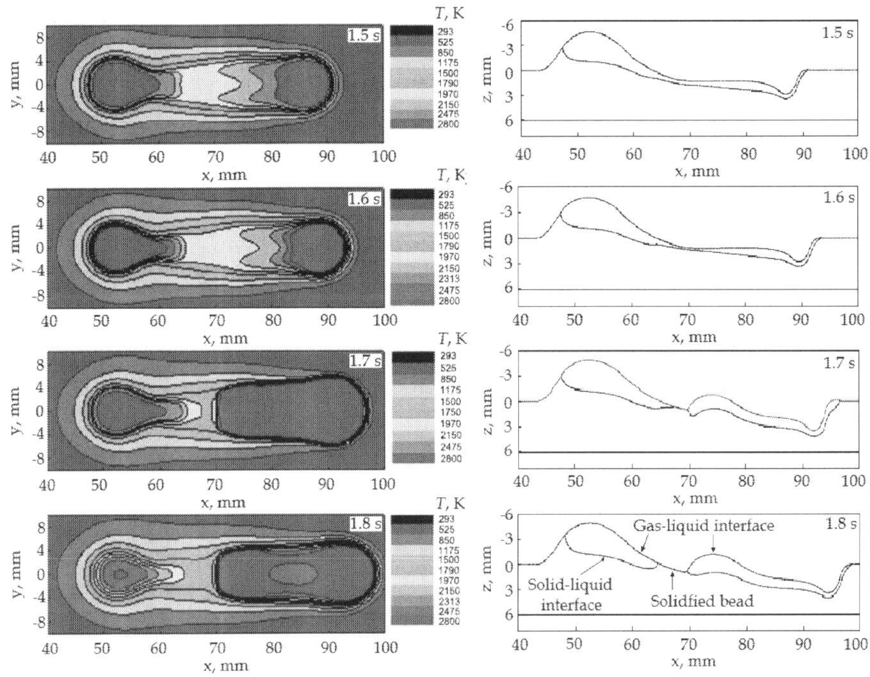

Fig. 10. Simulated temperature profiles on the top surface and shape evolution of longitudinal section of the pool of a low carbon steel sample welded for different times at welding speed 1.5 m/min, current 350 A, arc voltage 27 V, and sample thickness 6 mm; after Chen & Wu, 2009

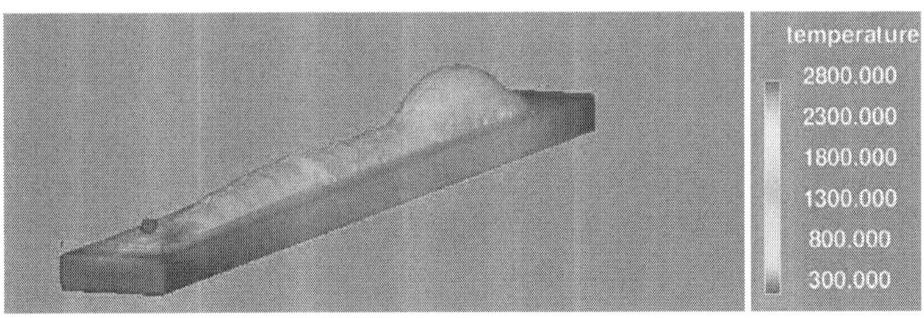

Fig. 11. Three-dimensional thermohydrodynamic simulation of a single hump in hybrid laser-GMA welding. The calculation time by software Flow-3D was 89 hours; after Cho & Farson, 2007

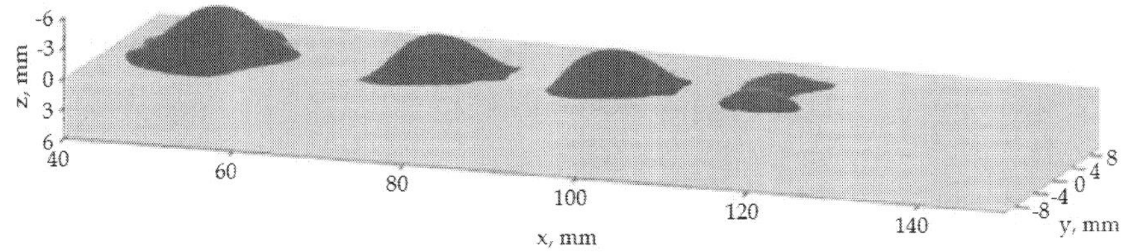

Fig. 12. Three-dimensional thermohydrostatic simulation of humps formation in arc welding. Calculation time was approximately in 100-1000 times less; after Chen & Wu, 2009

another that the buffer arc plays a role in preliminary heating of a surface of the basic metal. Both factors force liquid metal of a weld pool to disperse so that formation of a welded seam could be improved. It is to similarly hybrid laser-GMA welding where the leading laser beam can preliminary warm up a surface of the basic metal as an auxiliary heat source. Thus, change a fluid flow in a weld pool and the form of a way which suppresses the stooping platen of a weld. Models such novel processes while are unknown, but it is expected that they will appear in the near future.

4. Conclusion

Study of formation mechanisms of defects such as an undercut and humps is examined, and also thermohydrodynamic and thermohydrostatic models for simulation of corresponding defects formation are presented at arc welding. Thermohydrodynamic and thermohydrostatic approaches to construction of mathematical models of a weld pool with the specified formulation of boundary conditions are reconsidered. Formulations of stationary mathematical models of welding nonconsumable and consumable by electrodes, and also non-stationary model of a consumable electrode which allows reproducing formation of defects of type of an undercut and a burn-through are resulted. The solution of problems of search of parameters of a mode of the welding, providing defectless areas is illustrated by means of two examples of two-dimensional areas for GTAW and three-dimensional areas for GMAW.

5. References

Batchelor G. K. (1970). *An introduction to fluid dynamics*. Cambridge. University Press. 615p.

Bradstreet B.J. (1968). Effect of surface tension and metal flow on weld bead formation. Welding Journal. Vol. 47. No. 7, pp. 314 - 322.

Cao G., Yang Z. & Chen X. L. (2004). Three-dimensional simulation of transient GMA weld pool with free surface. Welding Journal, No. 6, pp. 169s - 176s.

Chen J., Wu C. S. (2009) Numerical simulation of forming process of humping bead in high speed GMAW. Acta Metallurgica Sinica, Vol.45, No.9, pp.1070-1076

Cho M. H. and Farson D. F. (2007). Simulation study of a hybrid process for the prevention of weld bead hump formation. Welding Journal, Vol. 86, No. 9, pp. 253s - 262 s.

Choi H. W., Farson D. F. & Cho M. H. (2007). Using a hybrid laser plus GMAW process for controlling the bead humping defect. Welding Journal, Vol. 86, No. 8, pp. 174s - 179 s.

Erokhin A.A.; Bukarov V. A. & Ishchenko U.S. (1972). Influence of electrode sharpening of the tungsten cathode on undercuts formation and gas cavities at welding. Automatic Welding, №5.

Gratzke U, Kapadia P D & Dowden J. (1992). Theoretical approach to the humping phenomenon in welding processes. J. Phys. D: Appl. Phys. Vol. 25, pp. 1640 - 1647.

Hu Z K. & Wu C S. Experimental investigation of forming process of humping bead in high speed MAG arc welding. Acta Metallurgica Sinica, 2008, Vol. 44, No. 12, pp. 1445-1449.

Hu J., Guo H. & Tsai H. L. (2008). Weld pool dynamics and the formation of ripples in 3D gas metal arc welding. Inter. J. of heat and mass transfer. Vol. 51, pp. 2537 - 2552.

Kumar A., DebRoy T. (2006). Toward a unified model to prevent humping defects in gas tungsten arc welding. Welding Journal. Vol. 86, No. 12, pp. 292s - 304 s.

Kou S.; Sun D. K. (1985). Fluid Flow and Weld Penetration in Stationary Arc Welds. Metallurgical transactions. A. Vol. 16A, pp. 203 - 213.

Landau S.D. & Lifshits E.M. (1958). *Statistical Physics*. Pergamon Press. London.

Landau L.D. & Lifshits E.M. (1959). *Fluid Mechanics*. Pergamon press. New York. 536 p.

Mendez P. F. & Eagar T. W. (2003). Penetration and defect formation in high-current arc welding. Welding Journal. Vol. 82, No 10, pp. 296s - 306s.

Myshkis A.D., Babckii V.G., Kopachevskii N.D., Slobozhanin L.A. & Tyuptsov A.D. (1987). Low-Gravity Fluid Mechanics. Springer-Verlag, Berlin, Heidelberg, New York, London, Paris, Tokyo. 583 p.

Nguyen T. C., Weckman D. C., Johnson D. A. & Kerr H. W. (2005). The humping phenomenon during high speed GMAW. Sci. and Techn. of Weld. and Joining. Vol. 10, No. 4, pp. 447 - 459.

Nguyen T. C., Weckman D. C., Johnson D. A. & Kerr H. W. (2006). High speed fusion weld bad defects. Sci. and Techn. of Weld. and Joining. Vol. 11, No.6, pp. 618-633.

Nomura H.; Sugitani Y. & Tsuji M. (1982). Behaviour of Molten Pool in Submerged Arc Welding. Observation by X-Ray fluoroscopy. Journal Japan Welding Society. Vol. 51, No. 9, pp. 43 - 51.

Ohji T., Nishiguchi K. (1983). Mathematical modelling of a molten pool in arc welding of thin plates. Technol. Rep. of Osaka Univ. Vol. 33, No. 1688, pp. 35 - 43.

Ohji T., Miyasaka F., Yamamoto T. & Tsuji Y. (2004). Mathematical model for MAG welding in a manufacturing environment. *Proc. of Int. Conf. on Mathematical Modelling and Information Technologies in Welding and Related Processes*. Katsiveli, Crimea, Ukraine, Sept. 2004. Kiev: Paton Welding Institute, pp. 205--209.

Ohji T., Onkubo A. & Nichiguchi K. (1992). Mathematical modelling of molten pool in arc welding. Mechanical Effects on Welding. Proceedings IUTAM Symposium (Lulea Sweden Jun 1991) Karlson L., Lindgren L.-E., Jonson M. eds. (Berlin, Springer-Verlag), pp. 207 - 214.

Paton B.E.; Mandelberg S.L. & Sidorenko B.G. (1971). Some Features of Weld Formation during Welding on the Raised Speeds. Automatic Welding, №8.

Pogorelov A.V. (1967). *Differential Geometry*. Groningen. Nurdhoff. 171p.

Rykalin N. N.; Uglov A. A. & Anishchenko L. M. (1985). *High-temperature technological processes: Thermalphysic basics*. Moscow: Nauka, 175 p.

Savage W. F., Nippes E. F. & Agusa A. (1979). Effect of arc force on defect formation in GTA welding. Welding Journal, Vol. 58, No. 7, pp. 212s - 224s.

Shimada W. & Hoshinouchi S.A. (1982). A study on bead formation by low pressure TIG arc and prevention of undercut bead. J. Jap. Weld. Soc. Vol. 51, No. 3, pp. 280–286.

Sudnik V. A. (1985). Digital and experimental temperature distribution in the weld zone when subjected to the effect of a defocused energy beam. *2nd Int Conf "Beam Technology"*, Düssedorf, DVS Verlag, 1985, pp. 158 - 161.

Sudnik V. A. (1991a). Welded joints quality prediction based numerical models of weld formation during fusion welding of thin-walled structures. Doctoral Dissertation. St.-Petersburg State Polytechnic University .

Sudnik V. A. (1991b). Research into fusion welding technologies based on physical-mathematical models. Welding and Cutting, No. 10, pp. E216 - E217, S. 588 - 590.

Sudnik V. A. & Erofeev V. A. (1986). *Calculations of Welding Processes by computer.* Tula. Tula State University. 100 p.

Sudnik V. (1997). Modelling of the MAG process for the pre-welding planning. Mathematical modelling of welding phenomena 3. ISBN 1 86125 010 X, Graz-Seggau. September 1995, pp. 791 - 816.

Sudnik V. A., Rybakov A. S. & Kurakov S. V. (1999). Numerical solution of the coupled problem of temperature and deformation fields of weld pool during arc welding. Computer technology in joining of materials: Trans. of Tula State University / Edited by V. A. Sudnik. Selected Proc. 2nd All-Russia Conf. Tula: pp. 97 - 109.

Sudnik V. A. & Mokrov O. A. (1999). Mathematical model and numerical simulation of GMA-welding of fillet weld in various spatial positions. Computer technology in joining of materials: Trans. of Tula State University. Edited by V. A. Sudnik. Selected Proc. 2nd All-Russia Conf. Tula: pp. 81 - 96.

Tchernyshov G. G.; Rybachuk A. M. & Kubarev V. F. (1979). About metal movement in a weld pool. Trans. of High Schools. Mechanical Engineering. No 3, pp. 134 - 138.

Wu, C. S., Zhang, M. X., Li, K. H. & Zhang, Y. M. (2007). Study on the process mechanism of high speed DE-GMAW. Acta Metall. Sinica, Vol. 43, No. 6, pp. 663–667.

Yamamoto T. & Shimada W. (1975). A study on bead formation in high speed TIG arc welding. In: Int. Symposium in Welding. Osaka. Japan. Paper 2-2-7, pp. 321 - 326

Yamauchi N. & Inaba Y. , Taka T. (1982). Formation mechanism of lack of fusion in MAG welding. J. Jap Welding Soc., Vol. 51, No. 10, pp. 843 - 849.

Zacharia T., Eraslan A. H. & Aidun D. K. (1988). Modelling of non-autogenuous welding. Weld. J. Vol. 67. № 1, pp. 18s - 27s.

Zhang, Y. M. & Jiang, M. (2004). Double electrodes improve GMAW heat input control. Weld. J., Vol. 83, No. 11, pp. 39–41.

12

Using Solid State Calorimetry for Measuring Gas Metal Arc Welding Efficiency

Stephan Egerland[1] and Paul Colegrove[2]
[1]*FRONIUS International GmbH,*
[2]*Cranfield University,*
[1]*Austria*
[2]*United Kingdom*

1. Introduction

The thermal profile of fusion welding or its heat input can cause degradation of the material properties, which is reflected in the microstructural changes, occurring in the heat affected zone (HAZ). Hence, quantifying the amount of thermal energy transferred from the welding arc to the workpiece is beneficial to understanding this phenomenon. High accuracy in determining the thermal weld process efficiency, improves the predictive ability of numerical models. Weld 'process efficiency' is also called 'efficiency', 'energy absorption' or 'heat transfer efficiency'. (AWS, 2001) defines "energy absorption" by the workpiece, regularly denoted by the Greek symbol η (eta), as the fraction of the "total energy supplied by the heat source", that is, the arc. Depending amongst others on material properties and heat source density, the final energy absorption can vary. According to (Lancaster, 1986) this relationship can be described by:

$$\eta = 1 - \frac{q_e + (1-n)q_p + mq_w}{EI} \quad (1)$$

η represents the thermal arc efficiency, q_e is the rate of heat transfer from the arc to the electrode in cal s^{-1}, n stands for the energy proportion radiated and convected from the arc column per unit time and transferred to the workpiece, q_p is the energy radiated and convected from the arc column per unit time in cal s^{-1}, m represents the proportion of anode energy radiated away from the workpiece, q_w is the arc heat fraction absorbed by the workpiece in cal s^{-1}. As E and I stand for voltage and current, respectively, representing particularly constant voltage welding processes, for advanced welding power supplies equation (1) can be written in a more general form as:

$$\eta = 1 - \frac{q_e + (1-n)q_p + mq_w}{q_a} \quad (2)$$

here q_a is the average instantaneous power from the welding process, being defined as:

$$q_a = \frac{1}{T}\int_0^T UIdt \tag{3}$$

with T representing the total welding time, and t the time. As q_e can be ignored with consumable electrode processes such as Gas Metal Arc Welding (GMAW) the expression may be written as (Lancaster, 1986):

$$\eta = 1 - \frac{(1-n)q_p + mq_w}{q_a} \tag{4}$$

The power input to the plate, q_i can be found simply by:

$$q_i = \eta q_a \tag{5}$$

Of the energy that is transferred to the workpiece, some will be used to melt the material in the fusion zone, while the remainder heats up the base material. Therefore it is useful to define the melting efficiency η_m according to (DuPont & Marder, 1995 and Fuerschbach & Eisler, 1999 and Eder, 2009) as the energy required to melt the fusion zone area divided by the energy input to the plate:

$$\eta_m = \frac{vA\rho\delta_h}{q_i} \tag{6}$$

v represents welding speed, A is fusion zone cross section; ρ is the density and δ_h is the melting enthalpy per unit mass which is given by:

$$\delta_h = \Delta h_f + \int_{T_r}^{T_m} c_p(T)dT \tag{7}$$

where Δh_f is for heat of fusion. c_p stands for the specific heat, as T, T_r and T_m represent absolute-, room- and melting temperature, respectively.

Welding calorimetry is used to measure the process efficiency through determining the energy transferred to the workpiece, as well, as to question physical aspects of heat and current flow distribution as studied e.g. by (Tsai and Eagar, 1985 and Lu and Kou, 1988). The authors used a calorimeter consisting of a split hollow water cooled copper Dee-anode (split-anode), developed by (Nestor, 1962). Current and voltage in autogenous gas tungsten arc welding (GTAW) and their affect on energy distribution and process efficiency were investigated. A study on efficiency of variable polarity plasma arc welding on AA 6061 aluminium alloy specimens was carried out by (Evans et al., 1998). The samples were "quickly placed into a calorimeter and the retained heat measured after the temperature of the water in the calorimeter stabilised (about 2 minutes)". However, no detailed information is provided concerning the time scatter between welding the sample and immersing it into the calorimeter. This fact has been taken into greater account by (Bosworth, 1991), using water calorimetry for ferrous parent material gas metal arc welding. Researching the effective heat input applying solid wire electrodes the delay between "cessation of welding to quenching of the sample was standardised at 15 s for all of the tests". A "maximum uncertainty of ± 5% to the (efficiency) value" was indicated and it was found an increasing voltage or arc length, respectively, decreased the efficiency. The method, reported by (Kou, 1987), involved GTAW on an aluminium tube ("if the workpiece is a pipe") which is continuously cooled with water. The temperature rise throughout welding was measured using "differential-thermocouples" and plotted over time. The energy input to the plate was then calculated by (Kou, 1987, 2003):

$$q_i t_{weld} = \int_0^\infty W c_p \left(T_{out} - T_{in}\right) dt \approx W c_p \int_0^\infty \left(T_{out} - T_{in}\right) dt \tag{8}$$

W is water mass flow rate and c_p is the specific heat of water. T_{in} and T_{out} represent the water inlet- and outlet temperature, as t and t_{weld} are for time and weld time, respectively. The *Seebeck* envelope calorimeter method uses a similar principle, however the weld is sealed in an insulated, water cooled box after welding and a temperature gradient layer is used to calculate the heat loss to the water. According to (Kou, 2003), knowing the gradient layer thickness L, its thermal conductivity k and the heat conducting area A_c allows to calculate the heat transfer from the heat source to the calorimeter as:

$$q_i t\text{weld} = \mathrm{A_c} \int_0^\infty k \frac{\Delta T}{L} dt \tag{9}$$

Seebeck welding calorimetry was particularly applied for studying gas tungsten- and plasma arc welding (Giedt et al. 1989 and Fuerschbach and Knorovsky, 1991). Conducting efficiency investigations on AISI 304 stainless steel coupons in a Seebeck calorimeter (Giedt et al., 1989) found ~ 80% process efficiency, confirmed to be "consistent with results from other calorimeter type measurements". The main issue with these methods of calculating the heat input to the weld is the time to undertake the experiment; which restricts its suitability for general application. It was shown e.g. by (Giedt et al., 1989) that "up to six hours was required for the workpiece to come to equilibrium with the constant-temperature cooling water." More rapid process efficiency measurement is possible, applying the *liquid nitrogen* calorimeter method, as used e.g. by (Joseph et al., 2003 and Pepe et al., 2011). The specimen, welded and immersed immediately into a Dewar filled with liquid nitrogen, vaporises a specific mass of nitrogen, Δm_n. Knowing the latent heat of vaporisation for liquid nitrogen c_n the energy to cool the welded sample to liquid nitrogen temperature, E_s, can be calculated:

$$E_s = \Delta m_n c_n \tag{10}$$

To enable the energy input to the sample to be calculated, two energy losses need to be considered: the energy loss from normal nitrogen vaporisation, E_n; and the energy required to cool the specimen from room temperature to liquid nitrogen temperature, E_a. Therefore the final expression for calculating the energy input to the specimen is:

$$q_i t_{weld} = E_s - E_n - E_a \tag{11}$$

A final method for measuring the process efficiency is that reported in (Cantin & Francis, 2005) who used a solid state calorimeter encased in an *insulated box*. To determine the process efficiency of aluminium gas tungsten arc welding, an appropriate weld specimen was welded within an insulated box. As for the other processes, the energy input to the plate was found by:

$$q_i t_{\mathrm{weld}} = m_{\mathrm{w}} \int_{T_0}^{T_e} c_{\mathrm{pw}}(T)\mathrm{d}T + m_{\mathrm{b}} \int_{T_0}^{T_e} c_{\mathrm{pb}}(T)\mathrm{d}T \qquad (12)$$

Here m_{w} and m_{b} represent the workpiece and backing bar mass, respectively, and c_{pw} as c_{pb} stand for their specific heat. T, T_0 and T_e are for temperature, initial temperature and equilibrium temperature, respectively. The method is similar to the Seebeck method in that the weld is contained within an insulated box after welding, however rather than waiting for the weld to cool back to room temperature, the final equilibrium temperature is calculated. The main advantage of the solid state calorimeter is a significant reduction in the measurement time.

Calorimetric measurements have been done on a variety of processes including Gas Tungsten Arc Welding (Fuerschbach and Knorovsky, 1991 and DuPont and Marder, 1995 and Giedt et al. 1989 and Cantin and Francis, 2005), Gas Metal Arc Welding (DuPont and Marder, 1995 and Joesph et al. 2003 and Pepe, 2010 and Bosworth, 1991), and Plasma Arc Welding (Fuerschbach and Knorovsky, 1991 and DuPont and Marder, 1995). The process efficiency for consumable electrode processes is generally about 10-20% higher than non-consumable processes (DuPont and Marder, 1995).

The process efficiency of GMAW which is the subject of this investigation vary. (DuPont and Marder, 1995 and Bosworth, 1991) who used water based calorimeters claimed that the efficiency could be between 80-90%. Joseph et al., 2003 who used a liquid nitrogen calorimeter and longer duration welds (up to 60 seconds) claimed that the value was closer to 70%. Also using a liquid nitrogen calorimeter (Pepe, 2010) found that the process efficiency varied between 78-88% for CMT welding. Although there doesn't appear to be any difference between CV and pulsed welding (Joseph et al. 2003 and Bosworth, 1991), two articles (Hsu and Soltis, 2002 and Bosworth, 1991) have reported that the efficiency with short circuiting or surface tension transfer modes is significantly higher (up to 95%). The latter (Bosworth, 1991) found that increasing arc voltage and therefore arc length reduced the efficiency, however interestingly (Cantin and Francis, 2005) found no such link with arc length in their investigation of GTAW. Finally, the arc efficiencies are increased when welding in a groove compared with bead on plate welds (Bosworth, 1991).

This chapter compares the process efficiency of pulsed GMAW with the Fronius Cold Metal Transfer (CMT) GMAW process. Pulsed GMAW may be classified as 'free flight' and, if appropriately adjusted, short circuit free. In comparison, CMT which was invented by (Hackl and Himmelbauer, 2005) is principally a 'short arc' process. The major difference to natural short circuit droplet transfer is CMT applies both a reproducible transient control of weld current and voltage, as well as mechanical support to the molten droplet detachment. These features are explained in Fig. 1.

The wire electrode is fed forward until short-circuiting with the liquid weld pool. Detected by the weld system, the wire is instantaneously retracted from the weld pool by reversing the feeding direction, and simultaneously decreasing weld current and voltage. The process has high process stability and reproducibility, and reduced thermal input to the parent material.

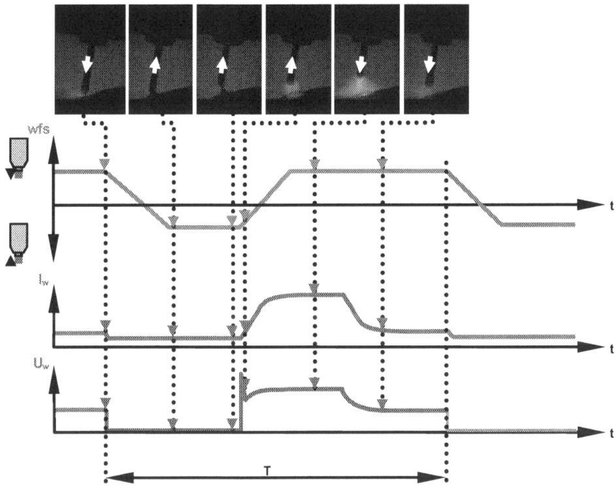

Fig. 1 Representative wire-feed speed (wfs), voltage (U_w) and current plots (I_w) vs. time (t) for the CMT process.

2. Experimental

2.1 Welding systems and experiments

GMAW-P and CMT were investigated. In order to simplify the experimental setup, a single welding system was chosen, capable of operating both processes. See Fig. 2, for configuration overview.

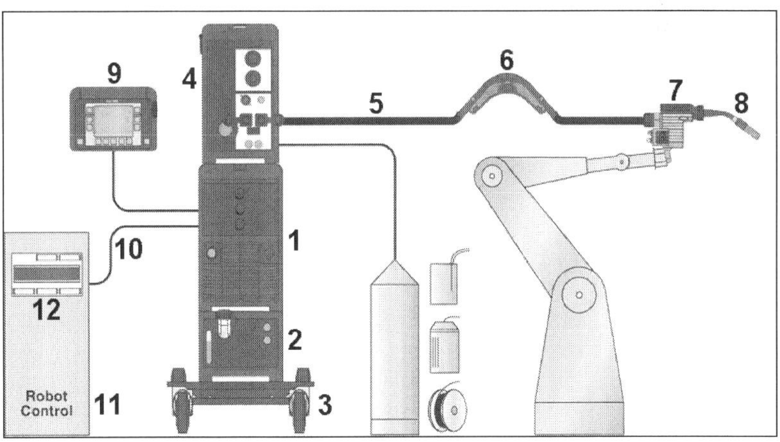

Fig. 2. Schematic of welding system configuration.

Note that items 1 - 12 in Fig. 2 are as follows:

1. Inverter Welding Power Source (FRONIUS TPS 4000 Type *)
2. Cooling Unit (FRONIUS FK 4000 R Type)
3. Trolley
4. 4-wheel drive wire feeding unit (FRONIUS VR 7000 CMT Type)
5. Wire Buffer hose package (water cooled 4.25 m - equipped with appropriate wire liner)
6. Wire buffer + torch hose package (1.2 m - equipped with appropriate wire liner)
7. Special CMT drive unit welding torch
8. Torch neck (36°/500A - equipped with appropriate wire liner and contact tip ∅ 1.0 mm)
9. Remote Control Unit (FRONIUS RCU 5000i Type)
10. Robot Control Cable
11. Robot Control
12. Robot-Power Source Interface

(*) CMT Release

For high reproducibility reasons, an industrial welding robot type ABB IRB 2400 + IRC 5 robot control and DEVICENET-robot interface was used. Welding current was measured by applying a Hall-effect current sensor (LEM™ shunt). A sense lead, connected to the torch neck (closely to the contact tip area) was used in order to obtain the voltage measurement. Current and voltage acquisition was carried out using a high-speed digital oscilloscope (*Tektronix DPO 4034*), adjusting a sampling rate of 25 kS s^{-1}. The power input from the welding process was calculated from equation (3).

Mild carbon steel S235 J2 (DIN EN 10025) was used for the experiments and Table 1 provides the chemical composition according to this standard. The material was sandblasted prior to welding and two different geometries were used for the welding: 250 x 50 x 5 mm (see Fig. 3 (a, b)) which was used for the bead on plate welds; and 250 x 50 x 12 mm (see Fig. 3 (c)) which was the square groove geometry and was meant to simulate welding in a narrow gap. Two of the square groove coupons were not sandblasted to evaluate the effect of surface condition on the process efficiency.

Grade	C max.	Si max.	Mn max.	P max.	S max.	Cu max.	N max.
S 235 J2	0.17	-	1.40	0.030	0.030	0.55	-

Table 1. Steel grade 'S 235 J2' chemical average composition in weight percent (acc. to EN 10025).

Solid filler wire, grade G3 Si1 (acc. to EN 440), nominal ø 1.0 mm, and shielding gas 82 $Ar/18CO_2$ (M21 acc. to EN 439) were used for the experiments. The shielding gas flow rate was 12 l min^{-1}. The contact tip to workpiece distance (CTWD) was 12 mm and the torch was positioned normal to the plate surface. A total of 12 experiments were done, which included:

- 3 x pulsed GMAW bead on plate
- 2 x pulsed GMAW square groove
- 3 x CMT bead on plate
- 2 x CMT square groove
- 2x CMT square groove (non-sandblasted)

In each case the average wire feed speed was 8.0 ± 0.04 m min^{-1} which was verified by measurement. The standard synergic line for each process was used and the welding speed was 0.6 m min^{-1}.

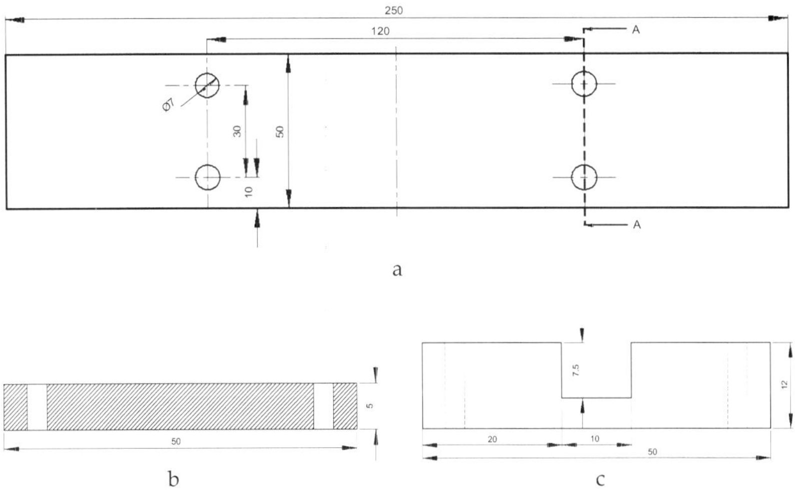

Fig. 3 (a) Bead on plate weld specimen dimensions and hole pattern; (b) Cross section (A-A) from (a); and (c) square groove geometry.

2.2 Insulated box calorimeter

Unlike the calorimeter reported in (Cantin & Francis, 2005), a large copper block is used to conduct and absorb the energy from the welding process. Fig. 4 schematically shows the design. The calorimeter was constructed as an insulated box, containing an 'electrolytic copper' block, which had a bolt for connecting to the power source work cable. The whole copper part weighed 5.90 kg including some small copper spacer plates which were used to ensure consistent contact between the sample and the copper block. The steel workpiece material was weighed before and after welding using a high accuracy scale (Type: *KERN EW*) for determining the weld metal deposition, as well as the mass of the workpiece (see equation (12)). The specimens, were fixed to the copper block using four steel screws. Their heat capacity was included in the calorimeter's total heat capacity.

Although this work applies equation (12), the specific heat of the copper and steel is assumed constant between the initial and equilibrium temperatures. The values of specific heat used for the analysis were 388 J kg^{-1} K^{-1} for the electrolytic copper and 484 J kg^{-1} K^{-1} for S235 steel (Holman, 1990).

Three thermocouples were attached to the copper block at the locations (denoted TC_start, TC_centre and TC_finish) shown in Fig. 4 and were recorded with an Agilent Type 34970A data logger. The copper block was fixed upon two insulating blocks of high strength polyamide-imide (PAI). To further reduce heat loss to the surroundings, the box is manufactured from a polyurethane (PU) polymer, completely laminated with self-adhesive aluminium foil. According to (BS EN 12524, 2000), PU shows low specific heat capacity (1.80 kJ kg^{-1} K^{-1}) and thermal conductivity (0.25 W m^{-1} K^{-1}). The whole calorimeter is fixed upon a low thermal conductive synthetic resin bonded paper plate (PERTINAX™), clamped to the welding turntable. The calorimeter is closed by a top cover (lid) of the same material as the insulated box, as schematically shown in Fig. 5 . Throughout welding, this lid is consistently

manually moved along the welding direction, most closely following the welding torch (see Δ_S in Fig. 5) for reducing radiation and heat losses to the widest possible extent.
A typical temperature vs. time plot from the calorimeter is shown in Fig. 6. The temperatures converge on a steady state value between ~ 200 s and ~ 300 s, depending on the welding conditions. By examining the slope after convergence it is possible to estimate the average heat loss from the calorimeter as a function of time. The steady state temperature reading includes this effect.

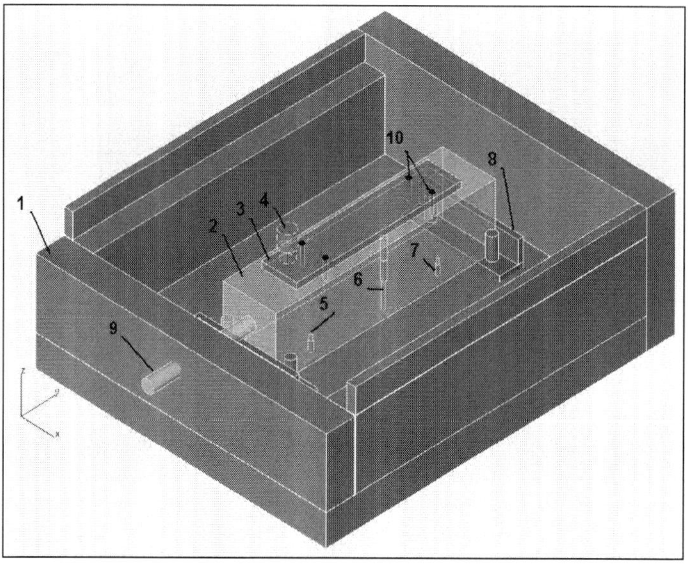

Fig. 4. Sketch showing the design of the insulated box calorimeter (note: lid not depicted in this figure).

Note that items 1 - 10 in Fig. 4 are as follows:
1. Insulated box (aluminium foil laminated polyurethane)
2. Copper block
3. Weld specimen
4. Welding torch
5. Thermocouple (TC_start)
6. Thermocouple (TC_centre)
7. Thermocouple (TC_finish)
8. Polyamide-Imide insulating block
9. Copper connection to work cable
10. Bolt holes

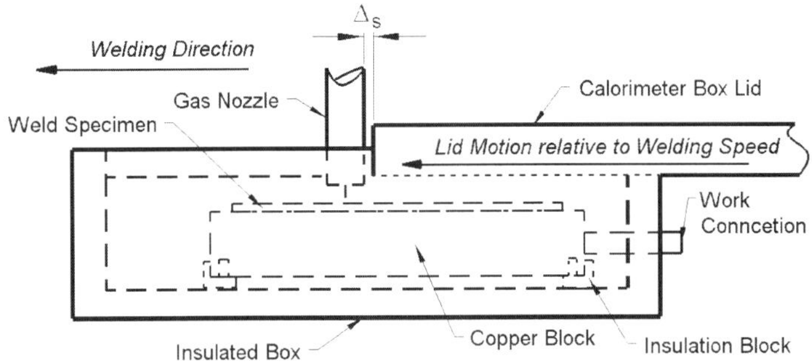

Fig. 5. Schematic showing the operation of the insulated box calorimeter.

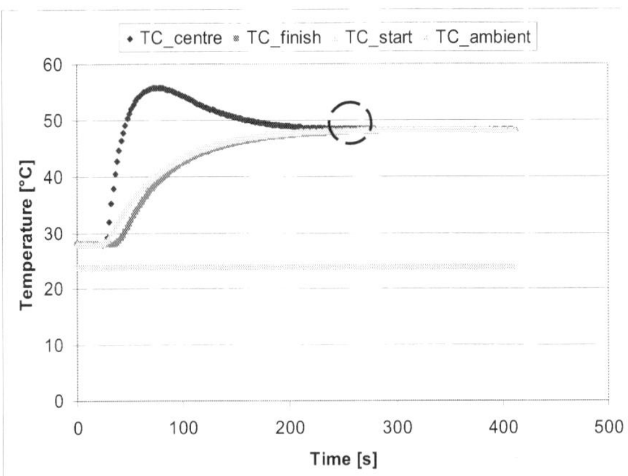

Fig. 6. Typical temperature vs. time plot. Note that the dashed circle shows when equilibrium is established.

3. Results

The average welding currents and voltages for the 5 conditions are shown in Fig. 7 (a). Pulsed GMAW-employs higher arc voltages but lower welding currents vs. CMT leading to comparably higher electrical average instantaneous power. The pulsed GMAW average instantaneous power was calculated approximately 1.6 higher than the corresponding CMT process for bead on plate geometry and 1.65 higher for square groove geometry.

Fig. 7 (b) shows the average instantaneous power, q_a, the measured average power delivered to the calorimeter, q_i and the corresponding process efficiency. CMT shows only marginally increased efficiencies vs. pulsed GMAW, when applied to BOP welding and virtually no difference when applied to the square groove geometry. Regardless of which process was used, energy losses were found considerably decreased when applying the square groove design. For GMAW-P, this configuration allowed for reducing energy losses by ~ 37%, thereby improving the process efficiency to ~ 87%. With CMT the square groove design could drop the energy losses by ~ 20%. The non-sandblasted groove surface condition was found to have no significant influence on thermal efficiency, compared with the sandblasted grooves.

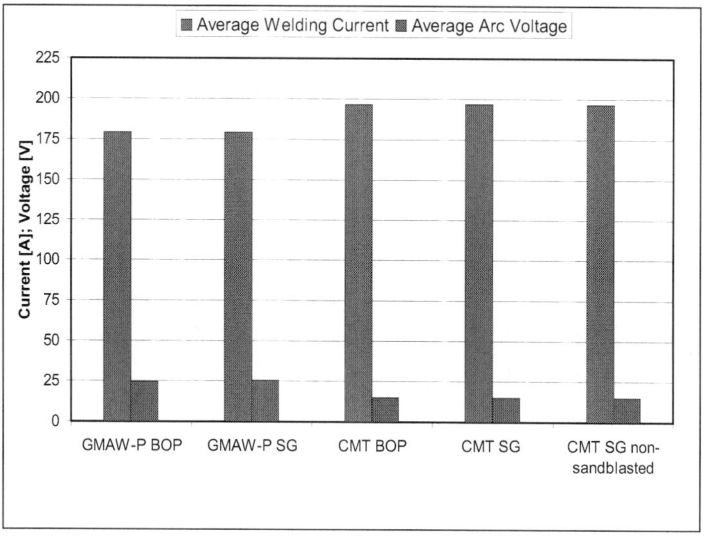

a

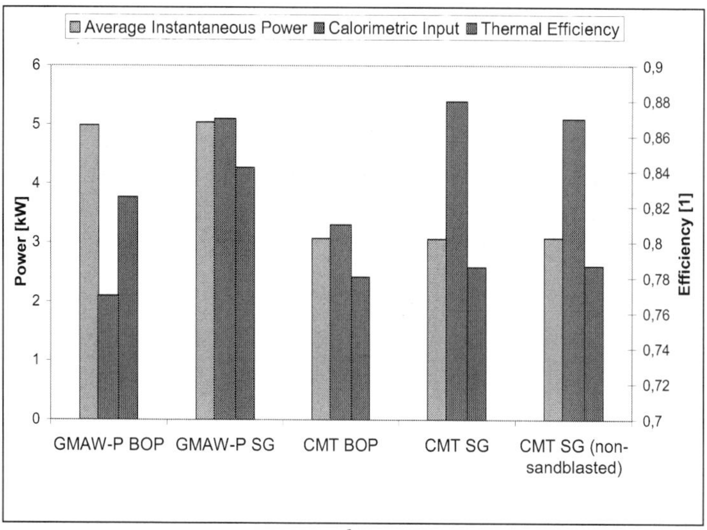

b

Fig. 7. (a) Average welding current and arc voltages; (b) average instantaneous power, calorimetric power input and process efficiency for the welded specimens.

The weld bead shapes for the GMAW and CMT processes are shown in Fig. 8. This illustrates how the lower heat input with the CMT process leads to a narrower, more reinforced weld bead and less penetration (and lower dilution) of the substrate.

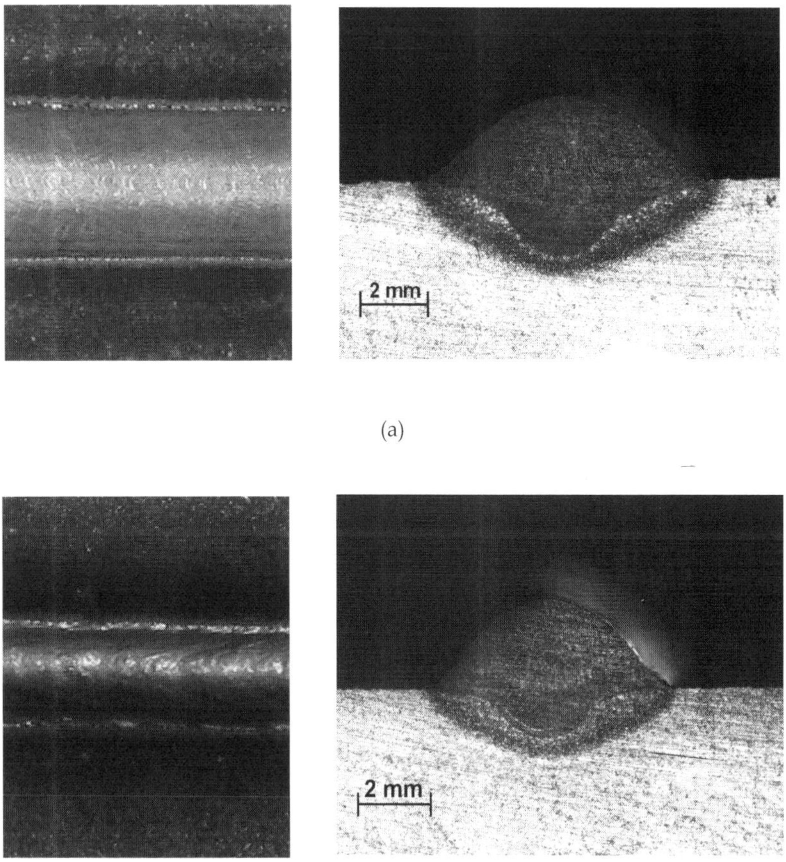

(a)

(b)

Fig. 8. Typical bead shape and transversal cross section for (a) pulsed GMAW and (b) CMT.

Although this fact is believed to be quite important, it seems yet to be often neglected in '*low energy*' GMAW process discussions. This calorimetric study could basically approve a lower thermal energy input to the base material using CMT, leading to significant changes in bead shape and penetration behaviour with constant given conditions. Lower heat input is known beneficial for enhancing the process window and joining lower wall thickness parts. *Only* focusing on the reduced energy input is believed, however, to neglect an important part of the whole physical process spectrum - inner and outer changes in the weld result.

4. Discussion

'*Controlled*' GMAW processes, such as CMT, are often considered capable of reducing the thermal energy input to the base material which has been demonstrated in this work. This work has shown that CMT supplies lower average voltage and different arc characteristics. The pulsed GMAW electrical parameters corresponding to the average wire feed speed of 8 m min^{-1} chosen in this study, produce an average instantaneous power which is similar to that from (Joseph et al., 2003) who used a wire feed speed of 7.62 m min^{-1}. However the process efficiency in this work (77%) was much higher than that measured by (Joseph et al. 2003) 60%. It is supposed that the different calorimetric principle (liquid nitrogen), strongly affected by the transmission time, required to immerse the sample into the Dewar, may provide an explanation for the varying result. For GMAW-P BOP a heat input of ~ 0.38 kJ mm^{-1} and ~ 0.2 kJ mm^{-1} for CMT BOP, respectively, was found in this study, which was in very good agreement with (Pepe and Yapp, 2008). It was supposed in the present work, both processes, due to their different characteristics, would show considerably distinct energy losses. Interestingly however, they were found to have almost similar thermal efficiencies. The small arc efficiency increase with CMT BOP is believed to confirm the results from (Hälsig et al., 2010) and (Eichhorn and Niederhoff, 1972) assessing arc length reduction or short circuit affliction, respectively, as "efficiency increasing". However, an efficiency of ~ 95% for short circuit arc welding as exceptionally stated by (Bosworth, 1991) with low deposition rate short circuit arc welding and ø 1.2 mm steel wire electrode could not be found in the present study; nevertheless reasonable good agreement with the efficiency results (~ 87% at ~ 4 kg h^{-1} deposition rate) could be proved for GMAW-P when welding in a square groove or "narrow gap" (Bosworth, 1991). Although the authors of this study acknowledge the influence of convection and/or vaporisation effects on welding efficiency, they unlike (DuPont and Marder, 1995) consider arc radiation rather the major reason for energy losses in GMAW. This is suggested due to the square groove results, which could prove the process efficiency to be significantly increased. It is believed the groove side walls are capable of capturing a considerable fraction of the energy regularly radiated away from the arc. As an interesting detail, the thermal efficiency values for CMT and GMAW-P become equal in magnitude when applying the square groove design, showing both processes to finally loose ~ 15% of their energy. This could indicate a stronger 'compensation effect' of particular groove configurations, e.g. square groove, for stronger radiating or higher performance processes. Besides changing arc radiation losses it might be assumed that also the remaining energy losses, such as convection, might be affected by specific groove designs. It is suggested however, that further investigation is needed in order to thoroughly explain the equivalence between GMAW-P and CMT when employing special groove configurations. Finally, the insignificance between sandblasted and non-sandblasted square groove conditions - having lower or higher side wall reflectivity, respectively - is suggested to be explainable due to either oxidation, generated by heat conduction in front of the arc, or general secondary physical importance in respect to the given conditions.

Calorimetry, as one assessment of this study, is considered an appropriate means in order to determine the interaction between arc and solid matter; confirming hereby the results from other researchers. An extensive amount of work in welding calorimetry has been conducted through the past decades supplying, however, quite different results joined to these experiments. It is considered likely that the noticeable spread in the results may arise from both "systematic and random errors" and especially the former can lead to "underestimates of the actual welding process" (Pepe et al. 2011). Nevertheless, Sievers and Schulz in (Kohlrausch, 1985) estimate - from a rather more physics viewpoint - even "low complexity" calorimetry methods as being adjustable within an accuracy scatter of ± 1%. The grade of accuracy again is the main parameter for the final choice of calorimeter method and -type, respectively. According to (Kohlrausch, 1985), reducing the uncertainty in measurement e.g. toward ± 0.1% requires to rise the experimental complexity "exponentially" by "one ore even more orders of magnitude". The insulated box calorimeter type, as used in this study, was applied for assessing two different approaches. First, gaining calorimetric data for an advanced or 'controlled' GMAW process (CMT), being unknown as yet. Secondly, if the calorimeter type, described in the present work, could provide an accuracy output similar to e.g. the Seebeck envelope calorimeter. As described GMAW-P efficiency results obtained in this study could show a reasonable good agreement with data from other researchers (DuPont and Marder, 1995 and Bosworth, 1991). The efficiency increase in short circuit arc welding, as stated by (Eichhorn and Niederhoff, 1972 and Hälsig et al., 2010) was also approved but was lower for the bead on plate welding conditions however, vs. the results stated by (Bosworth,1991). The CMT efficiency data are thus considered sufficiently accurate within the experimental and systematic scatter. As shown by (Pepe et al. 2011) an error of ± 1.5 for the insulated box calorimeter was comparable to the Seebeck envelope calorimeter as used in (Giedt et al. 1989), whereas an error of ~ 8% was found for the liquid nitrogen calorimeter as used in (Pepe et al, 2011). It is considered important to mention that the Seebeck envelope calorimeter efficiency measurements, as known from the literature (Giedt et al., 1989 and Fuerschbach and Knorovsky, 1991) are focused on autogenous gas tungsten- or plasma arc welding, respectively. It is also suggested important that, albeit a row of welding calorimetric investigation was carried out, the calorimeter types show a broad variety. This is considered to, at least in part, contribute to the scatter in the known efficiency data. Finally, the data, gained through this investigation, showed quite good agreement with both calculated and experimental results of (Sudnik et al., 2001). The authors have developed a mathematical GMAW-P model including the description of the heat source. Using a calorimeter type as with (Bosworth, 1991) for model verification applying different conditions, they could find a process efficiency of ~ 80% when adjusting a wire feed rate of 8 m min^{-1}.

It is believed that the welding calorimetry method, as used for this investigation, is capable of providing sufficient accuracy for measuring the process efficiency in much less time compared e.g. with the Seebeck envelope calorimeter type and with lower error compared with the liquid nitrogen calorimeter type.

5. Conclusions

An investigation on the accuracy and suitability of a self constructed solid state insulated box calorimeter for measuring and comparing the arc efficiency of controlled GMAW processes was conducted and the following conclusions were reached:

- The solid state insulated box calorimeter showed precise measurements with both processes applied, and little random error. That is, it could be approved suitable for detecting also slight performance variations with low performance or controlled GMAW processes such as CMT.
- At given experimental conditions and a wire feed speed of 8 m min^{-1}, pulsed gas metal arc welding showed approx. 2 kW higher arc power in average vs. the CMT process.
- The thermal efficiency with both processes was found slightly higher with CMT vs. GMAW-P when welding BOP, approving thereby the work of other researchers suggesting higher arc efficiencies for short circuit- or dip transfer.
- Applying a square groove joint design was found capable of reducing radiation losses, hereby increasing considerably the arc efficiency. Almost equivalent average thermal efficiencies were found with both processes when welding in a square groove joint.
- Further work seems necessary to explain this similarity of arc efficiency with both processes when applying the square groove joint design.
- The insulated box calorimeter could show reasonable good agreement with efficiency data as known from other researchers and is believed to be a reasonable technological alternative in welding calorimetry vs. the Seebeck envelope calorimeter (higher measurement times) or liquid nitrogen calorimeter (greater experimental error).

6. Acknowledgements

The authors should like to acknowledge the generous permission of FRONIUS International Wels Austria to use equipment and facilities. Special thanks shall belong to Dipl.-Ing. Mr Andreas Leonhartsberger, who has constructed and built the insulated box calorimeter, and to Mr Gerhard Miessbacher, both with FRONIUS International Research & Development, for providing unselfish assistance throughout this investigation.

7. References

Bosworth, M.R. (1991). Effective Heat Input in Pulsed Gas Metal Arc Welding with Solid Wire Electrodes, *Welding Journal*, Vol. 70, No. 5, pp. 111s-117s

British Standard EN 12524:2000 (2000). Building materials and products - Hygrothermal properties - Tabulated design values

Cantin, C.M.D. & Francis, J.A. (2005). Arc Power and Efficiency in Gas Tungsten Arc Welding, *Science and Technology of Welding and Joining*, Vol. 10, No. 2, 03/2004, pp. 200-210

DuPont, J.N. & Marder, A.R. (1995). Thermal Efficiency of Arc Welding Processes, *Welding Journal*, Vol. 74, No. 12, pp. 406-s - 416-s

Eder, A. (2009), Private Discussion (unpublished)

Eichhorn, F., & Niederhoff, K. (1972). Streckenenergie als Kenngröße des Wärmeeinbringens beim mechanisierten Lichtbogenschweißen (in German). *Schweißen und Schneiden*, Jg. 24, H. 10, pp. 399-403

Evans, D., Huang, D., McClure, J.C., & Nunes, A.C. (1998). Arc Efficiency of Plasma Arc Welds, *Welding Journal*, Vol. 77, No. 2, pp. 53s-58s

Fuerschbach, P.W. & Eisler, G.R. (1999). Effect of Very high Travel Speeds on Melting Efficiency in Laser Beam Welding, *SAE Transactions: Journal of Materials and Manufacturing*, Vol. 108, pp. 1-7

Fuerschbach, P.W. & Knorovsky, G.A. (1991). A Study of Melting Efficiency in Plasma Arc and Gas Tungsten Arc Welding, Welding Journal, Vol. 70, No. 11, pp. 287s-297s

Giedt. W.H., Tallerico, L.N. & Fuerschbach, P.W. (1989). GTA Welding Efficiency: Calorimetric and Temperature Field Measurement, Welding Journal, Vol. 68, No.1. pp. 28s-32s

Hackl, H. & Himmelbauer, K. (2005). The CMT-Process - A Revolution in Welding Technology, International Institute of Welding, *IIW Doc. No. XII-1875-05*

Hälsig, A., Kusch, M., Bürkner, G. & Matthes, K.-J. (2010). Bestimmung von Wirkungsgraden moderner Schutzgasschweißverfahren, IGF-No. 15.562B/DVS-No. 03.078 (in German), Investigation of the *Institute for production technology/ welding technology (Technical University of Chemnitz, Germany)*

Holman, J.P. (1990). *Heat Transfer* (7th edn.), McGraw-Hill, ISBN 0-07-909388-4, San Francisco (CA)

Hsu, C. & Soltis, P. (2002). Heat input comparison of STT vs. short-circuiting and pulsed GMAW vs. CV processes, *Proc. ASM Proceedings of the International Conference: Trends in Welding Research*, Pine Mountain, GA, USA, 15-19th April, 2002

Jenney, C.L., O'Brien, A. (Editors). (2001). *Welding Handbook - Welding Science and Technology*, American Welding Society, ISBN 0-87171-657-7, Miami (FL), USA

Joseph, A., Harwig, D., Farson, D.F. & Richardson, R. (2003). Measurement and calculation of arc power and heat transfer efficiency in pulsed gas metal arc welding, *Science and Technology of Welding and Joining*, Vol. 8, No. 6, 01/2003, pp. 400-406

Kohlrausch, F., Hahn, D. †, Wagner, S. (1985). *Kohlrausch Praktische Physik* (in German), B.G. Teubner, ISBN 3-519-13001-7, Stuttgart, Germany

Kou, S. (1987). Heat Flow during Welding, In: *Welding Metallurgy*, pp. 29-59, John Wiley & Sons, Inc., ISBN 0-471-84090-4, Hoboken (NJ)

Kou, S. (2003). Heat Flow in Welding, In: *Welding Metallurgy*, pp. 37-62, John Wiley & Sons, Inc. (2nd ed.), ISBN 0-471-43491-4, Hoboken (NJ)

Lancaster, J.F. (1986). *The Physics of Welding*, (2nd ed.), Pergamon Press, ISBN 0-08-030555-5, Oxford (UK)

Lu, M. & Kou, S. (1988). Power and Current Distributions in Gas Tungsten Arcs, *Welding Journal*, Vol. 67, No. 2, pp. 29s-32s

Nestor, O.H. (1962). Heat Intensity and Current Distributions at the Anode of High Current, Inert Arcs, *Journal of Applied Physics*, Vol. 33, No. 5, (May, 1962), pp. 1638-1648, ISSN 0021-8979

Pepe, N. & Yapp, D. (2008). Process Efficiency and Weld Quality for Pipe Root Welding, International Institute of Welding, *IIW Doc. No. XII-1951-08*

Pepe, N.C. (2010). Advances in Gas Metal Arc Welding and Application to Corrosion Resistant Alloy Pipes, *PhD Thesis*, Cranfield University

Pepe, N., Egerland, S., Colegrove, P., Yapp, D., Leonhartsberger, A. & Scotti, A. (2011). Measuring the Process Efficiency of Controlled Gas Metal Arc Welding Processes, *Science and Technology of Welding and Joining*, Vol. 16, No. 5, (July, 2011), pp. 412-417

Sudnik, V.A., Ivanov, A.V., & Dilthey, U. (2001): Mathematical model of a heat source in gas-shielded consumable electrode arc welding, *Welding International*, Vol. 15, No.2, pp.146-152

13

Chemical and Physical Properties of Fluxes for SAW of Low-Carbon Steels

Ana Ma. Paniagua-Mercado and Victor M. Lopez-Hirata
Instituto Politécnico Nacional (ESFM-ESIQIE)
Mexico

1. Introduction

The submerged-arc welding of steels has been used since 1930. It is well known that the mechanical properties of steel weldments depend on the chemical compositions of electrodes and fluxes. The development of welding electrodes has been based on practical experiences. The study of welding deposits by means of physical metallurgy permitted to develop electrodes and fluxes for SAW process of steels. In contrast, the use of chemical and physical properties of fluxes for the development of welding process started in the 70´s. (Shah, 1986). Most of the works concerning with the fluxes for SAW process have been focused on its effect on microstructure and mechanical properties. Likewise, it has been interesting the study of the thermochemical and electrochemical reactions that occur at the welding pool which are very important for the transferring of metallic elements to the welds. In addition to the stated above, the electrode coverings is also a very important aspect to obtain weld metal with good mechanical properties. The covering materials are fluxes which are composed of different mineral chemical compounds such as, oxides, fluorides and carbonates (Singer & Singer, 1979). The firing and sintering process of fluxes during welding electrode processing promotes chemical reactions and phase transformations of these minerals. All these factors determine the valence or electric charge of the different elements which are deposited to the weld metal. All the oxides from flux may contribute to the dissolution process of different metallic elements and oxygen in the welding pool. These metallic elements can react with oxygen to form oxide inclusions which can serve as nucleation sites for the formation of some benefit phases such as acicular ferrite during the welding process (Davis & Bailey 1991). These events may improve the mechanical strength and ductility of weld metal. Thus, the purpose of this chapter is to show the effect of different fluxes on the chemical composition, microstructure and mechanical properties of weld metals by the submerged-arc welding procedure.

2. Chemical and structural characterization of the crystalline phases in agglomerated fluxes for SAW

The weld metal chemistry is affected by the electrochemical reaction at the weld pool flux interface. The main characteristics of submerged-arc welds are a function of the fluxes and their physicochemical properties (Indacochea et al., 1989). Submerged-arc welding (SAW) fluxes are manufactured in three main forms; fused at temperatures exceeding 1400 $^{\circ}$ C, agglomerated from 400 to 900 ° C and sintered from 1000 to 1100 ° C from mineral constituents, (Jackson, 1982). Reactions, which occur during firing of a ceramic oxide, comprise phase transformations of minerals (Singer, 1963), reaction among the different phases (Allen, 1966), formation of crystal phases from the melt and formation of a melt. The phase transformation of the crystal phases predominates at the early stages of firing, before a melt is formed. The formation of a melt can be observed at temperatures as low as 920 °C, due to the presence of minor amounts of impurities such as, alkaline earth oxides (Dunham & Christian, 1984). In this way, the agglomerated fluxes can be studied by the chemical analysis of the crystal phases formed during the increase of temperature. This enables us to quantify types of ions and their distribution, which is useful to predict their behavior when they are dissociated in the submerged-arc welding process, (Indacochea et al., 1989). Thus, this section shows a study of the crystalline phases and the chemical characterization of the ions formed in agglomerated fluxes using Chemical Analysis, X-Ray Diffraction (XRD) and Differential Thermal Analysis (DTA). It is also intended to show the effect of ion content of flux on the behavior of submerged-arc welding.

2.1 Composition, preparation and characterization of fluxes

Three agglomerated welding fluxes were prepared, using mixtures of mineral oxides in different proportions (designated as A, B and C), as shown in Table 1. Each mixture was weighed, mixed in a vibratory camera for 30 minutes and subsequently pelletized in a laboratory machine of 80 cm diameter and 28 cm height at 40 rpm, using sodium silicate as the agglomerate agent. The average size of pellets was 5 mm. They were dried in a stove at 200 °C for 24 hours, fired for 3 hours. at 950 °C in a gas kiln at a heating rate of 50 °C/h, crushed and then screened to a 240 μm size. For comparison, a commercial agglomerated welding flux was also used, designated as T sample. The chemical analysis was determined by X-ray fluorescence. The XRD analysis was conducted in a diffractometer with a monochromated Cu Kα radiation. For all the samples, the DTA analysis was carried out in a Al_2O_3 crucible from 25 to 1350°C at a heating rate of 10°C/min. Alumina was used as standard reference for DTA analysis.

2.2 Chemical, structural and thermal characterization of fluxes

2.2.1 Chemical analysis

Table 2 shows the elemental chemical composition of the mixtures A, B and C, calculated with the data shown in Table 1 and the stequiometric formula of each compound. Tables 3 and 4 show the elemental chemical analysis and oxide compounds, respectively, for the fused fluxes A, B, C and T. By comparison of Table 1 with Table 3, it can be noticed that there was an increase in the percentage of Al_2O_3, SiO_2, and Na_2O. These oxides increased at the same proportion as they were present in the original mixtures. This can be attributed to the evaporation of impurities such as alkaline oxides with a low melting point that were present in the original mineral, (Dunham & Christian, 1984). The amount of CaO showed the highest decrease because it came from the $CaCO_3$ compound, in the original mixture, which is decomposed at about 900 °C (Singer, 1963). The Na_2O also is formed from Na_2CO_3, but its proportion is a little higher than the original one because the sodium silicate, Na_2SiO_4, was used for the agglomerating process. In order to be agglomerated, a higher amount of sodium silicate was added to the flux B than that added to the fluxes A and C. This fact explains the difference in Na_2O composition.

In the SAW process, the temperature of electric arc causes the dissociation of oxides and these remains as ions in the plasma, (Belton et al. 1963). The temperature in the welding pool reaches 1560-2300 °C. Christensen and Gjermundsen, (Christensen & Gjermundsen, 1962) calculated temperatures above 2500 °C for the welding pool in mild steels. Apold (Apold, 1962) suggested that the heat energy provided by the electric arc is concentrated on a circumference of 5mm diameter. According to the above information, most of the oxides are melted, but oxides with high melting points such as MgO (2500-2800 °C), CaO (2572 °C) and ZrO_2 (2720 °C) are not melted. That is, it is important that the chemical formulation of fluxes enables all the oxides to be melted in order to avoid the presence of inclusion in the weldment.

Compounds	Flux A	Flux B	Flux C
Al_2O_3	2.6	8.00	8.0
SiO_2	43.8	25.00	20.0
Fe_2O_3	6.0	2.0	9.0
K_2O	0.11	0.07	0.07
Na_2CO_3	0.09	1.0	0.20
$CaCO_3$	20.00	3.0	5.0
MnO	7.00	---	11.0
MgO	6.00	1.0	5.0
TiO_2	10.00	29.00	31.00

Table 1. Initial composition (wt. %) of fluxes.

Element	Flux A	Flux B	Flux C	Flux T
Al	3.85	12.17	12.00	9.71
Si	25.53	14.72	11.47	5.46
Fe	1.74	0.52	2.74	2.4
K	0.37	0.24	0.24	0.32
Na	1.05	10.4	2.36	0.9
Ca	8.1	1.21	1.88	4.85
Mn	4.11	-	6.38	9.2
Mg	2.8	0.37	2.49	4.47
C	0.46	0.53	0.03	0.08
S	0.08	0.028	0.06	0.016
Ti	5.3	15.5	16.7	21.6

Table 2. Elemental chemical analysis (wt. %) of fused fluxes.

Oxide	Flux A	Flux B	Flux C	Flux T
Al_2O_3	7.29	23.00	23.00	18.36
SiO_2	54.65	31.50	24.50	11.69
Fe_2O_3	2.48	0.74	3.92	3.43
K_2O	0.44	0.28	0.28	0.38
Na_2O	1.41	14.00	3.18	1.21
CaO	11.33	1.69	2.63	6.78
MnO	5.30	-	8.23	11.88
MgO	4.64	0.61	4.12	7.41
TiO_2	9.00	26.00	28.00	36.31

Table 3. Oxide compound (wt. %) of fused fluxes.

2.2.2 Structural analysis

Figures 1, 2, 3 and 4 shows the X-ray diffraction patterns of agglomerated and sintered fluxes. Here, it can be observed the presence of crystalline phases, such as Nepheline ($NaAlSiO_4$), Gismodine ($CaAl_2Si_2O_8.4H_20$) and Vesuvianite ($Ca_{19}Al_{II}Mg_2Si_{18}O_{69}(OH)_9$), which were formed at low temperature, (Berry & Mason, 1959). These are equilibrium phases and were formed due to the reaction at the sintered temperature between different ions of the original mixture with the sodium silicate, added during the agglomeration process. Other oxides, such as titanium oxide and manganese oxide showed practically no reaction with other reactants at these temperatures; however, they appeared with a different oxidation degree and crystalline structure. The silicon oxide, aluminum oxide and calcium oxide did not react at all.

The Nepheline phase can be formed by replacement of the half of silicon atoms with aluminum atoms in the sodium silicate compound, (Klein & Hurbult, 1999). The electric balance is kept by the addition of sodium ions in the interstitial sites of the hexagonal structure (Berry & Mason, 1959). Its quantification is shown in Table 4 and it was based on the X-ray diffraction peaks with the highest intensity (Cullity, 2003). Table 5 shows the quantification of ions calculated according to its disassociation reaction at the temperature of electric arc, as proposed by Davis and Bailey (Davis & Bailey, 1991).

The Gismodine phase can be melted at low temperatures and has a monoclinic structure and also appeared in the sintered fluxes. The quantification of compound and ions is shown in Tables 4 and 5, respectively.

The Vesuvianite phase has a low melting point and tetragonal crystalline structure. This could be formed by means of the replacement of aluminium with silicon. This phase has also a low melting temperature and a tetragonal structure. Its quantification is also shown in Tables 4 and 5.

The other detected phases were Quartz, Rutile, Anastase and γ-titanium oxide. Fluorite was only detected in the flux T. All these compounds showed practically no reaction at the sintering temperature. These were also stable at room temperature. Tables 4 and 5 show their quantification as ions and compounds, respectively.

The formation of different oxides and silicates observed in fluxes A, B and C, but not in flux T can be attributed to the addition of sodium silicate used in the agglomeration process of the former fluxes, as well as the high sintering temperature.

2.2.3 Thermal analysis

All oxides were stable during heating of DTA up to 1000 °C. Thus, the DTA curves are shown in the range of 1000-1350°c for fluxes A, B, C and T in Figs. 5, 6, 7 and 8, respectively. Most of the reactions observed in DTA curves were endothermic type. Some of the endothermic peaks are located at close temperature values for the different fluxes. For instance, an endothermic peak is located at 1147.8, 1121.4, 1122.3 and 1161.6 °C for the fluxes A, B, C and T, respectively. Another one is located at 1226.0, 1189.3, 1213.4 and 1228.6 °C for the same fluxes, respectively. Finally, a peak located at 1288.4, 1267.9 and 1286.6 °C for the fluxes B, C and T, respectively. The difference in temperature for these endothermic events can be attributed to the difference in chemical composition of fluxes.

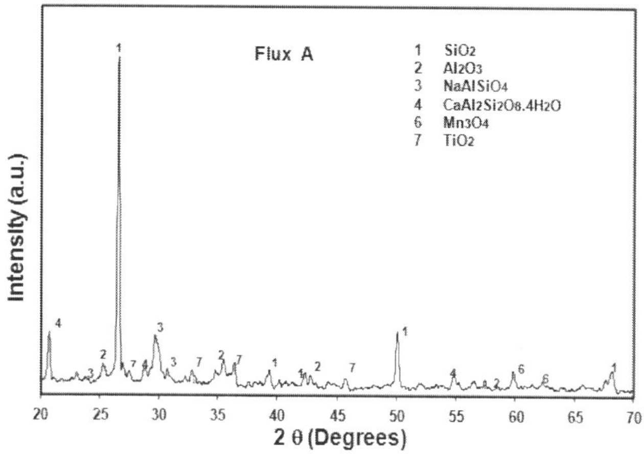

Fig. 1. XRD pattern of the flux A.

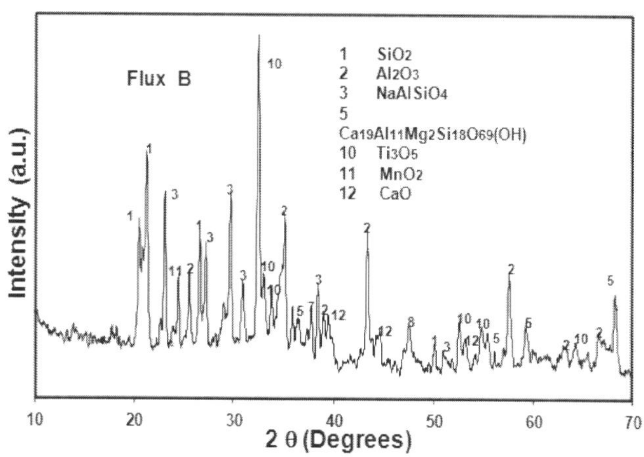

Fig. 2. XRD pattern of the flux B.

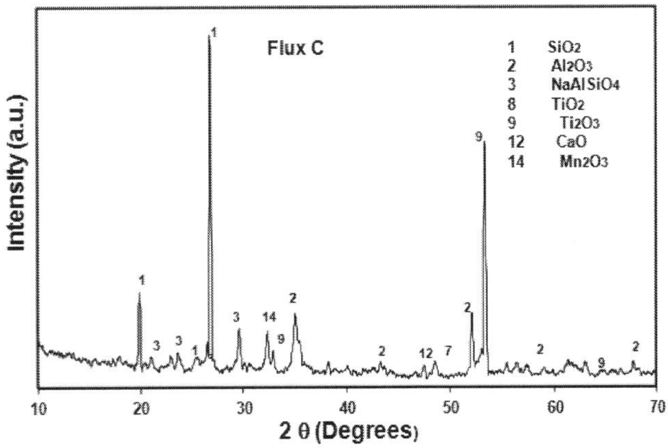

Fig. 3. XRD pattern of the flux C.

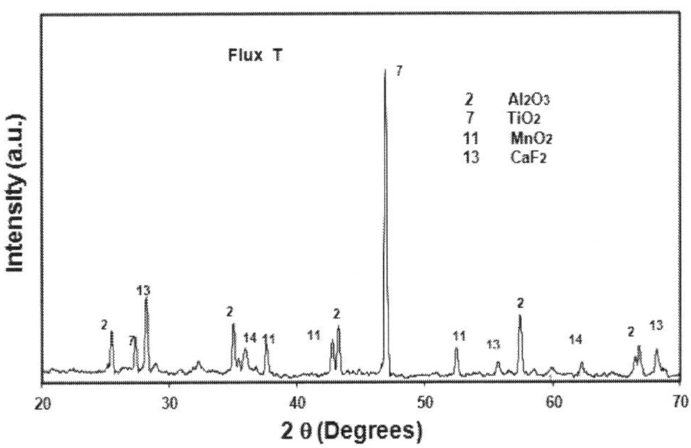

Fig. 4. XRD pattern of the flux T.

Name	Composition	A	B	C	T
Corundum	Al_2O_3	9.17	17.90	12.11	19.07
Quartz	SiO_2	51.10	17.05	18.38	6.90
Nepheline	$NaAlSiO_4$	12.65	19.95	17.12	-----
Gismondine	$CaAl_2Si_2O_8.4H_2O$	12.65	-----	-----	-----
Vesuvianite	$Ca_{19}Al_{11}Mg_2Si_{18}O_{69}(OH)_9$	-----	10.23	-----	-----
Hausmannite	Mn_3O_4 [$MnOMn_2O_3$]	5.24	3.80		6.10
Anatase	TiO_2	-----	6.95	3.60	53.03
Rutile	TiO_2	9.17	-----	-----	-------
γ-Titanium oxide	Ti_3O_5	-----	20.32	-----	-------
Titanium oxide	Ti_2O_3	-----	-----	25.65	--------
Manganese oxide	Mn_2O_3	----	-----	8.55	-------
Calcium oxide	CaO			3.56	
Fluorite	CaF_2	-----	-----	----	14.90

Table 4. Weight percent of compounds determined from XRD.

Ion	Flux A	Flux B	Flux C	**Flux T**
% Al^{+3}	10.51	14.28	11.01	10.09
% Si^{+4}	29.55	22.80	19.79	3.22
% Na^{+}	2.05	3.23	1.15	---
% Mn^{+2}	1.17	0.84	2.65	1.36
% Mn^{+3}	2.60	1.88	5.90	3.03
% Mn^{+4}				
% Ti^{+2}	5.5	6.95	2.15	31.79
% Ti^{+3}			17.08	
% $Ti^{5/3}$		13.05		
% Ca^{+2}	2.34	2.57	2.54	7.64
% Mg^{+2}		1.88		
% H^{+1}	0.12	0.03		
% F^{-1}				7.26
% O^{-2}	46.16	42.21	32.47	35.61

Table 5. Ion content (mol %) of fluxes.

The endothermic reactions corresponding to the above temperatures can be compared with the melting reaction of Gismodine, which has been reported to occur in a temperature range between 965 and 1082 °C. Quartz transformation from Cristobalite to Tridimite occurs between 1200 and 1650 °C. Na_2O sublimation takes place at about 1275 °C. Fluorite and

Vesuvianite melt at about 1200 °C. Nefeline melts at temperatures higher than 1248 °C. The melting of corundum, rutile, anastase, hausmannite and quartz was not detected because it occur above 1500 °C. The zigzag behavior of DTA curves reveals blistering or gassing of the glass formation (Gordon & Chu, 1966).

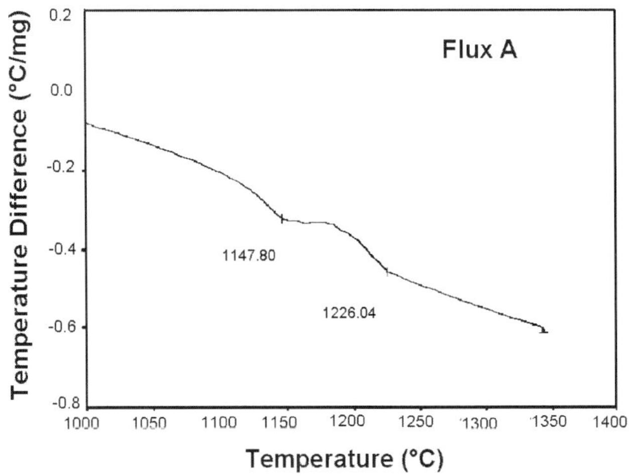

Fig. 5. DTA diagram of flux A.

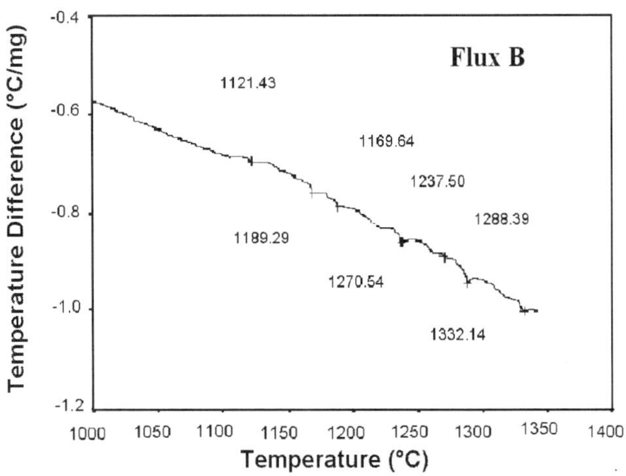

Fig. 6. DTA diagram of flux B.

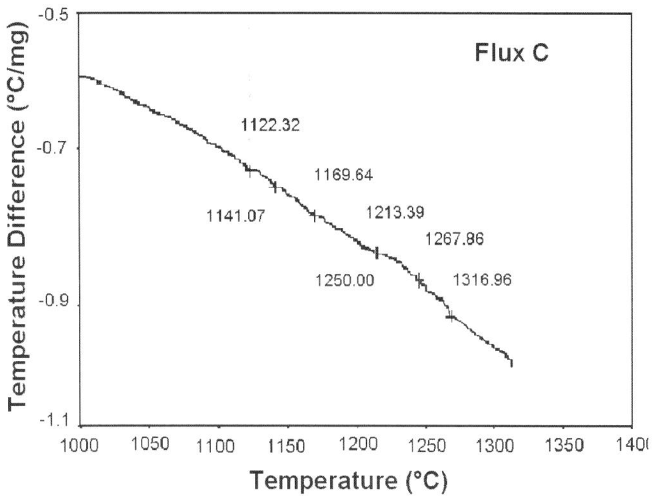

Fig. 7. DTA diagram of flux C.

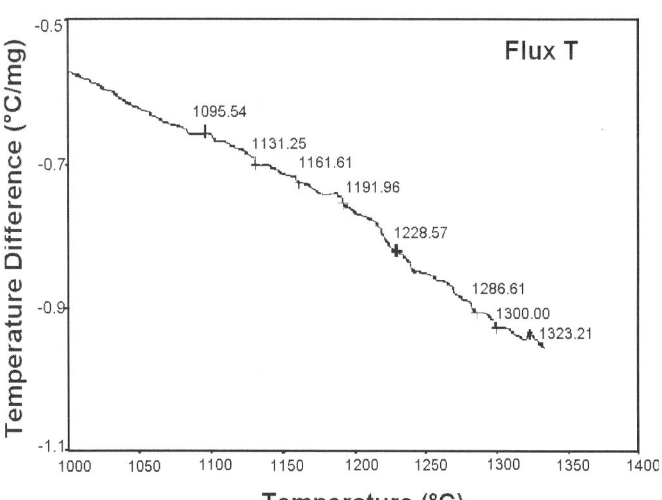

Fig. 8. DTA diagram of flux T.

2.3 Effect of crystalline phases and ion contents on the behavior of flux

Oxides and carbonates were used for manufacturing the different fluxes. These formed different crystalline phases after sintering. According to the X-ray diffraction analyses, silica (SiO_2), manganese oxide (MnO) and titanium oxide (TiO_2) were the compounds that reacted during the heating of fluxes.

The silica reacted to form anionic species such as silicates. Three compounds with this characteristic were identified in the sintered fluxes: Nepheline ($NaAlSiO_4$), Gismodine ($CaAl_2Si_2O_8.4H_2O$) and Vesuvianite ($Ca_{19}Al_{11}Mg_2Si_{18}O_{69}.(OH)_9$). The silicon electrodeposition in the welding cathodic pool is not as favored as the deposition of volatile calcium from flux. It has been found (Davis & Bailey, 1991) that a small amount of silicon can be formed electrochemically, but most of it is formed by means of calcium evaporation, which can reduce the SiO_2 in the flux.

It was found that the hausmannite (Mn_3O_4) was formed in the sintered fluxes. This is a spinel with electron valences, Mn^{+2} and Mn^{+3}. Thus, it is possible to have different reactions which permit the formation of several oxide compounds and oxide radicals from MnO, forming inclusions.

The ion quantification determined from the X-ray diffraction results enables us to estimate the amount of the ions from flux formed in the plasma of electric arc. These ions react with oxygen and the oxides will be deposited on the weld. The most important reactions between the electric arc and welding pool correspond to those where oxygen is involved.

Oxygen can react with any cationic component from the flux, as Na^{+1}, Ca^{+2}, Mg^{+2}, Al^{+3}, Si^{+4}, Fe^{+2}, Fe^{+3}, Mn^{+2}, Mn^{+3}, Mn^{+4}, Ti^{+2}, Ti^{+3}, $Ti^{+5/3}$ and probably SiO_4^{-4} from the silicates formed in the fluxes, the anions present are O^{-2} and F^{-1}.

It is possible to make a prediction of the reactions with these cations and anions. Calcium and magnesium are expected to react first with oxygen in the welding arc because its corresponding oxides have the largest negative formation free energy ΔG_f. Al^{+3} has the next highest value for the formation of a stable oxide. Si^{+4} and Titanium reacts readily with oxygen in the next order. Ti^{+2} almost react at the same time with Mg^{+2}, and then Ti^{+3} and Ti^{+4} react to form stable dioxides to give anions and cations with similar set of reactions. Manganese too exhibit variable valency and may form many oxides and oxide radicals when react with the oxygen in the next order Mn^{+2}, Mn^{+3} and Mn^{+4} in agreement with the corresponding ΔG_f values (Davis & Bailey, 1991).

The silicates formed will be more soluble in the slag than in the weld pool. Transfer of Mn, Si and Al from the flux to weld pool depends on the amounts in the flux (Davis & Bailey, 1991).

The oxygen can react with any available cationic species to give non-metallic inclusions.

The identification of phases in fluxes can be used to know the behavior of fluxes during their depositing process, as well as the effect of them on the mechanical properties of welds.

In the fluxes A, B, C, and T, the presence of corundum can be observed, which has an effect on the facility for slag removing (Jackson, 1982). The same effect has been observed for the presence of rutile in fluxes.

Calcium ions in fluxes increase the stability of electric arc (Butler & Jackson, 1967). These ions can come from either oxide or fluoride compounds. Significant calcium content was detected for fluxes A, C and T but not for flux B. Thus, the latter might present an unstable electric arc during welding.

It is also known that quartz and corundum increase the viscosity of fluxes, while the additions of manganese oxide, fluorite and titanium oxide reduce viscosity (Jackson, 1982).

The latter compounds were detected in the fluxes A, C, and T, but only a small content, may be as impurity, and were observed in the flux B. Thus, a problem of flux fluidity might be present in this flux.
The manganese oxide and quartz have been observed to have a beneficial effect on the mechanical properties of welds (Lancaster, 1999). Manganese oxide and quartz were included in the formulation of fluxes A and C, but not for flux B.
Titanium oxide has been observed to promote the formation of acicular ferrite, which is less susceptible to cracking (Evans, (1996). Fluxes C and T are expected to produce the better weld since they contain quartz, manganese oxide, titanium oxide, and the content of oxygen ions (calculated by X-ray diffraction) is smaller than A and B.

2.4 Influence of chemical composition on the microstructure and tensile properties of SAW welds

The mechanical properties of welds are determined by the microstructure developed during the submerged-arc welding process (Joarder et al. 1991). It was suggested that the acicular ferrite provided a good toughness and tensile strength to the welds because its fine size has a higher resistance to the crack propagation (Liu & Olson, 1986). Thus, it seems to be convenient to increase the volume fraction of ferrite in welds. Widmanstätten, equiaxial and acicular ferrite can be developed by the welding process. A method for promoting the formation of acicular ferrite consists of the additions of oxides into the flux, such as boron oxide, vanadium oxide and titanium oxide (Evans, 1996). The oxides in the flux may contribute to different metallic element dissolution and oxygen into the weld. These elements may react to form oxide inclusions, which are trapped into the weld and facilitate the nucleation of acicular ferrite during the weld cooling (Dowling et al. 1986) and (Vander et al. 1999). The main factors that determine the microstructure of a weld are: chemical composition, austenite grain size and cooling rate (Lancaster, 1980). The chemical composition comes from the composition of base metal, electrode and flux (Davis & Bailey, (1991). In order to increase the mechanical properties of welds for low-carbon steels, the selection of an appropriate flux composition plays a very important role to obtain a fine acicular ferrite, which has been shown to improve the properties in this type of welds. The objective of this work is to study the effect of different fluxes on the chemical composition, microstructure and tensile properties of weld applied on an AISI 1025 steel by the arc-submerged welding process.

2.4.1 Weldment preparation and mechanical tests

The chemical composition of fluxes used in this work are given in Table 6. It also includes the composition for a commercial flux to be used for comparison purposes. The compositions were selected in order to analyze the effects of SiO_2, MnO and TiO_2 on weld properties. The base metal was a 250 x 60 x 13 mm plate of an AISI 1025 steel, with a chemical composition as given in Table 7. The joint preparation was a 45° single V-groove. The welding fluxes were dried at 200 °C for 24 hours. The welding conditions were 600 A, 30V and a welding speed of 0.118 ms^{-1}, and kept constant for all cases. All welds were carried out in one pass. Tension specimens were extracted from welded plates and prepared according to the E8M-86 ASTM standard. Tension tests were conducted at room temperature and 3.33×10^{-4} s^{-1} in a universal machine and on 10 specimens for each flux. Vickers hardness measurements were carried out on the base metal, heat affected zone and weld site with a load of 10 kg. The chemical analysis of welds was performed by an X-ray fluorescence spectrometer. Weld samples were prepared metallographically and etched with Nital to be observed in a Scanning Electron Microscope (SEM) at 20 kV and a light microscope. Microanalysis were also carried out on the inclusions of welds using an EDX equipment attached to the SEM. Macroetching was also carried out with Nital reagent.

Flux	wt.% SiO_2	wt.% Fe_2O_3	wt.% MnO	wt.% MgO	wt.% K_2O	wt.% Na_2O	wt.% CaO	wt.% Al_2O_3	wt.% TiO_2
A	54.65	2.48	5.30	4.64	0.44	1.41	11.33	7.29	9.00
B	31.51	0.74	---	0.61	0.28	14.00	1.69	23.00	26.00
C	24.55	3.92	8.23	4.12	0.28	3.18	2.63	23.0	28.0
T	11.69	3.43	11.88	7.41	0.38	1.21	6.78	18.36	36.31

Table 6. Chemical composition of fluxes.

2.4.2 Chemical composition

As is well known, the reactions between liquid weld metal and fused flux in the SAW process are similar to those between molten metal and slag in the steel making process. As shown in Table 7, the manganese, silicon, calcium, titanium and aluminum contents of the weld metal increase as the content of the corresponding oxides for the initial flux also increases. It was reported (ASM, *Metals Handbook*, 1989) that a rapid pick up of manganese and silicon might occur until its corresponding oxide content of the flux was about, 10 and 40 %, respectively. The fluxes T and A have the highest contents of MnO and SiO2, respectively. The highest contents of manganese and silicon were also detected for the weld metal of fluxes T and A, respectively.

The lowest carbon content was detected for the weld metals of the fluxes B and C. This is attributed to the higher contents of Al_2O_3 and SiO2 for those fluxes, which give a higher content of O^{-2} ions that react with carbon (Davis & Bailey, 1991). Additionally, these fluxes also have the lowest content of metallic cations.

2.4.3 Mechanical properties

Table 8 shows the average values of tensile properties, yield strength (S_y), ultimate tensile strength (S_{uts}), elongation percentage and area reduction percentage. The Vickers hardness at the center of weld is also shown in this table.

The tensile strength and hardness of a steel can be related to the equivalent carbon, given by the following equation (Zhang et al. 1999):

$$C_{equivalent} = C + Mn/6 + Si/24 + Ni/40 + Cr/5 + Mo/4 + V/4 \tag{1}$$

Where C, Mn, Si, Ni, Mo and V represent the metallic content, expressed as a percentage. As expected, the highest and lowest tensile strength and hardness were obtained for the weld metal with the highest and lowest equivalent carbon, respectively. It was reported, (Lancaster, 1980) that an equivalent carbon higher than 0.45 had a high susceptibility to cold cracking after welding. Besides, the formation of martensite is facilitated during cooling of welds. The equivalent carbon of welds for all fluxes is lower than 0.45. The highest toughness, detected by the largest area under the engineering stress-strain curve, corresponded to the weld metals of fluxes T and C. This can be attributed to their higher content of manganese, which is known to improve the steel toughness (ASM, *Metals Handbook*, 1989). Table 8 also shows the yield and ultimate tensile strengths calculated using a computer program based on a non-linear multiparameter regression of the strength versus the composition (Material Algorithms Project, 1999). The calculated values are higher than the values determined in this work. The differences can be attributed to the fact that the regression program considers neither the cooling condition of the weld nor the microstructure constituents. Nevertheless, the highest and lowest calculated yield and ultimate tensile strengths also corresponded to the welds for fluxes T and B, respectively.

Element (wt. %)	Base Metal	Electrode	Weld (Flux A)	Weld (Flux B)	Weld (Flux C)	Weld (Flux T)
C	0.248	0.082	0.124	0.103	0.098	0.171
Mn	0.785	1.312	0.640	0.710	1.000	1.380
Si	0.218	0.762	0.610	0.300	0.470	0.530
P	0.024	----	0.008	0.008	0.10	0.019
S	0.037	0.008	0.0044	0.034	0.046	0.033
Mo	0.016	---	0.006	0.006	0.006	0.012
Ni	0.118	----	0.030	0.030	0.030	0.090
Cu	0.345	----	0.175	0.130	0.134	0.285
Sn	0.048	----	0.004	0.004	0.004	0.022
Al	----	----	0.004	0.027	0.001	0.010
Ti	0.002	----	----	---	0.020	0.023
As	----	----	0.004	0.004	0.006	0.007
Ca	----	----	0.0031	----	0.002	0.0007
Cr	0.0054	----	----	----	----	0.090
V	0.004	----	----	----	----	----

Table 7. Chemical composition of steel, electrode and welds.

Flux	S_y (MPa)	S_{uts} (Mpa)	Elongation (%)	Area Reduction (%)	Vickers Hardness	$C_{equivalent}$	Calc. S_y (MPa)	Calc. S_{uts} (MPa)
A	208	410	22	47	167	0.28	446	572
B	197	418	23	51	163	0.24	407	507
C	330	551	19	36	179	0.30	565	583
T	345	569	20	43	192	0.43	586	662

Table 8. Tensile properties, hardness an equivalent carbon of welds for the different fluxes.

2.4.4 Macrostructure and microstructure of welds

The macrographs of the as-deposited region of welds corresponding to the fluxes A, B, C and T show a dendrite structure. The width of the heat affected zone (HAZ) is about 0.4 mm for most of the welds. In the case of HAZ, the ferrite mean grain size is about 5-9 μm. Figures 9 (a)-(d) and (e)-(h) show the light and SEM micrographs, respectively, of the welds corresponding to the fluxes A, B, C and T. Table 9 summarized the microconstituents and its volume percentage observed in the welds corresponding to the fluxes A, B, C and T. Pearlite and equiaxial ferrite are observed for all welds. No Widmanstätten ferrite is detected for all

welds. The acicular ferrite was only detected for the welds corresponding to the fluxes C and T. This can be attributed to the TiO_2 content of flux because this type of oxides favored the nucleation of acicular ferrite at the interface between austenite matrix and inclusion (Zhang et al. 1999) and (Babu et al. 1999). Table 9 also shows the calculated microstructure constituents using a computer program based on the weld composition and welding conditions (current, voltage and welding speed), (Material Algorithms Project, 1999). According to the calculated results, no Widmanstätten ferrite is predicted to be formed for these weld compositions, which is in agreement with the results described above. However, the calculated results indicated no presence of Pearlite, which can be observed clearly for all welds in Fig. 9. The calculated microstructure results also show the increase in the volume percentage of bainite + acicular ferrite for the fluxes C and T, compared with microstructure results for fluxes A and B. This fact shows a good agreement with the presence of acicular ferrite observed for the welds corresponding to fluxes C and T.
The volume percentage of inclusions is show in Table 10 for the welds corresponding to fluxes A, B, C, and T. The lowest and highest contents of inclusions are detected for the welds of fluxes C and T, and those of fluxes A and B, respectively. Table 10 also shows the content of oxygen ions, calculated from the flux composition. The lower the content of oxygen ions, the lower inclusion content in the weld steel is resulted. The elongation and area reduction percentages are shown to have a dependence on with the inclusion volume percentage. This seems to be reasonable since the ductility is drastically decreased with the increase of inclusions in steels (Liu & Olson, 1986).

Weld	Equiaxial Ferrite (%)	Widmast. Ferrite (%)	Pearlite (%)	Acicular Ferrite (%)	Calculated Widmanst. Ferrite (%)	Calculated Ferrite (%)	Calculated Acic. Ferrite+Bainite (%)
Flux A	83	0	17	0	0	1.85	98.15
Flux B	88	0	12	0	0	1.97	98.03
Flux C	69	0	20	11	0	1.86	98.14
Flux T	68	0	8	24	0	1.24	98.76

Table 9. Microstructure of welds.

Weld	O^{-2} ion %	Basicity Index	Inclusion vol. %
Flux A	46.16	0.35	24.40
Flux B	46.21	0.30	21.60
Flux C	32.47	0.32	15.00
Flux T	35.61	0.60	15.11

Table 10. Content of inclusions.

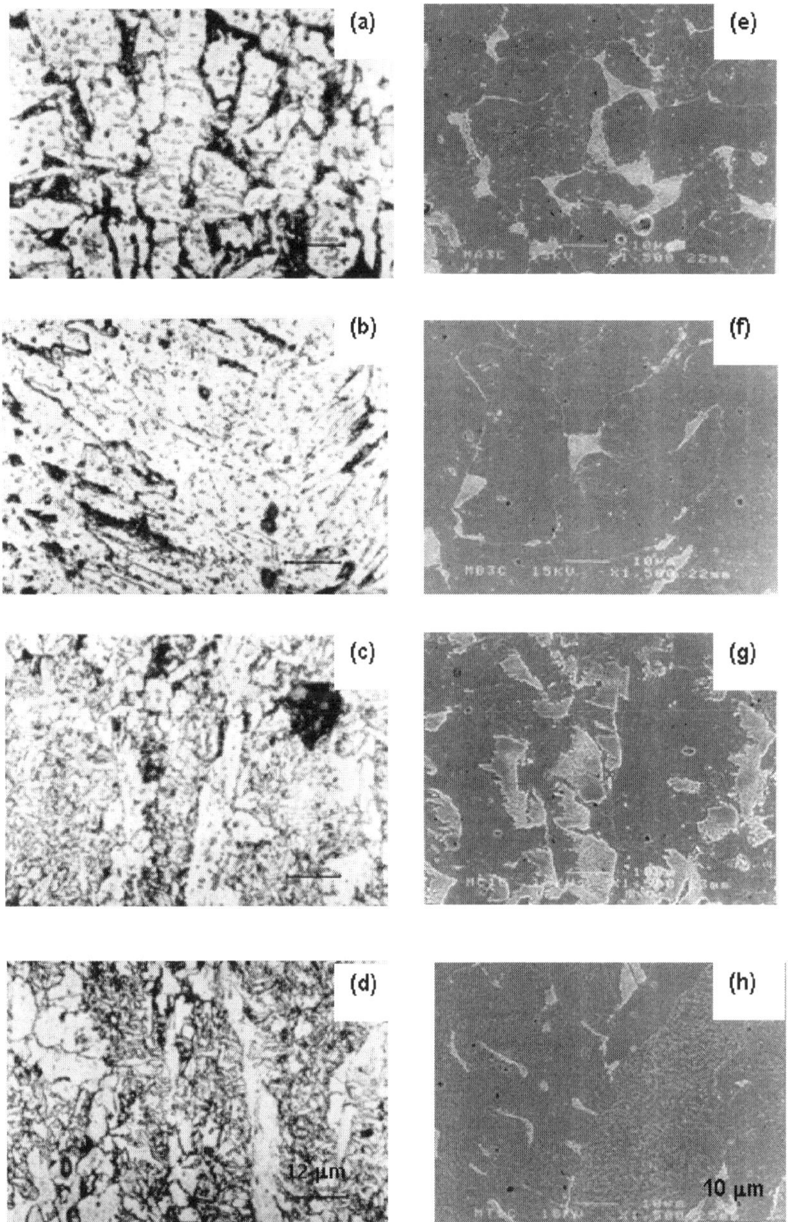

Fig. 9. Light and SEM micrographs of welds for (a) and (e) flux A, (b) and (f) flux B, (c) and (g) flux C, and (d) and (h) flux T, respectively.

The microanalysis on inclusions indicated that most of inclusions are aluminum and silicon oxides, manganese and calcium sulfides; however, the presence of round and bright titanium oxide inclusions is observed only for the welds of fluxes C and T, Fig. 10. This type of inclusions were mainly composed of Ti, Al, Mn and Si. This fact suggests that the acicular ferrite seems to be nucleated at the interface between austenite matrix and titanium oxide inclusions welds. We believe that almost no TiO_2 inclusions are formed for welds of fluxes A and B because these fluxes have the lowest content of MnO. MnO was reported (Davis & Bailey, 1991), to react with SiO_2 to form silicomanganates. However, the absence of MnO causes SiO_2 to react with TiO_2 to form silicotitanates. Thus, this may explain the absence of acicular ferrite for welds corresponding to fluxes A and B. The titanium-containing welds corresponding to fluxes C and T have the highest yield and ultimate tensile strengths. This fact is in agreement with the formation of acicular ferrite observed for these welds.

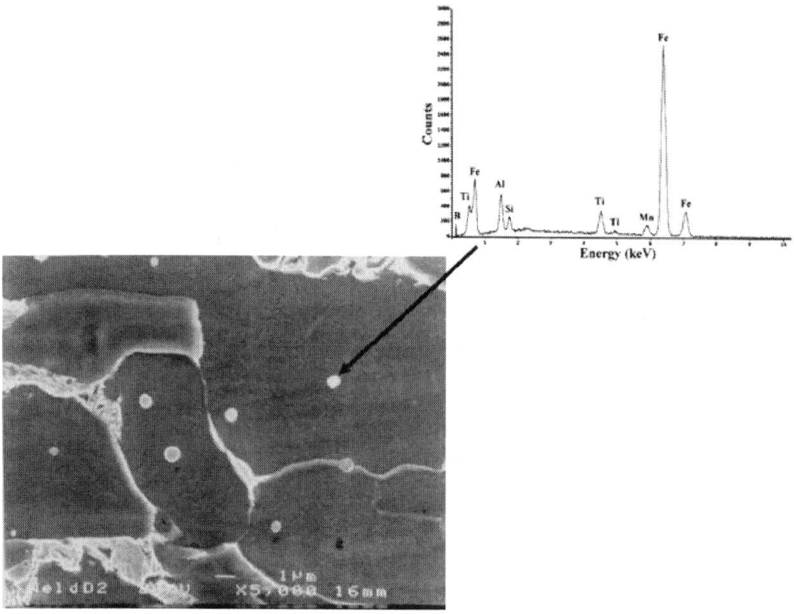

Fig. 10. SEM micrograph of Ti-Inclusions and it corresponding EDS spectrum.

3. Conclusion

The determination of phase in fluxes enables us to identify the different type of oxides and radicals formed during sintering of the initial materials. This quantification makes possible know what anions and cations will be present in the electric arc. The most reactive reacts quickly in the weld pool and might be either absorbed in the slag or retained in the weld as inclusions. In summary, this work shows the importance of the selection for flux composition in order to improve the mechanical properties of steel welds.

4. Acknowledgment

The authors wish to thank financial support from SIP-IPN-CONACYT.

5. References

Allen, A.W. (1966). Optical Microscopy in Ceramic Engineering, *Proceedings Of the Third Berkeley International Materials Conference* (6), p.p. 123-142. ISBN 471-28720-2 New York, United States.

Apold, A. (1962) *Carbon Oxidation in the Weld Pool.* Pergamon Press, the Macmillan Co., ISBN-10: 0080096921; ISBN-13: 978-0080096926, New York, United States.

ASM, 1989 *Metals Handbook*, vol. 8, Mechanical Testing 9th. Ed. pp. 39–74. ISBN: 087170389. Metals Park, OH, United States.

Babu, S.S. David, S.A. & Vitek, J.M. (1999) Thermo-chemical-mechanical effects on microstructure development in low-alloy steel welds elements, in: *Proceedings of the International Conference on Solid–Solid Phase Transformations'99 (JIMIC-3),* June, Kyoto, Japan.

Belton G.R; Moore, T. J. & Tankins E. S. (1963). Slag-metal reactions in submerged -arc welding. *Welding Journal Research. Suppl. Research.* 68 (3). pp. 289-290. ISSN: 00432296.

Belton G.R; Moore, T. J. & Tankins E. S. (1963). Slag-metal reactions in submerged -arc welding. *Welding Journal Research. Suppl. Research.* Vol. 68, No. 3, pp. 289-290. ISSN: 00432296.

Berry L. G. & Mason B. (1959) *Mineralogy,* Freeman (Ed.), pp. 150-155, ISBN: 13:978-0716702351, New York, USA.

Butler, C. A. & Jackson, C. E. (1967). Submerged-arc welding characteristics of the CaO-TiO2-SiO2, System, *Welding Journal Supplement Research.* Vol, 46 No, 10, pp. 448-456. ISSN: 00432296.

Christensen, N. & Gjermundsen, k.(1962) Measurements of Temperature Outside and in Weld Pool in Submerged Arc Welding. *U.S. Department of Army, Europea Research Office, Report No. 273091.*

Cullity, B. D., (2003) *Elements of X-ray Diffraction,* Addison Wesley(Ed.), 3a. Ed., p.p. 407-417. ISBN:13: 9780201011746, ISBN: 0201011743.

Davis, M.L.E. & Bailey N. (1991). Evidence from Inclusion Chemistry of Element Transfer during Submerged Arc Welding, *Journal Welding Research, Welding Research Supp,* Vol.70, No. 2, p.p. 58. ISSN: 00432296.

Davis, M.L. & Bailey, N. (1991) Evidence from inclusion chemistry of element transfer during submerged arc welding, *Welding Journal Supplement Research* Vol. 70 No. 2, 57- 61. ISSN: 00432296.

Davis, M.L.; & Bailey, N. (1991). Evidence from Inclusions Chemistry of Elements Transfer During Submerged Arc Welding. *Welding Journal, Supplement Research,* February, p.p. Vol.70, No. 2, pp. 61- 65, ISSN: 00432296.

Dowling, J.M. Corbett, H.W. & Kerr, (1986) Inclusion phases and the nucleation of acicular ferrite in submerged-arc welds in high-strength low alloy steels, *Metallurgical Transactions.* A Vol.17, No.9, 1613–1618. ISSN:0360-2133.

Dunham, W.B. & Christian, A. (1984). *Process Mineralogy of Ceramic Materials,* Elsevier Science (Ed), p.p. 20-48, ISBN: 0-444-00963-9, ISBN: 0-444-00963-9, New York, United States.

Evans, G.M. (1996) Microstructure and properties of ferritic steel welds containing Ti and B, *Welding Journal Supplement Research*.Vol. 75, No. 8, pp. 251- 254. . ISSN: 00432296.

Evans, G.M. (1996) Microstructure and Properties of Ferritic Steel Welds Cointaining Ti and B. *Welding Journal, Supplement Research*, August, p.p. 251-259. ISSN: 00432296.

Gordon, & Chu, P.K. (1966) Microstructure of Complex Ceramics. *Proceedings of the Third Berkeley International Materials Conference,* (6), p.p. 828-861. Berkeley Conference, CA, July . ISBN: 471-28720-2 USA New York United States.

Indacochea, J.E., Blander, M. & Shah, S. (1989). Submerged Arc Welding: Evidence for Electrochemical Effects on the Weld Pool, *Journal Welding Research Welding Supp.*, No. 3, p.p. 77-82, ISSN: 00432296.

Jackson, C.E., (1982). Submerged- Arc Welding Fluxes and Relations among Process Variables, *Metals Handbook, ASM, Metals*, p.p. 74-78. ISBN: 9780871706546, Park, Ohio, United States.

Joarder, A. Saha, S.C. & Ghose, A.K. (1991) Study of submerged arc weld metal and heat-affected zone microstructures of a plain carbon steel, *Welding Journal Supplement Research* Vol. 70 No. 6, pp. 141–146. ISSN: 00432296.

Klein, C. & Hurbult, C. S. (1999) *Manual of Mineralogy*, Wiley, (Ed.), 566-567. ISBN-10: 0471821829, ISBN-13: 978-0471821823, New York, United States.

Lancaster, J.F. (1999) *Metallurgy of Welding*, Alden Press Ltd., London, 1980, pp. 25-50, Woodhead publication, 6a. (Ed.) p.p. 110-177. ISBN: 0046690042 / 0-04-669004-2, Hertfordshire, HRT, United Kingdom.

Liu, S. & Olson, D.L. (1986) The role of inclusions in controlling HSLA steel weld microstructures, *Welding Journal Supplement Research* Vol.65, No. 6, pp. 139–141, ISSN: 00432296.

Liu, S. & Olson, D.L. (1986). The role of inclusions in controlling HSLA steel weld microstructures, *Welding Journal Supplement Research* Vol. 65, No.6, pp. 144- 149, ISSN: 00432296.

Material Algorithms Project, (1999) MAP: Perpetual Library of Computer Programs, Available from www.msm.cam.ac.uk/map/mapmain.html.

Redwine R. H. & Conrad M. A., (1966), Microstructures developed in crystallized glass-ceramics in: *Proceedings in the Third Berkeley International Materials Conference*, No. 40, pp. 900- 922, Berkeley Conference, CA, July . ISBN: 471-28720-2 USA New York.

Shah, S. (1986). *The Influence of Flux Composition on the Microstructure and Mechanical Properties of Low-carbon Steel Weld Metal*, M.S. Thesis, University of Illinois, Chicago Ill. United States.

Singer, F.; & Singer, S. (1979) Cerámica Industrial , Enciclopedia de la Química Industrial Tomo 9, Vol. 1, ed. URMO, S.A (Ed.), 236-257, ISBN 84-314 -0177-X, Bilbao, España.

Vander Eijk, Grong, C. O. Hjelen, J. (1999) Quantification of inclusions stimulated ferrite nucleation in wrought steel using the SEM EBSD technique, in: *Proceedings of the International Conference on Solid–Solid Phase Transformations'99 (JIMIC-3), pp.* 351, *Kyoto, Japan*, June.

Zhang, M. He, K. & Edmons, D.V. (1999) Formation of acicular ferrite in C–Mn steels promoted by vanadium alloying elements, in: *Proceedings of the International Conference on Solid–Solid Phase Transformations' 99 (JIMIC-3), Kyoto, Japan*, June.

14

Arc Welding Health Effects, Fume Formation Mechanisms, and Characterization Methods

Matthew Gonser and Theodore Hogan
Northern Illinois University
College of Engineering & Engineering Technology
USA

1. Introduction

Welding fume health effects, fume formation, and characterization are reviewed. The applicability of several collection and characterization methods is discussed. Detailed work on Flux-Cored Arc Welding (FCAW) consumables is presented along with current trends in welding fume particle characterization.

2. Health effects of welding fume exposure

Worker exposure to welding fume is most often associated with acute and chronic lung damage, but there are a number of other health concerns. While exposure to specific gases and metals are associated with certain disease outcomes (such as hexavalent chrome and lung cancer), health effects have also been well documented for total welding fume exposure.

2.1 Routes of exposure

Workers can be exposed by inhaling, ingesting, and coming into skin contact with the fume. All three can be important contributors to disease outcome. Inhalation is the primary, but not only, route of exposure. Welding worker exposures are usually measured in the breathing zone. To evaluate actual exposure, the filter or other sampling media should be placed under the welding helmet (sample adapters are available). As the worker lifts up and puts down the helmet, the device collects a more representative sample. Particle size affects respirability, but virtually all welding fume is in the respirable range. However, it's important to note that the mass collected in typical exposure measurements is not all metal fume. A study found that worker welding fume exposure was 25-55% of the mass collected was metal fume, with the balance of the metal mass collected was due grinding and spattering (Linden & Surakka, 2009).

Workers can also be exposed to welding-related metals through ingestion and skin contact. This needs to be taken into account when evaluating worker exposure as welders eating with dirty hands or eating or drinking contaminated food/liquids can ingest a significant dose. This route is important because lung cancer has been associated with human consumption of drinking water containing high levels of arsenic and chromium. Additionally, a number of metals (including beryllium, chromium and cobalt), can directly affect the skin (irritation and allergic impacts) or be absorbed through the skin and cause lung damage and other health effects. Skin absorption is enhanced by small particle size and by cuts or other damage to the skin. Surface wipe sampling has been used to evaluate fume distribution in facilities (Nygren, 2006). Biomonitoring of blood and urine can be used to evaluate total exposures.

2.2 Acute health effects

Short term exposures to welding fumes can cause shortness of breath, irritation to eye, nose and throat, and other nonspecific effects such as headache and nausea. These effects are similar to those from exposures to just ozone and/or nitrogen dioxide, important components of welding fumes, but metal fume components also contribute to lung effects (Antonini et al, 2004).

Welding-related asthma, which has been related in the past to fluoride exposures, can also be seen in non-fluoride aluminum welding exposures (Vandenplas et al, 1998). Metal Fume Fever, defined in one study as "having at least two symptoms of fever, feelings of flu, general malaise, chills, dry cough, metallic taste, and shortness of breath, occurring at the beginning of the working week, 3-10 hours after exposure to welding fumes", may also be an indicator of welding-related asthma (El-Zein et al, 2003). In addition to fluoride and aluminum, it is thought that chromium, nickel and/or molybdenum are some welding fume components that may contribute to asthma (Hannu et al, 2007).

There is some suggestion that Manual Metal Arc welding may result in acute decreased lung function compared to other processes (Leonard et al, 2010). Generation of Reactive Oxygen Species (ROS) is theorized as one mechanism for welding fume acute adverse health effects, (Taylor et al, 2003) with stainless steels containing chrome and nickel producing more reactive oxygen species (ROS) than mild steel (Leonard et al, 2010).

2.3 Chronic health effects

Primary concerns with chronic exposures include chronic bronchitis (a component lf COPD), and lung cancer (discussed in more detail in chromium section) (Christensen et al, 2008). Welding fume exposures have also been associated with:

- Susceptibility to infections (Palmer et al, 2009).
- Decrease in semen quality and other adverse reproductive effects (welding-related heat and electromagnetic field exposures may also be contributors) (Meeker et al, 2008).
- Sino-nasal cancer (d'Errico et al, 2009).
- Sarcoid-like (immune response) lung disease (Kelleher et al, 2000).
- Peripheral Artery Disease, particularly cadmium (Navas-Acien, 2005).
- Cardiotoxicity (Fang et al, 2010)

A potential mechanism for cardiovascular disease, acute systemic inflammatory response, has been found in welding-exposed workers (Kim et al, 2005). Siderosis, once thought to be an iron fume related "benign pneumoconiosis" without health effects, has been associated with pulmonary fibrosis. (McCormick et al, 2008).

Inorganic dust pneumonias: the metal-related parenchymal disorders

Liberating coatings on surfaces being welded (such as the well-known risk of metal fume fever from zinc coating) is an important source of welder exposure. While not a metal exposure, heating polyurethane-containing coatings to as little as 150°C can release free isocyanates, potentially resulting in severe acute lung damage and debilitating respiratory sensitization. This can occur even without visible smoke generation, as even grinding and power sanding to remove the coatings before welding can be sufficient to release isocyanates.

2.4 Health effects of elements

2.4.1 Manganese

Chronic manganese exposure has long been associated with central nervous system (CNS) effects which are similar in nature to Parkinsonism. There is considerable controversy to whether manganese exposure can cause Parkinsonism itself, or whether Manganism is a separate disease entity. Either way, manganese can cause a degeneration of CNS function that gets progressively worse after symptoms first appear. (American Conference of Governmental Industrial Hygienists, 2011b). Note that Vanadium exposure may also be related to Parkinsonism. (Afeseh Ngwa et al, 2009).

2.4.2 Beryllium

Beryllium is recognized as a known human lung carcinogen (U.S. Department of Health and Human Services, 2011). Workers can become sensitized to beryllium, exhibiting an asymptomatic immune response (Committee on Beryllium Alloy Exposures, 2007). Sensitization can lead to chronic beryllium disease, a granulomatous disease primarily affecting the lungs. Beryllium contact with skin may play a role in sensitization, so this should be considered in addition to fume levels when evaluating and controlling exposure. (Committee on Beryllium Alloy Exposures, 2007)

2.4.3 Cadmium

Cadmium is a known human lung carcinogen (U.S. Department of Health and Human Services, 2011). High short-term exposures can result in chemical pneumonitis (lung symptoms similar to pneumonia). Sometimes an acute overexposure to cadmium may cause death. Chronic exposure can cause kidney damage and osteoporosis. (Bernard, 2008) (Nawrot et al, 2010). Cadmium exposure may also increase the risk of prostate and bladder cancer. (Yi-Chun Chen, 2009) (Kellen et al, 2007)

2.4.4 Chromium

DNA damage in welders has been associated with hexavalent chromium exposure. (Sellappa et al, 2010). This is consistent with the classification of hexavalent chromium as a human lung carcinogen. (U.S. Department of Health and Human Services, 2011) This risk may be independent of smoking. (Halasova, 2005). The chromium valence state appears to be critical, and with water soluble compounds being more carcinogenic. Unfortunately, chromium welding exposures can be difficult to control. An exposure survey by National Institute for Occupational Safety and Health (NIOSH) found "Most operations judged to be moderately difficult to control...involve joining and cutting metals with relatively high chromium content." (Blade, 2007)

2.4.5 Nickel

There is considerable ongoing research on the effects on toxicity of nickel speciation, solubility and particle size. Ni (II), which is water soluble, is recognized by IARC as a "known" human carcinogen causing lung, nasal, and sinus cancers. (IARC, 2011). Nickel metal and other nickel alloy materials are classified as a "possible" human carcinogen by

IARC. However, The US Report on Carcinogens states that metallic Nickel is "reasonably anticipated to be a human carcinogen, and classifies Nickel compounds as known human carcinogens that can cause both lung and nasal cancer (Report on Carcinogens, 2011).

2.4.6 Aluminum

Chronic aluminum exposure is associated with Alzheimer's disease, although concerns have been raised about the strength of the epidemiology studies. (Ferriara, 2008). (Santibáñez, 2007). A recent review identified mechanisms of how aluminum may contribute to the formation of Amyloid proteins in the brain, a marker of Alzheimer's Disease. (Kawahara, 2011).

2.4.7 Other welding-related metal carcinogens

Both lead and lead compounds are also "reasonably anticipated to be human carcinogens" (12th Report on Carcinogens). Arsenic is a recognized human carcinogen, implicated in both lung and bladder cancer (IARC, 12th Report on Carcinogens, Marshall et al, 2007). Antimony is "possibly carcinogenic to humans", but the studies suggested that lung cancer may be associated with the arsenic co-exposures (McCallum, 2005).

2.5 Threshold limit values

The Threshold Limit Values (TLVs(r)), published by the American Conference of Governmental Industrial Hygienists (ACGIH), provide guidance for controlling exposures (Table 1). The TLV for each material has formal documentation that should be reviewed before using a TLV to understand the bases and limitations of the recommended control levels. (American Conference of Governmental Industrial Hygienists, 2011a) A common misunderstanding (perhaps because they are called "Thresholds") is that keeping worker exposures below TLV(r) concentrations will protect all workers against "all" health effects. TLV(s) are based on protecting against "known" health effects, and some workers exposed to sub-TLV(r) concentrations can still experience those known adverse effects (and other unknown effects). TLVs are periodically revised as new information emerges. Using the TLV(r) as a control value without these considerations can put workers at risk.

2.6 Welding fume

Toxic substances produced during welding include heavy metals, ozone, carbon monoxide, carbon dioxide, and nitrogen oxides. Ozone, or O_3, is a strong oxidant that generates reactive oxygen species in tissue, which can cause DNA damage. Ozone is produced within 30 seconds during welding, depending on the arc power. The highest O_3 levels, 195 parts per billion (ppb) occurred during the welding operation (Liu et al, 2007). Welding fume is generated primarily by fusion welding processes, such as arc welding and to a lesser extent laser beam welding. Fume formation as a result of other processes, such as friction welding, a solid state process, are generally very minimal. Welding fume is produced when the filler material or electrode is melted and vaporized in the presence of a high temperature heat source. The combination of the molten metal with the compounds in the flux or electrode coating can cause chemical reactions that can change the composition of the fume particles. Fume particle morphology and composition are therefore a product of the electrode composition and shielding atmosphere. The materials typically found in welding fume include aluminum, beryllium, cadmium oxides, chromium, copper, fluorides, iron oxide, lead, manganese, molybdenum, nickel, vanadium, and zinc oxides. Welding also produces gases, which can contain carbon monoxide, fluorine, hydrogen fluoride, nitrogen oxide, and ozone. All welding processes produce fume, but most fumes are produced via arc welding process such as Shielded Metal Arc Welding (SMAW), Flux-Cored Arc Welding (FCAW), and Gas Metal Arc Welding (GMAW). High heat from an electric arc that is maintained between the electrode and workpiece is used to melt and fuse the metal at the joint. The heat from the arc vaporizes a portion of the electrode in the air, thereby creating a weld plume.

Substance	Threshold Limit Value-8 Hour Time Weighted Average
Aluminum metal and insoluble compounds	1 mg/m^3
Antimony and compounds, as Sb	0.5 mg/m^3
Arsenic and inorganic arsenic compounds, as As	0.01 mg/m^3 (A1)
Beryllium and compounds as Be	0.00005 mg/m^3 (A1)
Cadmium	0.01 mg/m^3 (A2)
compounds, as Cd	0.002 mg/m^3 (A2)
Chromium and inorganic compounds, as Cr	
Metal and Cr III compounds	0.5 mg/m^3 (A4)
Water-soluble Cr VI compounds	0.05 mg/m^3 (A1)
Insoluble Cr VI compounds	0.01 mg/m^3 (A1)
Cobalt and inorganic compounds, as Co	0.02 mg/m^3 (A3)
Iron Oxide	5 mg/m^3 (A4)
Lead and inorganic lead compounds, as Pb	0.05 mg/m^3 (A3)
Manganese and inorganic compounds, as Mn	0.2 mg/m^3 *
Nickel, as Ni	
Elemental	1.5 mg/m^3 (A5)
Soluble inorganic compounds	0.1 mg/m^3 (A4)
Insoluble inorganic compounds	0.2 mg/m^3 (A1)
Zinc oxide	2 mg/m^3

A1: Confirmed Human Carcinogen
A2: Suspected Human Carcinogen
A3: Confirmed Animal Carcinogen with Unknown Relevance to Humans
A4: Not Classifiable as a Human Carcinogen
A5: Not Suspected as a Human Carcinogen
* ACGIH issued a notice of intended change - TLV may be lowered

Table 1. 2011 Threshold limit values (TLV) for selected metals.

2.6.1 Fume formation mechanisms

There are several mechanisms by which welding fume can originate. More than 90% of the particulates in welding fume is formed by vaporization of the electrode filler metal in addition to the core and coatings (Brown, 1997). Particles form when the metal vapor condenses. About 1% of the electrode condenses into metal oxide nanoparticles which aggregate together to form particle agglomerates (Haider, 1999). Welding fume can be present in different morphologies. Individual spherical particles less than 20 nm are formed by vapor condensation, while aggregates of 20 nm particles are formed by the collision of primary particles. Welding fume particle size can be divided into three groups: ultrafine ($0.01<d<0.1\mu m$), fine ($0.1<d<2.5\mu m$), and coarse ($d >2.5\mu m$) (Jenkins & Eager, 2003).

Inert atmospheres, such as those used in GTAW, facilitate fume formation through vaporization and condensation of elemental material with some light oxidation. Fume formed from FCAW and SMAW results from vaporization, condensation, and subsequent oxidation of elemental and lower oxide species, and the vaporization/condensation of oxide and fluoride flux species and compounds. The chemical composition of fume is, therefore, highly dependent on flux composition since a significant amount of low melting point components are contained within the flux and wind up in the fume particles themselves (Heille & Hill, 1975). Micro and nano-particles of fume are comprised of metal oxides and unoxidized metals and compounds, such as fluorides and chlorides (Konarski et al, 2003). There are five main forms that the metal from the electrode can assume. These include metal droplet, metal vapor, primary metal oxides, volatile element vapor, and volatile metal oxides.

Figure 1 shows the various methods of fume formation. Metal alloy droplet expulsion occurs when heating of the liquid metal occurs rapidly. What results is small droplets being forced off of the parent metal droplet. When this metal droplet cools and solidifies outside the presence of oxygen, as in an inert argon or helium atmosphere, mostly metallic particles with a slight oxide layer will occur. In cases where inert gas shielding is not used, such as with SMAW or self-shielded FCAW, the liquid droplet is in direct contact with oxygen and will readily oxidize. That the welding process require flux or an electrode coating is not a prerequisite for oxidation. For inert processes, such as GMAW or GTAW, oxygen may be incorporated from the surrounding atmosphere or may be an inherent component in the gas mixture, as in $Ar\text{-}O_2$ mixtures or even $Ar\text{-}CO_2$ mixtures. If heating of the liquid metal is sufficient, vaporization of the metal will occur. If this metal vapor condenses in the presence of oxygen then primary metal oxides can form, the composition of which depends on the metal composition. This type of fume formation is referred to as unfractionated (Gray et al, 1982). In some cases volatile elements within the liquid metal can cause selective evaporation of these elements. Selective evaporation occurs in elements that have the highest vapor pressure and the lowest vaporization temperatures. Elements with higher vapor pressure are more prevalent in fume, and thus produce higher amounts of fume (Kobayashi et al, 1983). High vapor pressures for manganese and iron explain why oxides that contain these two elements are formed by the vaporization method of fume formation. The main steps or phases in fume formation are listed below:

- Metal droplet expulsion
- Vaporization
- Condensation
- Oxidation (fractionated and unfractionated)
- Agglomeration

These aerosols are formed primarily through the nucleation of vapors followed by competing growth mechanisms such as condensation and coagulation. High temperature metal vapors transform into primary particles via homogeneous nucleation. Condensation increases particle growth due to added metal vapors. Coagulation then results in agglomeration through particle collisions. (Zimmer, 2002).

2.7 Fume characterization

Fume characterization is generally utilized to determine the composition, structure, size, and distribution of welding fume so that permissible exposure levels can be determined. It is also used to determine how the welding fume behaves in the industrial environment. Since the chemical composition of the welding consumable controls what kinds of materials and compounds are released in the fume particles, it is sometimes possible to predict whether hexavalent chromium will form in any appreciable amount, or whether certain oxides will form that may form that could be deleterious to the welder's health. Figure 2 shows how Mn and Cr content in fume varies with consumable composition. The characterization of welding fume depends on what the investigator hopes to determine. Some forms of characterization are rudimentary and require a fume hood and glass fiber filters, as with fume generation rate collections, while others may require precision collection techniques

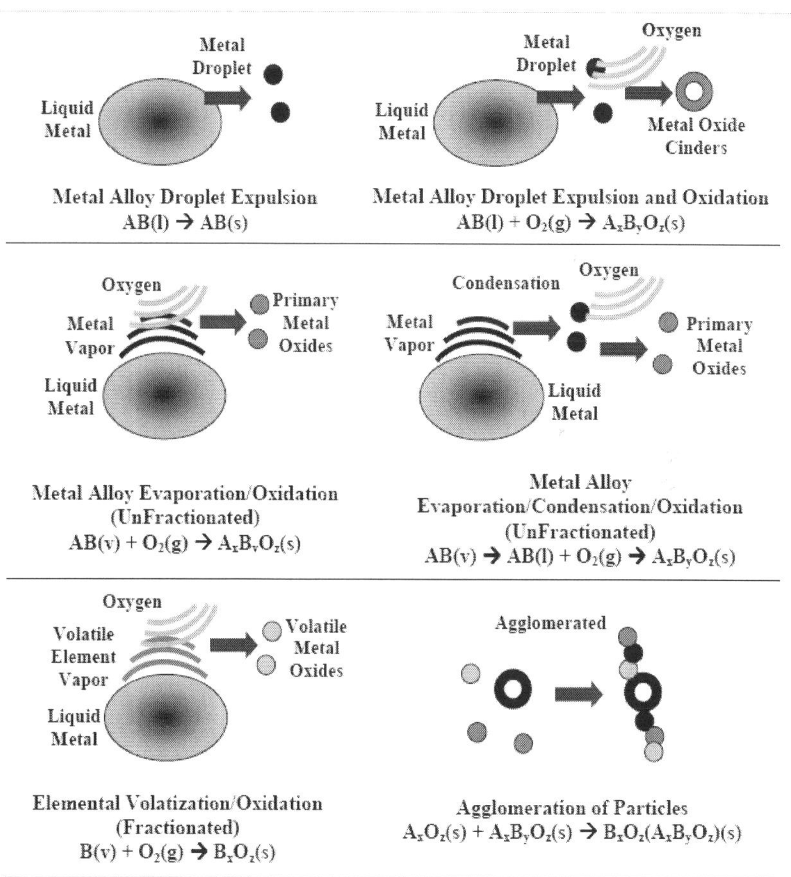

Fig. 1. Fume formation mechanisms. (Gray et al, 1982).

in concert with the latest in advanced characterization technology, such as electrical low pressure impactors and secondary ion mass spectrometers, respectively. Fume characterization can be broken down into two main categories: physical and chemical characterization. Due to the complexity and wide range of fume particles present in a typical weld plume, several techniques may be required to fully characterize the fume, as indicated in Table 2. This may, in turn, require several different fume collection techniques in order to identify different particles based on their size and morphology.

2.7.1 Physical characterization

The physical aspect of fume characterization includes such metrics as mass and number distribution and fume generation rate, which is also referred to as fume formation rate. Mass and number distributions can be quantified using cascading impactors or electrical low pressure impactors. The goal of physical characterization is to determine mass, number, weight, and fume generation rate. Also, the morphology of the fume, which is quantified by analyzing the size and shape of the fume particles, is included in this category.

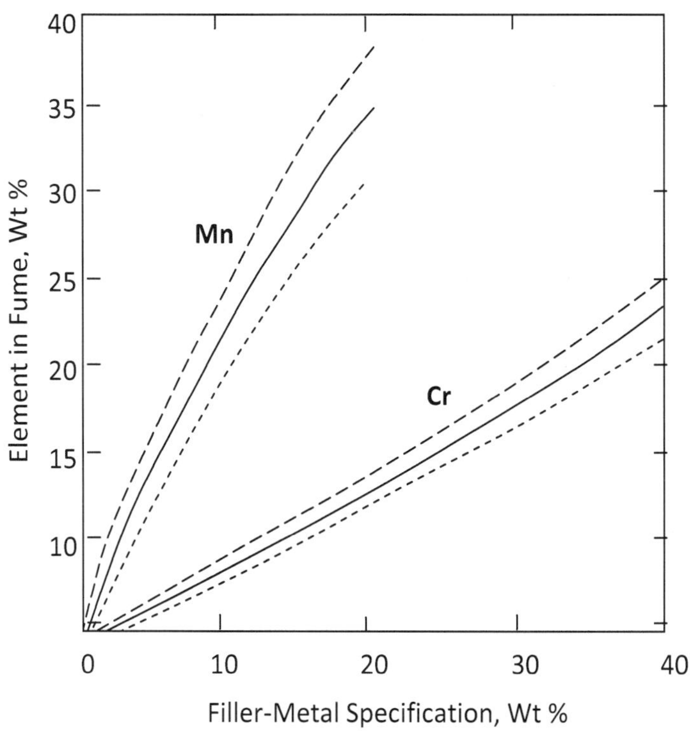

Fig. 2. Mn and Cr content in SMAW fume as a function of consumable composition (ACGIH, 2002).

The fume generation rate (FGR), sometimes referred to as fume formation rate (FFR), of a consumable is measured by collecting fume inside a fume hood, such as those designed by AWS standards. The total weight of the fume is determined by subtracting the weight of the filter from the final weight (with fume attached to the filter). Fume generation rate is generally determined using ANSI/AWS F1.2 method. The basic design of the fume hood used in this method is shown in Figure 3. Transfer mode can have an effect on fume formation rate. It has been demonstrated, using FCAW, that the surface tension transfer (STT) mode reduced the fume generation rate by 40-50% compared to continuous wave current mode. This reduction in fume formation was directly related to the amount of heat input in the electrode. The fume generation rate rises gradually as voltage and wire feed speed increased as the transfer mode changes from short-circuit to globular (Srinivasan & Balasubramanian, 2011).

Different shielding gas mixtures can be used with the GMAW process, which in turn, has an effect on fume generation rate. In Ar-based mixtures, increasing CO_2 had a greater impact than raising O_2. When O_2 was increased in ternary mixtures, the fume generation rate increased for Ar-5%CO_2, but not with Ar-12%CO_2. CO_2 additions in Ar-based shielding gases. A 100%CO_2 mixture produced the highest fume generation rate due to the globular transfer mode and increased spatter. Shielding gas mixture does not seem to have an effect on fume particle composition as the particles in all cases were characterized as $(Fe,Mn)_3O_4$ (Carpenter et al, 2009).

Fume generation rates rise as power is increased when the short-circuit transfer mode is implemented. Fume generation rates peak during the globular transfer mode and then drop as the transfer mode shifts to the spray type. Streaming transfer causes FGR to increase further (Quimby & Ulrich, 1999). Fume generation rates tend to increase with heat input for a given electrode. Fume generation rates also increase as the cross-sectional area of the electrode increases. Fume generation rates (FGR) and concentrations of total chromium and hexavalent chromium have been quantified for stainless steel FCAW electrodes using methods recommended by the American Welding Society, ICP-MS (NIOSH Method 7300) and ion chromatography (modified NIOSH Method 7604), respectively. The FGRs ranged from 189-344, 389-698, and 682-1157 mg/min at low, optimal, and high heat input, respectively. The ranges of total chromium FGR were 3.83-8.27, 12.75-37.25, and 38.79-76.46 mg/min at low, optimal, and high heat input, respectively. The hexavalent chromium range was 0.46-2.89, 0.76-6.28, and 1.70-11.21 mg/min at low, optimal, and high heat input, respectively. FGR, total chromium, and hexavalent chromium all increase with weld heat input (Yoon et al, 2003).

Gravimetric impactors and low pressure impactors can be used to collect fume used to quantify the mass distribution of the fume. The mass distribution of fume generated by a welding consumable is typically determined by collecting fume on an aluminum or polymeric substrate, whereby the weight of the fume is determined by weighing the substrate before and after the collection process. The mass of the fume is determined by subtracting the initial weight of the substrate or filter. Similarly to mass distribution an impactor is used to collect the fume used for the quantification of the number distribution. The impactors used can be gravimetric or low pressure, but they will generally have some sort of particle counting system incorporated into the design. In an electrical low pressure impactor, or ELPI, the particles drawn into the unit by suction are positively charged by the instrument. As these particles contact the electrometers, the system counts each individual particle so that a number distribution, from the ultrafines to the coarse particles, can be counted.

Characterization Method	Size Range (micron)	Detection Limit	Comments
Size Distributions			
Impactors	0.1 - 20		
Aerodyamic Particle Sizer	0.1 - 25		
SEM	0.5 - 50		
TEM	0.001 - 1		
Elemental Composition			
X-ray fluorescence (XRF)	bulk	100 ppm	*z > 10
Atomic absorption spectroscopy	bulk	10 ppm	z > 10
SEM-XEDS	1 - 50	0.10%	z > 10
Wavelength dispersive spectroscopy (WDS)	1 - 50	0.10%	z > 4
TEM-XEDS	0.01 - 0.5	0.10%	z > 5
Secondary ion mass spectroscopy (SIMS)	> 5	10 ppm	light elements
Auger electron spectroscopy (AES)	> 0.1	0.10%	z > 3
X-ray photoelectron spectroscopy (XPS)	> 5	0.10%	
Chemical Speciation			
X-ray diffraction (XRD)	bulk		
X-ray photoelectron spectroscopy (XPS)	bulk		
TEM Selected area diffraction (SAD)	0.3		

Table 2. Characterization methods used for analyzing welding fume particles (Jenkins & Eager, 2005b).

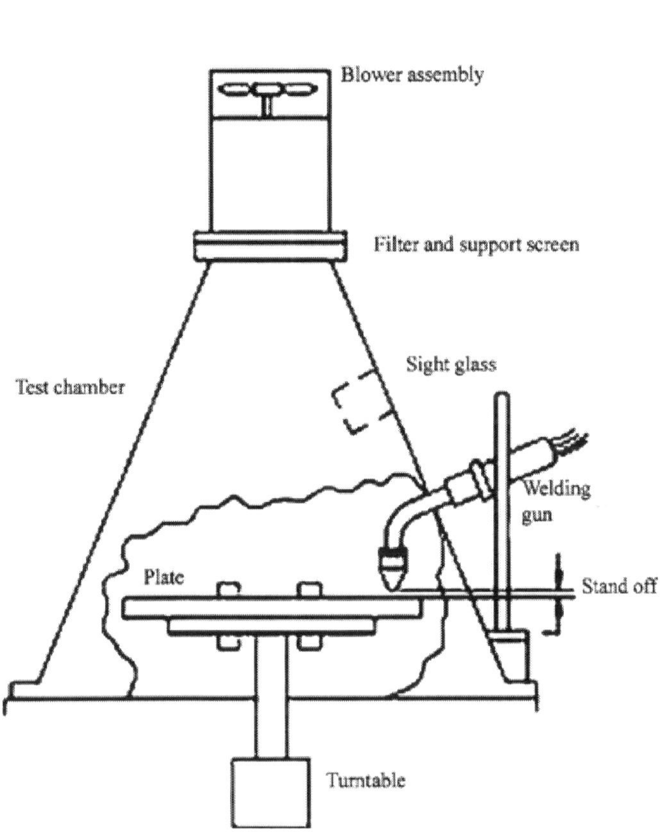

Fig. 3. Fume Hood Assembly (Srinivasan & Balasubramanian, 2011).

Cascade impactors, as shown in Figure 4, can be used to separate fume particles so as to determine their size distribution. The particle size distribution of GMAW fume typically consists of particle agglomerates smaller than one micrometer, meaning most of the fume is respirable. Less than 10% of the fume by weight can be microspatter. FCAW fume containes about 30% by weight of microspatter. This suggests that a smaller fraction of the FCAW fume is respirable compared to GMAW fume (Jenkins et al, 2005). There is a link between spatter and fume in arc welding. Spatter does not significantly contribute to the total amount of fume formed during arc welding. It is postulated that the proportional increase of fume and spatter formation is caused by the welding process variables (Jenkins & Eager, 2005a). Fume particle size distributions ($dN/dlogd_p$) can aslo be measured using a scanning mobility particle sizer (Zimmer, 2002).

Particle number distributions showed that E6010 and E7018 consumables produced fume where 95% of the particles were smaller in diameter than 0.3 um. For E308-16, 95% of the fume consisted of particles that were less than 0.6 µm in diameter. Most of the mass of the fume particles was larger than the respirable size range (>0.1 um). The mass of particles in the ultrafine range represented less than 2% of the total mass of the fume (Sowards et al,

2008). The number and mass distributions are shown for a series of FCAW consumables in Figure 5. Morphological analysis of particles via Scanning Electron Microscopy (SEM) can be used to reveal particle structures: spherical, irregular, and agglomerate (Figure 6). Another type of particle, rectangular, is generally present when magnesium is present. It is inferred that, based on X-Ray diffraction (XRD) results, the particles are MgO.
Exposure to welding fume can be quantified based on surface area and number concentration. Different measurement equipment can be used to characterize the total particle number concentration. These may consist of condensation particle counters or CPC, scanning mobility particle size or SMPS, and electrical low pressure impactor or ELPI. The CPC device can identify particle emission sources. The range of ultrafine particle number concentration can be detected by SMPS and ELPI. The ELPI is useful in that ultrafine sizes can be collected for further chemical analysis techniques. The specific surface area of the aerosols can be calculated using gas adsorption analysis, however ultrafine particles cannot be distinguished from particles with non-ultrafine sizes (Brouwer et al, 2004).

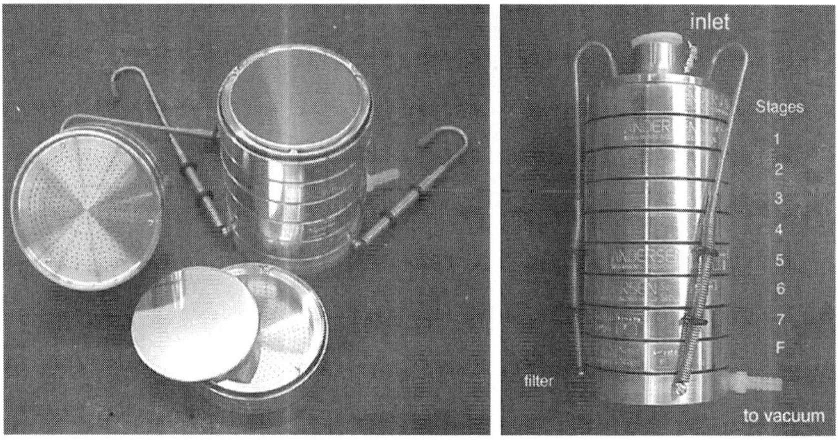

Fig. 4. Andersen Impactor as used by Jenkins. (Jenkins et al, 2005).

2.7.2 Chemical characterization

The other category of fume characterization is chemical and structural analysis. These include analysis of the chemical composition and crystal structure of fume particles. It is also concerned with the variation in composition with depth from the outer surfaces of the particles, as in X-Ray Photoelectron Spectroscopy or with Secondary Ion Mass Spectrometry. The crystalline phases present in fume can be identified with X-Ray diffraction (XRD) for fume collected in bulk, as in a fume hood. To identify crystalline compounds in individual particles, Transmission Electron Microscopy is generally performed. Particles are collected on a carbon-coated copper or gold grid and analyzed individually by selected area diffraction (SAD). Chemical composition can also be estimated in the TEM using X-Ray energy dispersive spectroscopy (XEDS).
X-Ray Flourescence and Wavelength Dispersive Spectroscopy can be used to determine stoichiometry and light elements, respectively. To determine elemental composition and stoichiometry, X-Ray Fluorescence can be implemented. This technique uses high-energy X-

rays to bombard a material. Upon bombardment, the material emits fluorescent X-rays. The characteristic X-ray radiation is detected by a Si(Li) detector. The amount of soluble hexavalent chromium, or Cr(VI), can be measured using ISO3613. Total chromium is determined using Inductively coupled plasma mass spectrometry or ICP-MS. This technique can also be used to determine the concentration of metals and non-metals in fume.

Various chemical analysis techniques can be implemented to characterize welding fume particles. The typical method of analyzing the elemental composition of welding fume is by the molar fraction of the metal cations, which can be done with SEM-XEDS and TEM-XEDS techniques. XRD is well suited for determining the phase composition of the welding fume. Welding fume collected from the GMAW process was predominately magnetite where approximately 10% of the cations were Mn. Welding fume collected from the SMAW process contains both a complex alkali-alkali earth fluoride phase and an oxide spinel phase. The metal cation fraction of the SMAW fume was 27% Fe, 10% Mn, 10% S, 28% K, and 25% Ca (Jenkins & Eager, 2005b).

Chemical analysis can be conducted with multiple techniques, which include energy-dispersive X-ray fluorescence spectroscopy (EDXRF), micro-Raman spectroscopy (MRS), and electron probe microanalysis (EPMA). In the fine fractions (0.25-0.5 μm), particles of irregular shape are rich in Fe, Si, Na, and C. In the 0.5-2.0um fractions, the particles are spherical with various types of Fe-rich particles that contain either Si, Mn, or both. MRS showed the presence of Fe-containing compounds such as hematite and goethite in all size fractions (Worobiec et al, 2007).

Changes in composition can have a drastic effect on fume composition. Fume from wires containing 1% Zn contained far less Cr(VI) than those produced by the wires that had no Zn. In addition, the use of 18V as opposed to 21V reduced the amount of Cr(VI) in the fume, most likely due to the lower FGR associated with a lower voltage (Dennis et al, 1996).

Previous work showed that reduction of sodium and potassium in SMAW led to reductions in Cr(VI) concentrations in fume as well as a reduction in fume generation rate. Lithium was used to replace potassium in a self-shielded FCAW electrode. This resulted in the reduction of Cr(VI) and fume generation rate. Another wire was made with 1% Zn resulted in reduction in Cr(VI). This wire resulted in greater than 98% reduction in Cr(VI) compared to the control wire when the shielding gas contained no oxygen. In the presence of oxygen, the 1% Zn increased the Cr(VI) generation rate higher than the control electrode. This study really underlined the importance of proper shielding gas usage (Dennis et al, 2002).

A colorimetric method for extracting and measuring manganese in welding fume was developed. It utilized ultrasonic extraction with an acidic hydrogen peroxide solution to extract welding fume collected on polyvinyl chloride filters. Absorbance measurements were made using a portable spectrophotometer. The method detection limit was 5.2 ug filter-1, while the limit of quantification was 17 ug filter-1. When the results are above the limit of quantification, the manganese masses are equivalent to those measured by the International Organization for Standardization Method 15202-2, which uses a strong acid digestion along with analysis by inductively coupled plasma optical emission spectrometry (Marcy & Drake).

One study utilized several characterization techniques to quantify number distribution, mass distribution, composition, structure, and morphology. FGR was dependent on heat input. XRD showed that flux composition had a profound effect on the phase structure of the fume. XRD results for common consumables include the following: E6010 (Fe_3O_4 or magnetite. Peak shifts suggest that Mn and Si substituted for Fe in the Fe_3O_4 structure. E7018: contained additional peaks for NaF and CaF_2. XRD of E308-16 showed Fe_3O_4 and K_2MO_4, where M accounts for Fe, Mn, Ni, and Cr, and NaF. (Sowards et al, 2008)

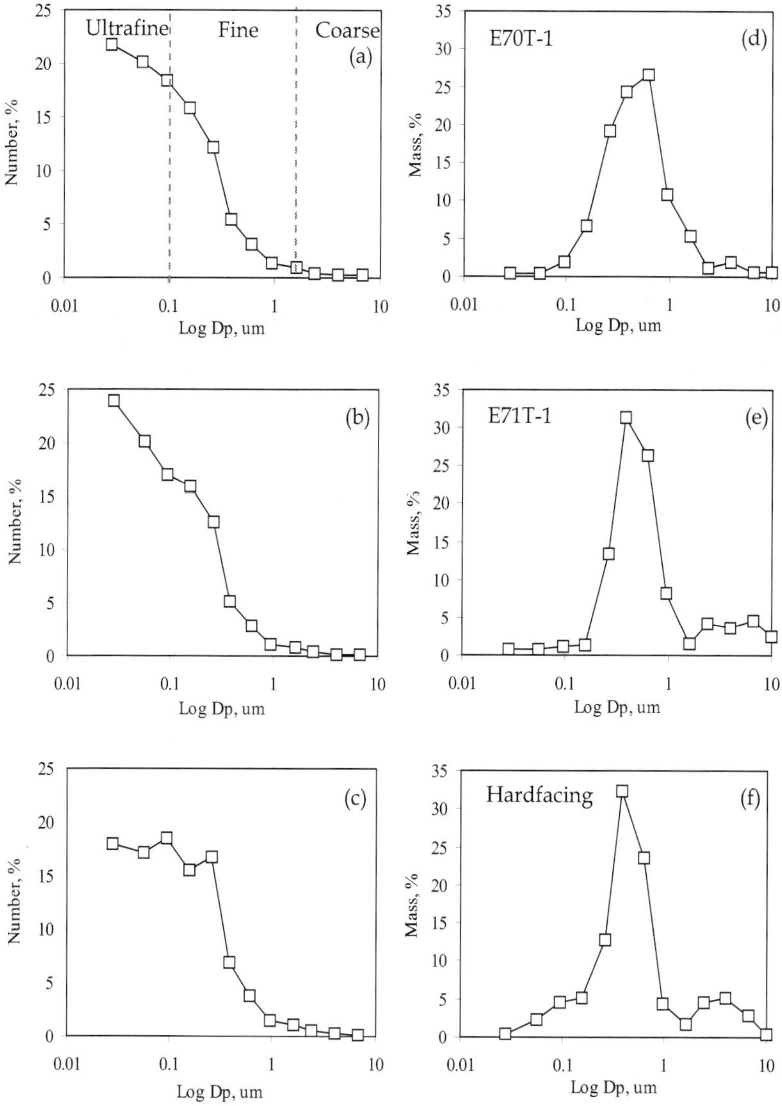

Fig. 5. Number (left) and Mass (right) distributions for FCAW fume.

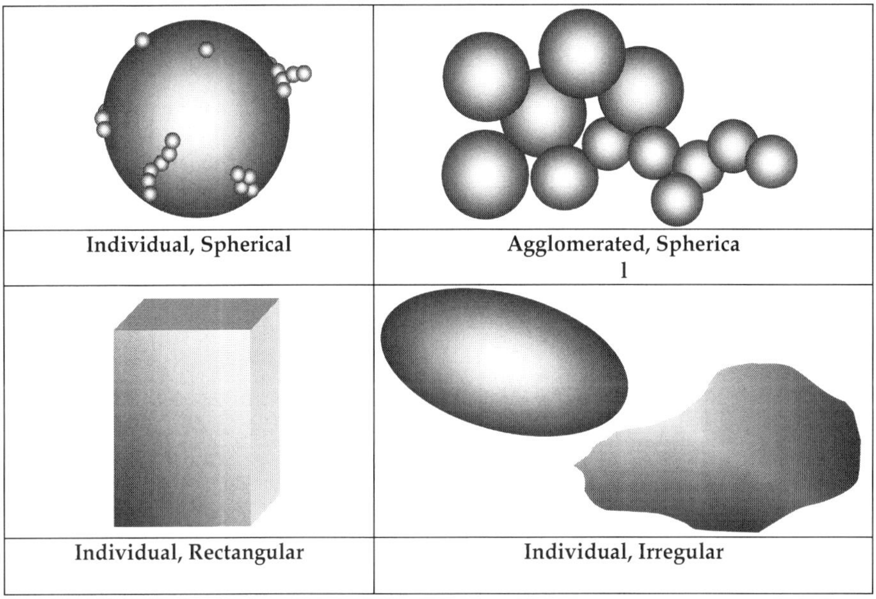

Fig. 6. General forms of fume morphology found in welding plumes.

Fume particles generated by SMAW can range in size from several nm to several µm. Therefore, multiple imaging and chemical analysis techniques were used to characterize them. SEM, TEM, and HR-TEM were used to characterize individual and bulk fume on each ELPI stage. SEM was used to characterize fume particles larger than 0.3 µm. The largest percentage of particles on all stages of the ELPI were agglomerates of spherical particles. Fume particles agglomerate together due to either charging or sintering. TEM and HR-TEM were used to characterize nano-sized particles. Most of the ultrafine particles had a crystalline structure with some having a core-shell morphology. Chemical analysis was done mainly with XEDS (both SEM and TEM) and XPS techniques. SEM-XEDS was used to analyze particles larger than 0.3 µm, while TEM-XEDS was used to analyze particles lower than 0.3 µm. TEM SAD showed that the ultrafine particles were predominantly meta oxides of the $(M,Fe)_3O_4$ type, where M stands for metals such as Mn and Cr that may substitute in the magnetite lattice. XPS confirmed the core-shell morphology by partial depth profiling and revealed the valence states for Fe and Mn as +2 and +3 as they are found mainly in the $(M,Fe)_3O_4$ compound. (Sowards et al, 2010) An XRD spectrum for E70T1-1 is shown in Figure 7. The XRD spectrum is used to show all of the crystalline phases or compounds present I the fume. TEM results show phase identification of individual particles and agglomerates in Table 3 for a series of FCAW consumables.

Secondary Ion Mass Spectrometry can be used to determine the elemental, isotopic, or molecular composition at the surface of the fume particles. Welding fume particles can be analyzed using low energy ion erosion. This can expose differences in fume particle structure and morphology. Fume particles can exhibit a core-shell morphology, which can be revealed using the low energy ion erosion technique (Figure 8). Differences in the process can have an effect on the chemical structure of the particles. For instance, particles

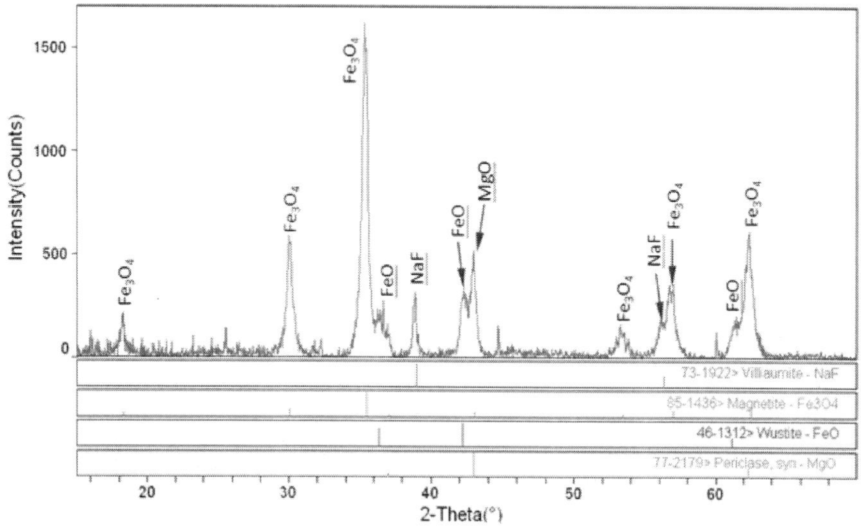

Fig. 7. XRD spectrum for E71T-1 fume.

formed during EBW contain a shell that is enriched in oxygen, fluorine, chlorine, and potassium, where the core is composed mainly of iron, chromium, and manganese. The GTAW particles are more oxidized than the EBW particles. The shells of the GTAW particles are composed mainly of fluorine and chlorine. The SMAW particles are more heavily oxidized than the other two processes. The shells of the particles produced with SMAW are rich in chlorides, fluorides, and potassium while the core is composed of iron, chromium, and manganese oxides. The differences in the shell composition are attributed to differences in atmosphere. Where the EBW process uses low pressure gases, the GTAW process uses an argon shield, and SMAW uses TiO_2 flux vapors (Konarski et al, 2003).

Sequential leaching is a method used to extract Ni and Mn species from welding fumes. Welding fumes were sampled with fixed-point sampling and the use of Higgins-Dewell cyclones. Ammonium citrate was used to dissolve soluble Ni, while a 50:1 methanol-bromine solution was used to dissolve metallic Ni. A 0.01 M ammonium acetate was used for soluble Mn, while a 25% acetic acid was used to dissolve Mn^0 and Mn^{2+}. A 0.5% hydroxylammonium chloride in 25% acetic acid was used for Mn^{3+} and Mn^{4+}. Insoluble Ni and Mn were determined after microwave-assisted digestion with concentrated HNO_3, HCl, and HF. Samples were analyzed by ICP quadruple mass spectrometry and ICP atomic emission spectrometry. Different welding processes and consumables were tested using the aforementioned techniques. The highest quantity of Mn in GMAW fumes was insoluble Mn where corrosion resistant steel was 46% and unalloyed steel was 35%. MMAW fumes contained mainly soluble Mn, Mn^0, and Mn^{2+}, while corrosion resistant steel contained Mn^{3+} and unalloyed steel contained Mn4+. GMAW fumes were rich in oxidic Ni, while the Ni compounds generated by MMAW are of soluble form. XRD was conducted and revealed that GMAW fumes were magnetite ($FeFe_2O_4$). MMAW fume contained $KCaF_3$-CaF_2 with some magnetite and jakobsite ($MnFe_2O_4$). (Berlinger et al, 2009)

MMAW of chromium nickel steel or high-alloyed steel generate chromium (VI) compounds such as chromium trioxide and chromates. Nickel oxides are generated when welding nickel and nickel base alloys during MIG welding. These include NiO, NiO_2, and Ni_2O_3. (Fachausschuss, 2008) Most aluminum welding operations involve the use of unalloyed or partially alloyed aluminum consumables. Ozone is produced by the effect of UV light emanated by the arc on the air. The use of these consumables produces fume that is mostly comprised of aluminum oxide. Aluminum-silicon and aluminum-magnesium alloys are the most commonly used aluminum alloy consumables.

Consumable	**Type**	**Size Range, nm**	**Compounds**
E70T-1	Agglomerates, spherical	30-230	$(Fe,Mn)_3O_4$ and SiO_2
E71T-1	Spherical, some square MgO	30-120	$(Mn,Fe)_3O_4$; MgO; small MgO particles
Hardfacing	Agglomerates, spherical	a) 30-70 b) 70-150	a) Al_2O_3, MgO and CaF_2 b) MgO and $(Mn,Fe)_3O_4$

Table 3. TEM results for fume particle type, size range, and compound as identified by SAD.

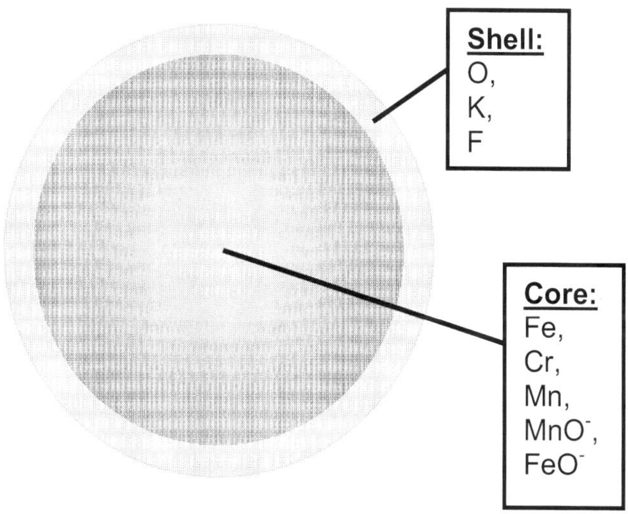

Fig. 8. Core-shell morphology of SMAW fume particles. (Konarski et al, 2003).

Proper ventilation systems: hoods, roof vents, high-speed intake and exhaust fans are used to capture fumes and gases in the direct vicinity of the welder. To capture welding fumes in large areas, using downdraft worktables is used. Outdoor fans can be used to direct plumes away from field workers. Working upwind from the welding operation can also decrease exposure. Respirators can be used to further protect the welder from exposure. These devices should be certified by NIOSH and practice or use of said device should adhere to OSHA's Respiratory Protection Standard. Proper training should be implemented to teach welders and non-welders alike of the dangers of welding fume exposure. The correct use of respirators, avoidance of weld plumes, implementation of engineering controls that minimize exposure, and other safety practices that mitigate exposure levels should be implemented in any workplace. In addition, posting warnings in welding areas is a good way to inform employees and visitors of the potential dangers. Familiarity with ANSI Z49.1, Safety in Welding, Cutting, and Allied Processes should be readily available at the facility. Industrial hygiene monitoring plans, such as area monitors and personnel should be in place or available in all welding work areas and industrial hygienists should be present to monitor welders for exposures and potential exposures. To reduce welding fumes it is advisable to sue low-fume welding rodes and alternative welding methods, such as SMAW, which generates lower amounts of fume than FCAW.

3. Conclusion

Based on the review of the literature, the health effects associated with welding warrant the thorough analysis of aerosol particles that are produced as a byproduct of the process. Many techniques, both physical and chemical in scope, are available that adequately characterize the fume particles. Physical characterization methods, such as mass or number distributions, as well as particle sizers, can be used to determine how much of the welding fume is in the respirable range. Chemical analysis can be used to determine how electrode composition affects fume composition. One of the more complex welding fumes, that which is generated by FCAW consumables, shows that over fifty percent of the fume, by number, is in the respirable range. By mass, nearly ninety percent of the fume is respirable. Fume particle produced via FCAW are typically magnetite, with some manganese substituting in the magnetite lattice. Other compounds form depending on the elements present in the flux.

4. References

ACGIH, American Conference of Governmental Industrial Hygienists (2011a) Threshold limit values for chemical substances and physical agents and biological exposure indices. Cincinnati, OH; American Industrial Hygiene Association

ACGIH, American Conference of Governmental Industrial Hygienists (2011b) Documentation of the Threshold Limit Value: Manganese and Inorganic Compounds, Cincinnati, OH.

Afeseh Ngwa, H.; Kanthasamy, A., Anantharam, V., Song, C., Witte, T., Houk, R. & Kanthasamy, A.G. (2009) Vanadium induces dopaminergic neurotoxicity via protein kinase Cdelta dependent oxidative signaling mechanisms: relevance to etiopathogenesis of Parkinson's disease. Toxicol Appl Pharmacol. Oct 15;240(2):273-85. Epub 2009 Jul 29.

Antonini, J.M.; Taylor, M.D., Zimmer A.T. & Roberts, J.R. (2004). Pulmonary Responses to Welding Fumes: Role of Metal Constituents. *Journal of Toxicological Environmental Health.* Vol.67, No.3. pp. 233-249.

Berlinger, B.; Náray, M., Sajó, I. & Záray, G. (2009) Critical Evaluation of Sequential Leaching Procedures for the Determination of Ni and Mn Species in Welding Fumes. Ann. Occup. Hyg, Vol.53, No.4, pp. 333-340.

Bernard, A. (2008) Cadmium & its adverse effects on human health. *Indian J Med Res 128.* October 2008, pp 557-564

Blade, L.M. (2007) Hexavalent Chromium Exposures and Exposure-Control Technologies in American Enterprise: Results of a NIOSH Field Research Study. *Journal of Occupational and Environmental Hygiene*, 4: 596–618

Brouwer, D.H.; Gijsbers, J.H.J. & Lurvink, M.W.M. (2004) Personal Exposure to Ultrafine Particles in the Workplace: Exploring Sampling Techniques and Strategies. *Ann. Occup. Hyg.* Vol.48, No.5, pp. 439-453.

Brown, K.L. (1997) Environmental aspects of fume in air and water. Villepinte: International Institute of Welding Document; Doc. CV111 1804-97.

Carpenter, K.R.; Monaghan, B.J. & Norrish, J. (2009) Influence of shielding gas on fume formation rate for GMAW of plain carbon steel. *Trends in Welding Research, Proceedings of the 8th International Conference*. Pp. 436-442.

Christensen, S.W.; Bonde, J.P. & Omland, O. (2008) A prospective study of decline in lung function in relation to welding emissions. *J Occup Med Toxicol*. Vol.26. pp. 3-6.

Committee on Beryllium Alloy Exposures. (2007) Health Effects of Beryllium Exposure: A Literature Review, Committee on Toxicology, National Academies of Science, ISBN: 0-309-11168-4, 108 pages

Dennis, J.H.; French, M.J., Hewitt, P.J., Mortazavi, S.B. & redding, A.J. (1996) Reduction of Hexavalent Chromium Concentration in Fumes from Metal Cored Arc Welding by Addition of Reactive Metals. *Ann. Occup. Hyg.* Vol. 40, No.3, pp. 339-344.

Dennis, J.H.; French, M.J., Hewitt, P.J., Mortazavi, S.B. & Redding, C.A.J. (2002) Control of occupational exposure to hexavalent chromium and ozone in tubular wire arc welding processes by replacement of potassium by lithium or addition of zinc. Ann. Occup. Hyg., Vol. 46, No.1, pp. 33042.

d'Errico, A.; Pasian, S., Baratti, A., Zanelli, R., Alfonzo, S., Gilardi, L., Beatrice, F., Bena, A. & Costa, G. (2009) A case-control study on occupational risk factors for sino-nasal cancer/ *Occup Environ Med*. Jul;66(7):448-55.

El-Zein, M.; Malo, J.L., Infante-Rivard, C. & Gautrin, D. (2003). Prevalence and Association of Welding Related Systemic and Respiratory Symptoms in Welders. *Occupational and Environmental Medicine*. Vol.60. pp. 655-661.

Fachausschuss (2008) Chromium(VI) compounds or nickel oxides in welding and allied processes. pp. 1-3.

Fang, S.C; Eisen, E.A., Cavallari, J.M., Mittleman, M.A., & Christiani, D.C. (2010) Circulating adhesion molecules after short-term exposure to particulate matter among welders. *Occup Environ Med*. Jan;67(1):11-6.

Ferrriara, P. (2008) Aluminum as a Risk Factor for Alzheimer's Disease, Rev Latino-am Enfermagem. Janeiro-Fevereiro; 16(1):151-7

Gray, C.N., Hewitt, P.J., Dare, P.R.M. (1982) New approach would help control weld fumes at source Part two: MIG Fumes. Welding and Metal Fabrication, October, p393-397.

Haider, J. (1999) An Analysis of Heat Transfer and Fume Production in Gas Metal Arc Welding. III. Journal of Applied Physics, 85(7), 3448-3459.

Halasova, E. (2005) Lung Cancer in Relation to Occupational and Environmental Exposure and Smoking. *Neoplasma* 52(4)

Hannu, T.; Piipari, R., Tuppurainen, M., Nordman, H. & Tuomi, T. (2007). Occupational Asthma Due to Manual Metal-Arc Welding of Special Stainless Steels. *European Respiratory Journal*. Vol.29. pp. 85-90.

Heile, R.F., Hill, D.C. (1975) Particulate fume generation in arc welding processes. Welding Journal, July, p201s-210s.

International Agency for Research on Cancer, (2011) Agents Classified by the IARC Monographs, Volumes 1–100

Jenkins N.T., Eager, T.W. (2003) Chemistry of Airborne Particles from Metallurgical Processing. Ph.D. Dissertation: Materials Science and Engineering, Massachusetts Institute of Technology, Cambridge, MA.

Jenkins, N.T. & Eager, W. (2005a) Fume formation from spatter oxidation during arc welding. *Science and Technology of Welding and Joining*. Vol.10, No.5. pp. 537-543

Jenkins, N.T. & Eager, W. (2005b) Chemical analysis of welding fume particles. *Welding Journal*, June. Pp. 87s-93s.

Jenkins, N.T.; Pierce, W.M.-G. & Eager, T.W. (2005) Particle Size Distribution of Gas Metal and Flux Cored Arc Welding Fumes. *Welding Journal*. Oct. pp. 156s-163s

Kawahara, M. (2011) Link between Aluminum and the Pathogenesis of Alzheimer's Disease: The Integration of the Aluminum and Amyloid Cascade Hypotheses., International Journal of Alzheimer's Disease Volume, Article ID 276393, 17 pages doi:10.4061/2011/276393

Kelleher, P.; Pacheco, K. & Newman, LS. (2000) Inorganic Dust Pneumonias: The Metal-Related Parenchymal Disorders. *Environ Health Perspect.* Aug;108 Suppl 4:685-96.

Kellen, E.; Zeegers, M.P., Hond, E.D. & Buntinx, F. (2007) Blood cadmium may be associated with bladder carcinogenesis: the Belgian case-control study on bladder cancer. *Cancer Detect Prev.* 2007;31(1):77-82

Kim, J.Y.; Chen, J.C., Boyce, P.D., Christiani, D.C. (2005) Exposure to welding fumes is associated with acute systemic inflammatory responses. *Occup Environ Med.* Mar;62(3):157-63.

Kobayashi, M.; Maki, S., Hashimoto, Y., Suga, T. (1983) Investigations on chemical composition of welding fumes. Welding Journal, July, p190s-196s.

Konarski, P.; Iwanejko, I. & Cwil, M. (2003) Core-shell morphology of welding fume micro- and nanoparticles. *Vacuum.* Vol. 70. Pp. 385-389.

Leonard, S.S.; Chen, B.T., Stone, S.G,. Schwegler-Berry, D., Kenyon, A.J., Frazer, D. & Antonini, J.M. (2010) Comparison of stainless and mild steel welding fumes in generation of reactive oxygen species. *Particle Fibre and Technology.* Nov.3. pp. 7-32.

Linden, G. & Surakka, J. (2009). A Headset-Mounted Mini Sampler for Measuring Exposure to Welding Aerosol in the Breathing Zone, Annals of Occupational Hygiene, Vol.53, No.2, pp. 99-106.

Liu, H.H.; Wu, Y.C. & Chen, H.L. (2007) Production of Ozone and Reactive Oxygen Species After Welding. Arch Environ Contam Toxicol. Vol.53, pp. 513-518.

Marcy, A.D. & Drake, P.L. Development of a field method for measuring manganese in welding fume.

Marshall, G; Ferreccio, C., Yuan, Y., Bates, M.N., Steinmaus, C., Selvin, S., Liaw, J. & Smith, A.H. (2007) Fifty-year study of lung and bladder cancer mortality in Chile related to arsenic in drinking water., *J Natl Cancer Inst.* Jun 20;99(12):920-8

McCallum, R.I. (2005) Occupational exposure to antimony compounds. *J Environ Monit.* Dec;7(12):1245-5

McCormick, L.M.; Goddard, M.& Mahadeva, R. (2008) Pulmonary fibrosis secondary to siderosis causing symptomatic respiratory disease: a case report. *J Med Case Reports.* Aug 5;2:257.

Meeker, J.D; Rossano, M.G., Protas, B., Diamond, M.P., Puscheck, E., Daly, D., Paneth, N. & Wirth, J.J. (2008) Cadmium, lead, and other metals in relation to semen quality: human evidence for molybdenum as a male reproductive toxicant. Environ Health Perspect. Nov. Vol. 116. No.11. pp. 1473-9.

Navas-Acien, A. (2005) Metals in Urine and Peripheral Arterial Disease. *Environ Health Perspect.* 113:164–169.

Nawrot, T.; Geusens, P., Nulens, TS. & Nemery, B. (2010) Occupational cadmium exposure and calcium excretion, bone density, and osteoporosis in men. *J Bone Miner Res.* Jun;25(6):1441-5.

Nygren, O. (2006). Wipe Sampling as a Tool for Monitoring Aerosol Deposition in Workplaces. *Journal of Environmental Monitoring.* Vol.8, pp. 49-52.

Palmer, K.T.; Cullinan, P., Rice, S., Brown, T. & Coggon, D. (2009). Mortality from infectious pneumonia in metal workers: a comparison with deaths from asthma in occupations exposed to respiratory sensitisers. *Thorax.* Nov. Vol.64. No.11. pp. 983-6.

Quimby, B.J. & Ulrich, G.D. (1999) Fume formation rates in Gas Metal Arc Welding. *Welding Journal,* April. pp. 142s-149s

Report on Carcinogens, Twelfth Edition (2011) U.S. Department of Health and Human Services, Public Health Service, National Toxicology Program, http://ntp.niehs.nih.gov/ntp/roc/twelfth/roc12.pdf

Santibáñez, M.; Bolumar, F. & García, A.M. (2007) Occupational risk factors in Alzheimer's disease: a review assessing the quality of published epidemiological studies. *Occup Environ Med.* Nov;64(11):723-32.

Sellappa, S., Prathyumnan, S., Keyan, K.S., Joseph, S., Vasudevan, B.S. & Sasikala, K. (2010) Evaluation of DNA damage induction and repair inhibition in welders exposed to hexavalent chromium. *Asian Pac J Cancer Prev.* 2010;11(1):95-100.

Sowards, J.W.; Lippold, J.C., Dickinson, D.W. & Ramirez, A.J. (2008) Characterization of Welding Fume from SMAW Electrodes - Part I. Welding Journal, Vol. 87, pp. 106s-112s.

Sowards, J.W.; Ramirez, A.J., Dickinson, D.W. & Lippold, J.C. (2010) Characterization of Welding Fume from SMAW Electrodes - Part II. Welding Journal, Vol. 89, pp. 82s-90s.

Srinivasan, K & Balasubramanian, V. (2011) Effect of Surface Tension Metal Transfer on Fume Formation Rate During FCAW of HSLA Steel. *Int J Adv Manuf Technol.* DOI 10.1007/s00170-011-3182-0.

Taylor, M.D.; Roberts, J.R., Leonard, S.S., Shi, X. & Antonini, J.M. (2003) Effects of Welding fumes of Differing Composition and Solubility on Free Radical Production and Acute Lung Injury and Inflammation in Rats. *Toxicological Sciences.* Vol.75. pp. 181-19.

Threshold Limit Values for chemical substances and physical agents and biological exposure indices. American Conference of Governmental Industrial Hygienists, Cincinnati, OH, 2011

U.S. Department of Health and Human Services. (2011) Report on Carcinogens, Twelfth Edition. Public Health Service, National Toxicology Program, http://ntp.niehs.nih.gov/ntp/roc/twelfth/roc12.pdf

Vandenplas, O.; Delwich, J.P., Vanbilsem, M.L, Joly, J. & Roosels, D. (1998). Occupational Asthma Caused by Aluminum Welding. *European Respiratory Journal*. Vol.11. pp.1182-1184.

Worobiec, A.; Stefanik, E.A., Kiro, S., Oprya, M., Bekshaev, A., Spolnik, Z., Potgieter-Vermaak, S.S., Ennan, A. & Van Grieken, R. (2007) Comprehensive microanalytical study of welding aerosols with X-ray and Raman based methods. *X-Ray Spectrometry*. Vol.36, pp. 328-335.

Yi-Chun Chen. (2009) Cadmium burden and the risk and phenotype of prostate cancer. BMC Cancer, 9:429 doi:10.1186/1471-2407-9-429

Yoon, C.S.; Paik, N.W.& Kim, J.H. (2003) Fume Generation and Content of Total Chromium and Hexavalent Chromium in Flux-cored Arc Welding. *Ann. Occup. Hyg*. Vol.47, No.8, pp. 671-680.

Zimmer, A.T. (2002) The influence of metallurgy on the formation of welding aerosols. J. Environ. Monit., Vol.4, pp. 628-632.